NOBEL LECTURES

PHYSICS

2006 – 2010

Nobel Lectures

Including Presentation Speeches
and Laureates' Biographies

Physics

Chemistry

Physiology or Medicine

Literature

Economic Sciences

Nobel Lectures

Including Presentation Speeches and Laureates' Biographies

PHYSICS

2006 – 2010

Editor

Lars Brink

Chalmers University of Technology, Sweden

World Scientific

NEW JERSEY • LONDON • SINGAPORE • BEIJING • SHANGHAI • HONG KONG • TAIPEI • CHENNAI

Published by

World Scientific Publishing Co. Pte. Ltd.
5 Toh Tuck Link, Singapore 596224
USA office: 27 Warren Street, Suite 401-402, Hackensack, NJ 07601
UK office: 57 Shelton Street, Covent Garden, London WC2H 9HE

NOBEL LECTURES IN PHYSICS (2006–2010)

ISBN 978-981-4612-67-8
ISBN 978-981-4612-68-5 (pbk)

Printed in Singapore by Mainland Press Pte Ltd.

PREFACE

Alfred Nobel was one of the greatest inventors, engineers, entrepreneurs and industrialists of the 19th century. His most profitable invention was dynamite that he patented in 1867. His most remarkable and lasting invention though was his founding of the Nobel Prizes. In his will from 1895 he stipulated that most of his wealth should be handled such that the interests from it should be divided into five equal shares which should be used for yearly Prizes in Physics, Chemistry, Physiology or Medicine, Literature and Peace to "those who, during the preceding year, shall have conferred the greatest benefit on mankind." Alfred Nobel died in 1896 and in 1901 an organization, the Nobel Foundation, was in place to hand out the first Prizes. Apart from some war years the Prizes have been given every year since then.

The only real change that happened to the will was that it was realized that the sentence "...during the preceding year..." could not be implemented literally. It is instead interpreted to mean that a discovery should have had renewed interest in recent years if it is of older date.

For the Physics Prize, which Nobel ranked as the first one, he stipulated that it should be given "to the person who shall have made the most important discovery or invention within the field of physics." This is the criterion that the Nobel Committee has had to work with since the beginning. It is not to the most deserving person, it is not for a lifetime achievement, nor is it for the clearest explanation of a physical phenomenon. It is to the discoverer of a physical phenomenon that changed our views, or to the inventor of a new physical process that gave enormous benefits to either science at large or to the public. This was an ingenious focus for the Prize. You do not need to be Albert Einstein to get the Prize but it helps. The discoverers are our scientific heroes and they are eligible for the Prize.

In this volume we will meet the laureates from the five years 2006–2010. It contains the presentation speeches of the Nobel Committee members given at the ceremony. It further contains the biographies and the Nobel lectures of all the laureates. We will learn to know the laureates, follow their lives up to their great discoveries or inventions and also follow them after these events that led to a Nobel Prize. We can learn how the great minds work and we will see that Nobel laureates are also human beings of flesh and blood. At the same time the Nobel lectures give us detailed explanations of the phenomena for which the laureates have been awarded.

The 2006 Nobel Prize in physics honored the discovery of the small anisotropy of the cosmic microwave background radiation that was discovered by the COBE satellite, and the Prize was given to two of its leaders, John C. Mather and George F. Smooth. Everyone knows that our universe is not homogeneous. We only have to look at ourselves and at the night sky on a clear night. Even so Albert Einstein assumed without any experimental indications already in 1917 that on large enough scales the universe should be homogeneous. This was finally verified in 1964 by Penzias and Wilson, who discovered the cosmic microwave background radiation that is the remnant of the Big Bang that we can detect today. This rendered them the Nobel Prize in 1978. It was a remarkable discovery to find that the background radiation seemingly have the same temperature whichever direction it comes form. However, there must be some anisotropy in order for galaxies to form at some time in the history of the universe. The COBE mission that started its planning in the late 1970's and was launched in 1989 finally found this anisotropy in the black body radiation which has a temperature of 2.7° above the absolute zero. The deviation is 1 part in 105, i.e. in the fifth digit, but this is enough for our universe to develop the structures we now see. In the memoirs and the Nobel lectures the reader can follow all the travails and brainstorming needed to successfully detect such a small but important discrepancy.

The 2007 Nobel Prize honored an important discovery that was also a great invention, namely the "Giant Magnetoresistance." The Prize was shared by Albert Fert and Peter Grünberg. It has long been known that the electric resistance of materials such as iron may be influenced by a magnetic field. The British physicist Lord Kelvin demonstrated this already in 1857. The resistance diminishes along the lines of magnetization when a magnetic field is applied to a magnetic conductor. If the magnetic field is applied across the conductor the resistance increases instead. This (anisotropic) magnetoresistance was used early on for example for reading out data. When the new techniques to produce material at the nanometer scale were developed, new possibilities arose. Both laureates realized that by sandwiching layers of non-magnetic material between layers of magnetic ones, electrons with different spin directions would scatter differently leading to differences in the resistance, when a magnetic field is switched on. The real discovery was that the effect was huge and this led directly to the invention of new miniaturized read-out heads revolutionizing techniques for retrieving data from hard disks. The discovery also played a major rôle in various magnetic sensors as well as for the development of a new generation of electronics called spintronics. In the memoirs the reader can follow the development of this new exciting field and also get an understanding of some of the enormous progress that has occurred in the IT-world.

The 2008 Nobel Prize was a shared prize with a common denominator, broken symmetries. One half was awarded to Yoichiro Nambu "for the discovery of the mechanism of spontaneous broken symmetry in subatomic physics" and the other half jointly to Makoto Kobayashi and Toshihide Maskawa "for the discovery of the origin of the broken symmetry which predicts the existence of at least three families of quarks in Nature."

It had long been known that physical systems could have ground states which have a smaller symmetry than the underlying equations that determine the systems. This came to be called "spontaneous symmetry breaking." Ferromagnetism is a typical example. The equations are rotationally invariant but the ground state is not. Yoichiro Nambu first explained the Bardeen-Cooper-Schrieffer theory of superconductivity (Nobel Prize 1972) as a theory where the electromagnetic gauge symmetry is spontaneously broken and proceeded to take this knowledge over to pion physics and the strong nuclear forces. This explained the low mass of the pion and is a background to our understanding of the strong nuclear forces. The relevant work by Nambu was made around 1960 but the consequences of his work on spontaneous symmetry breaking has been important ever since and is a pillar of modern fundamental physics.

The work by Kobayashi and Maskawa addressed a different issue of broken symmetries, namely the so-called CP-invariance. One of the most remarkable discoveries of the 20th century was that there is a difference between left and right in Nature. The so-called parity invariance P or mirror symmetry is indeed broken in the weak interactions. This led to the Nobel Prize in 1957 to Chen Ning Yang and Tsung-Dao Lee. However, if one also combines P with the charge symmetry C into CP-invariance it was believed to be a true symmetry. In 1964 James W. Cronin and Val L. Fitch discovered that also this symmetry could be violated, which rendered them the Nobel Prize in 1980. With the advent of the Standard Model for Particle Physics from 1971 on there did not seem to be any possibility to build in this symmetry violation in the theories for the weak interactions without destroying the model's good predictions. In 1972 Makoto Kobayashi and Toshihide Maskawa examined the possibility of introducing more fundamental particles — quarks — and found that if there are six different quarks, the theory can indeed break this symmetry. It was a daring assumption, since at that time the everyday world seemingly needed only two different kinds of quarks, whereas three had been discovered. Over the years the other three quarks were discovered and it was shown that Nature is indeed following the ideas of Kobayashi and Maskawa. It had in fact been pointed out by Andrei Sacharov (Nobel Peace Prize in 1975) that we need a violation of the CP-invariance in order to understand why our world consists of matter and no antimatter. The theory of Kobayashi and Maskawa does not explain the whole violation that is necessary but shows that it is possible to have a violation.

In 2009 it was time for a genuine invention prize. One half of the Nobel Prize was awarded to Charles Kuen Kao for "for groundbreaking achievements concerning the transmission of light in fibers for optical communication" and the other half to Willard S. Boyle and George E. Smith for "for the invention of an imaging semiconductor circuit — the CCD sensor."

We have all grown up with copper fibers leading the electric currents and when the telephones started to be used, they also utilized such fibers to lead the information in terms of weak electric currents. During the 1960's the young Chinese-British engineer Charles Kao started to experiment with transmitting light through glass fibers in order to send information. In the beginning the fibers were not transparent enough so the signal would disappear after a very short distance. (Think about how much you see through a thick glass.) In order to succeed one had to produce the purest glass in the world. In 1971 Corning Glass Works in the USA, a glass manufacturer with over 100 years experience, produced a 1-kilometer long optical fiber. This was the signal that it was possible to use glass fibers. Since then this technique has completely taken over. It has led to cheap telephone calls between any two places on earth, to internet, to sending emails also without any costs, also leading to spams. The modern communication society is totally dependent on the optical fiber. With the old system with for example satellites transmitting information between different parts of the world, we were used to delays in the transmission. Now all information flows on earth at the speed of light. Since light is used the frequencies are also much higher and one can pack much more information in each channel than before, which also explains how most of the present world can be connected. The biography and the Nobel lecture by Charles Kao are remarkably interesting to read. They show what one can achieve with persistence and hard work and belief in one's ideas.

The invention of the CCD-camera was done in one discussion between the two laureates one day in 1969 in the Bell Laboratories. It uses the photoelectric effect for which Albert Einstein got the 1921 Nobel Prize. The digital image sensor, CCD, is made out of silicon. The size of a stamp, the silicon plate holds millions of photocells sensitive to light. The effect occurs when light hits the silicon plate and knocks out electrons in the photocells. The liberated electrons are gathered in the cells which become small wells for them. The larger the amount of light, the larger the number of electrons is that fills these wells. The engineering problem was now to read out these numbers. An ingenious shifting of the wells so that each original well can be read off in a certain order does this. With this information the picture can be reconstructed. This is a very different technique from the old one using films that had to be developed. It has revolutionized the possibilities to take pictures. It started with TV cameras and since more than ten years electronic digital cameras occur not only as ordinary cameras but most telephones have them. They are everywhere. This is really what Big Brother should have used. It does not cost anything essentially to take a picture, and with the optical fiber technique it can be sent anywhere in the world at the velocity of light. The CCD-cameras have also played an enormous role for many science disciplines, such as astronomy where one camera can follow hundreds of thousands of stars for years. Also in medicine the use of small CCD-cameras have revolutionized the techniques to see inside the body. There are also other similar techniques to the CCD-concept that are used these days and they have also contributed much to our modern world of communications. For this prize it is interesting to read how an invention of this kind suddenly is ripe to be found.

In 2010 the Nobel Prize was awarded to two Russian-British scientists, Andre Geim and Konstantin Novoselov "for groundbreaking experiments regarding the two-dimensional material graphene." We know that carbon exists in many forms on Earth. It is very important for life and is the base for the DNA. The most common form of carbon is graphite, which consists of stacked sheets of carbon with a hexagonal structure. Under high pressure diamond is formed, which is a metastable form of carbon. A new form of molecular carbon is the so-called fullerenes. The most common one, called C60, contains 60 carbon atoms and looks like a football (soccer ball) made up from 20 hexagons and 12 pentagons which allow the surface to form a sphere. The discovery of fullerenes was awarded the Nobel Prize in Chemistry in 1996. Also a related quasi-one-dimensional form of carbon, carbon nanotubes, has been known for quite some time. However, it was commonly believed that a one-dimensional layer could not be a stable material. In 2004 the laureates managed to isolate such a one-dimensional sheet. Using scotch tape to pull layers from graphite they managed to reach one layer. This came to be called graphene. They then succeeded to characterize its physical properties, and indeed it has remarkable properties. It has opened up a new world of possibilities for innovative electronics. It is practically transparent and a good conductor, and it is now a race around the world to find novel applications for it. Also for these laureates it is very instructive and interesting to read their biographies and Nobel lectures. Especially their description of "Friday night experiments," where they tried odd ideas can teach many aspiring scientist the importance of playing and to have fun in science.

I am convinced that aspiring young scientists as well as more advanced ones but also the interested public will learn a lot from and appreciate the geniuses of these narrations.

Lars Brink

CONTENTS

Physics 2006

John C. Mather and George F. Smoot

"for their discovery of the blackbody form and anisotropy of the cosmic microwave background radiation"

THE NOBEL PRIZE IN PHYSICS

Speech by Professor Per Carlson of the Royal Swedish Academy of Sciences.
Translation of the Swedish text.

Your Majesties, Your Royal Highnesses, Ladies and Gentlemen,

Gaze into the sky on a clear, starlit night and feel the fascination of the cosmic spectacle it offers. Fifty years ago astronomers used earth-bound telescopes to explore space. The instruments then in use enabled us to see stars and galaxies so far away that their light had taken some hundreds of millions of years to reach us. This year's Nobel Prize in Physics is for another kind of astronomy, in which the instruments on the COBE satellite were used to observe radiation that was sent to us about 13 billion years ago, from another, much earlier era in the development of the universe.

At that time the character of the Universe had just altered. Previously it had been a dense, hot and impenetrable soup of electrons, protons and radiation. The temperature was so high and the Universe so dense that the radiation was swallowed up by the soup, just as light is veiled by fog. But the Universe is expanding, which means that its density and the temperature are sinking, and at the same time the energy of the radiation is lessening – which is the same as saying its wavelength is rising. The radiation that had previously been swallowed up by the soup was released – like when the fog lifts. The temperature had then sunk to 3,000 degrees and it is estimated that the Universe was about 380,000 years old. The released radiation could continue its long, unimpeded journey through the Universe.

During the radiation's 13 billion-year journey the Universe has expanded mightily and its wavelength has extended one thousandfold, at the same time the temperature has dropped from 3,000 to about three degrees above absolute zero. This cold background radiation fills the Universe and is a relic from its earliest stages, but it is invisible to the eye. The extended waves can be observed in the microwave area, with wavelengths of a few millimetres. It is this original radiation that the COBE satellite observed.

John Mather was responsible for the instruments on board COBE that were able to determine with great precision the temperature of the background radiation and also confirm that this spectrum had the characteristic blackbody form, characterising the hot uniform state of the early Universe.

George Smoot was in charge of the instruments that were searching for very small variations – about one per 100,000 – in the temperature of the background radiation in various directions. These variations would provide the seeds of the structures, in the shape of stars and galaxies that evolved from the original hot, uniform soup. After collecting data for several years, it was possible to show that the small variations sought for really existed. The first step towards understanding the development of structures in the Universe had been taken.

Dr Mather, Professor Smoot,

Cosmology has become a precision science and your ground-breaking research laid the foundation for that. With your carefully calibrated instruments you have shown the cosmic microwave background radiation to follow very precisely a blackbody form. The in-depth analysis of the radiometer data has shown the presence of the long sought small temperature anisotropies, the seed of the structures in the Universe that we observe today. In your successful experiments you have used space-based instruments on board the COBE satellite. We are now all together at sea level in Stockholm, and on behalf of the Royal Swedish Academy of Sciences it is my privilege and pleasure to congratulate you for your outstanding work and I now ask you to step forward to receive your Nobel Prizes from the hands of His Majesty the King.

John C Mather

JOHN C. MATHER

I was born on Aug. 7, 1946 in Roanoke, Virginia, a small city near Blacksburg where my father was a young faculty member at the school now called Virginia Tech. For some generations, my family on both sides has been populated with scientists and school teachers. My father, Robert E. Mather, was starting his research career in animal husbandry and statistics, specializing in dairy cattle breeding and feeding, having just received his PhD from the University of Wisconsin. My mother, Martha Cromwell Mather, was teaching high school French. When I was about a year and a half old, my parents moved to the Rutgers Agricultural Experiment Station, also known as the Dairy Research Station, in Sussex County, New Jersey. This is a very peaceful farming area, in the rolling foothills of the Appalachians, and our house was just a mile from Appalachian Trail. James Turner, a stockbroker from Montclair, New Jersey, and originally Scottish from Glasgow, founded the experiment station in 1931 I believe. So my childhood from age 1 to 15 was spent in that house, at the top of a long hill, looking over a valley filled with fields and farms and forests. My earliest memory is of a medical procedure in the hospital in Sussex – I had my tonsils out when I was about 2.5 years old, and they gave me ice cream. Our home looked out at a barn where 20 bulls were kept, and they were the sires of half the cows of northern New Jersey. Within a few yards of the house was an office building where the analysis of the experiments was done. Down the hill past the barn was a laboratory with calorimeters, chemical analysis equipment, liquid nitrogen tanks for keeping semen frozen until it was needed, some radioisotope equipment for studying metabolism, and so forth. As I understand it, my dad was responsible for developing a test that measured the protein content of milk, and thus indirectly for the re-optimization of the dairy industry to produce more protein and less butter. Later on, he became one of the early users of computers, and had the milk production records of 10,000 cows on punched IBM cards.

I attended Wantage Consolidated Elementary School near Sussex, N.J., which had about 600 students, and was established in the 1930's so that the one-room schoolhouses in the area could be closed. So I rode a big yellow school bus to school, along with many other students, most of them children of independent farmers. Their farm labors made them very strong, and our school athletes excelled at baseball and wrestling, but I was not among the athletes. I was much more interested in reading everything that came my way, hiding a book behind the desk while the other students learned from class. My interest in science started quite early. My earliest school recollection, from age 6, is actually of mathematics, realizing that one could fill an entire page with digits and never come to the largest possible number, so I saw what was meant by infinity. I recall that my parents took my sister and me a few times to the American Museum of Natural History in New York, after a journey of at least two hours, which to me seemed forever. The first time, Mars was very close to the Earth, and there was great excitement about whether the canals could be seen. My father bought a small telescope from Sears Roebuck, but of course it could not show the canals, and Mars was extremely small even with the telescope. He also bought a book *Astronomy Made Simple,* which got me started. My parents also enjoyed reading aloud from various books, including biographies of Darwin and Galileo. I was fascinated with the museum displays of the sequence of different kinds of skeletons of fishes, showing their changes through time, so evolution was quite the obvious conclusion even to a child. I liked thinking about geology and hunting for fossils in the roadside streams, as I built little dams of mud and pebbles. I didn't really appreciate what science was about yet, but it sounded very exciting, and a little dangerous in case one discovered things that were not consistent with previous knowledge, especially religious thought. In later years, I occasionally dreamed of being in court, defending the right to teach evolution in the schools. My mother's father, Hobart Cromwell, was a bacteriologist with Abbott Laboratories in suburban Chicago. I never got to know him well, as he died very young, but he was always a heroic figure in our family, wise and gentle and intelligent by reputation, with the courage to fight against the McCarthyites.

By the time I was in fourth grade (age = grade + 5 years) I was already pretty sure I liked scientific and engineering things, including electronics. For Christmas I got a one-tube radio kit, and then I saved my allowance for a 5-tube shortwave Heathkit radio that I put together so I could listen to exotic languages and broadcasts from far-away places. Around that time, the IGY (International Geophysical Year) was starting up, and at our spot in New Jersey it was marked by a fabulous bright aurora, the only one I've every seen. I think 4th grade also marked my first entry into the school Science Fair, and I submitted 4 little projects. Hardly anyone else at the school was much interested in science at the time, but I had one friend who walked several miles to come visit us at our house and talk about these things. We did have a Bookmobile, a traveling library from the County that visited the farms every couple of weeks, and I borrowed as much as I could. I started reading about optics, and I saved my allowance and ordered some lenses from Edmund Scientific and assembled small refractor telescopes. One summer my parents sent me off to a summer camp in the Poconos, a place that stood out because it had a science program. Another summer, they sent me to a day camp with the high school science teacher in Newton, New Jersey, Ben Cummings. With him, we climbed a hillside near High Point State Park and came back with bags of trilobites. And one year, I wanted to do a science fair project with a "robot" that I designed with vacuum tubes and remote controls. It never worked but I got a lot of experience, and now looking back on it I recognize that my parents contributed a substantial research grant when I was only about 11. So I had a lot of opportunity to learn science, even in our very rural setting.

When I finished 8th grade, it was time to go to high school, and my parents decided to send me to Newton High School, where they thought we would get the best available education in our area. That turned out well for me, and I had some excellent teachers in science, math, and English that I really liked. I took biology in 9th grade, chemistry in 10th grade, and physics in 11th grade. I was very fortunate to have the opportunity to go away to summer schools. After 10th grade I went to Assumption College for about 10 weeks to learn about the foundations of mathematics, and after 11th grade I went to Cornell University for a summer physics program. That was truly extraordinary for me, with an introduction to quantum mechanics, special relativity, optics, nuclear physics, and cosmology. Coming back from these programs, having done fairly well, I was convinced that I could have a future in science, and I was very glad to have a head start relative to others of my age. The National Science Foundation sponsored these summer programs, and they certainly did a wonderful thing for us. I competed in the nationwide math contest and placed 7th in New Jersey, I think, and in a statewide physics contest I placed first. With all this success I was feeling pretty good, but my parents reminded me frequently that I would still have to work hard in college, since I had been a big fish in a little pond, and I didn't know what was yet to come in the big world.

I never got very excited about dairy cattle, but my father did help me learn two important things, statistics and calculus. For one of my science fair projects, I had 8 baby rats that I kept in cages under the table in the kitchen, and I fed them various diets to see what they needed. My mother must have had immense patience with me to allow me to have my experiment there. My dad showed me how to design a Greco-Latin square for the experiment, and how to do the analysis of variance. The answer: dog food and vitamins are good, and corn flakes alone are inadequate. One summer, he returned to college to learn calculus himself, and when he got home I borrowed his textbook and studied it instead of taking an advanced science course in high school. This was another way of getting a little ahead of my cohort, and when I got to college it was a good thing.

I chose Swarthmore for college, largely because the atmosphere felt good and the faculty promised a complete education in physics. I tried without much success to learn a little of the humanities and the arts, but even passing the courses in art history and music history was a challenge. In those courses I understood what other folks felt when they saw me doing so well in physics; I knew it was hopeless to compete on that territory, but I persevered and even took up piano lessons again, with enjoyment but not skill or talent. I jumped a little ahead again, skipping the second half of freshman physics and diving right into sophomore physics. I got a lot of special attention from the faculty there and really appreciated it. I was in the honors program, with four seminars in math and four in physics and two in astronomy. When it was time to graduate, David Wilkinson, a young professor at Princeton, was one of my honors examiners. He asked me a question about everyday effects of relativity, and I said that magnetism was a relativistic effect of electron motion, or something like that. Years later, David was a founding member of the COBE team.

For graduate school I chose Princeton, and was making plans to go there, when a friend Ted Chang, who was my friend at Assumption College for a high school summer, sent me a photo of himself sitting on the fountain in Berkeley in January, wearing short sleeves. He sent me an application form for a summer job, and I went. As it happened, my job was at the Lawrence Berkeley Laboratory, working with Henry Frisch on control electronics for a spark chamber. Henry's father was a physicist too, and taught my co-Nobelist George Smoot. I liked Berkeley, and changed my mind about Princeton. Being a little churlish, I wrote to Princeton and told them that I was withdrawing because they had no women students. I was fortunate that my fellowship from the National Science Foundation was portable, so switching was easy.

At Berkeley, I found a big old brown shingle house to share, and my rent was very low. The household was organized by John Hauptman, another physics student, and held about 8 other people. Roy Torbert, now a space physicist at UNH, was a member of our little group. One household member, Richard Rotblatt, was a former architecture student and nuclear reactor piping engineer, and he was also an excellent chef. Now he's an accomplished wine maker as well. For a while my best friend from high school lived with us after returning from Vietnam. This old house could tell many stories of the times, with people of all sorts moving in and out.

At first, I thought I wanted to be an elementary particle physicist like my hero Richard Feynman. I took my courses for two years, during which my faith in my future was being challenged by the Vietnam War and by the Peoples' Park demonstrations that rocked the city. Governor Ronald Reagan's helicopters tear-gassed the campus, and people were seriously injured, but I tried to stay out of this trouble and had little sympathy with either side. Because I was very nearsighted, I was not drafted to be a soldier, so I didn't have to really deal with the great issue of principle that involved so many of my classmates. For a while I considered studying the law, in order to defend the country from the Government of the day, but when I read the course catalog I couldn't imagine studying those subjects. Now as a long-time Government employee my perspective has changed a bit. I also talked with my plasma physics professor about developing nuclear fusion power for the good of humanity, but he seemed to think this would be an extremely long and difficult project, as it has turned out to be.

So in 1970 I was looking for a thesis project, and interviewed with various faculty members. I found that Paul Richards was working with Charles Townes and a young postdoc Michael Werner to start up projects on the newly discovered Cosmic Microwave Background Radiation. I liked all three of them immediately, as well as the proposed experiment, and I started right in. It was a new world for me, much more tangible than years of books and classes and late nights in the library. The first project was to build a small far infrared spectrometer to take to the Barcroft station on White Mountain in eastern California, where the University was studying physiology at high altitudes. This project worked out well, but was of course limited in accuracy by the interference of the Earth's atmosphere. We were able to set some interesting limits on the CMBR intensity, and we got a few publications from them. Then, Paul went off to England on a sabbatical and came back with a concept for a new experiment, a balloon-borne far infrared interferometer to measure the CMBR spectrum. He explained it to the graduate students, and David Woody and I started trying to work out and build the design. This was the beginning of a baptism by fire, in the art of building instruments that would work in remote and hostile locations. It was a time to learn something of almost every area of engineering, from mechanical to optical to cryogenics to electronics. I'm afraid that my skill was stronger in understanding than it was in implementation, and it's a true story that the antenna on the balloon payload fell off while it was on the launch pad. It was my solder joint that failed. Fortunately this fault was noticed, and the payload was launched successfully.

However, it also was true that we had gotten tired of testing, and our instrument did not work, for three different reasons. It was an awful feeling, one that stayed with me for the rest of my life, and it was one of those ways of learning what one does not want to learn. Murphy's law had been proven one more time. Also, I wanted to finish my thesis, and had already lined up a job in New York as a postdoc with Pat Thaddeus. So Paul agreed, and my thesis described the ground-based work and the design for the balloon instrument. David Woody designed a test chamber for the payload, a cubical box of Styrofoam and plywood, and filled it with dry ice. He found out why the instrument had failed, fixed it, and made it work right for the second flight, the summer after I left Berkeley for New York. He analyzed and published the data and got his thesis out of the project too. Now he's a radio astronomer at CalTech, designing antennas for the new ALMA observatory in Chile.

With Pat Thaddeus at the Goddard Institute for Space Studies, a part of NASA housed in a building adjacent to Columbia University, I was hoping to go into a new field of study. I thought that my work on the background radiation was awfully difficult, and it was going to be hard to do much better with balloons. I suppose I was reacting too much to the failure of the instrument. At any rate, I arrived in New York at the end of January 1974, only days before the last possible allowed date, and I started theoretical and observational work on naturally occurring SiO masers. I learned how to build a microwave receiver with brilliant machinists and technicians, and I took it off to McDonald Observatory in Texas and to the Navy's Maryland Point observatory on the Potomac. We did observe the SiO emission at 43 GHz, which had never been seen before in space, and I made a little progress writing a giant Fortran program on the IBM 360 computer, but it never came to anything, and years later I threw many boxes of IBM cards into the trash, finally admitting defeat.

However, in the summer after I arrived, my trajectory took another abrupt turn when NASA issued Announcements of Opportunity 6 and 7, for Scout and Delta-launched satellite missions. My optimism was returning, and when Pat asked for ideas, I cheerfully asserted that my thesis experiment would have worked a lot better in space. He suggested that I call up Rainer Weiss, David Wilkinson, and Michael Hauser, and with their colleagues Dirk Muehlner and Bob Silverberg together we conceived of the new mission. It would have four instruments, a far IR interferometer to measure the CMBR spectrum, two instruments to measure its anisotropy (difference in brightness in different directions), and an instrument to hunt for the diffuse IR background from the first galaxies. Then, the balloon payload flew successfully after David Woody fixed it, and things were looking up. We sent in our proposal, typed by hand on real typewriters, and I at any rate thought that our odds of success were very low. None of us had any prior experience with space missions, and none of us knew that there would be about 150 other proposals, or that two of those (from JPL and Berkeley) would be direct competitors for ours. However, NASA was interested after all. There was already a negotiation with the Netherlands and the UK to build the Infrared Astronomical Satellite (IRAS). Ball Aerospace built the US part of the IRAS. So, the first expression of interest in our idea was to see whether the spectrometer could be miniaturized and given a ride to space as some part of the IRAS. I got a little money to study it, and I presented my concept to the IRAS science team at a meeting near Amsterdam. It went over with a resounding thud, for good reasons. I am very glad I never had to build this version of the instrument I had conceived, but I did learn a lot about what could be done, and I learned about the IRAS mission, which was to have a large helium cryostat much like the one we needed for the COBE.

So in the fall of 1976, NASA decided formally to study our concept, but not just with our team. Nancy Boggess, the Program Scientist at NASA Headquarters for infrared astronomy, appointed four members of our original team (Hauser, Weiss, Wilkinson, and me), along with George Smoot of UC Berkeley and Sam Gulkis of JPL, to form a Mission Definition Science Team. Anticipating this event, Mike Hauser had offered me a job at Goddard Space Flight Center in Greenbelt, Maryland, so I was already in place as a proper civil servant. We were assigned a manager, Martin Donohoe, and we were to compete with about 11 other missions that were also viewed as promising. Our little team elected a Chairman (Rainer Weiss), and three Principal Investigators (me, Mike Hauser, and George Smoot), and NASA assigned me the job of Study Scientist, to work with the engineering team to make this project happen. So this was the beginning of the COBE satellite project. Mike Hauser, who had hired me, was my main mentor, and I have learned to seek his advice whenever times are tough. Among all my colleagues, Mike is my greatest hero and example to follow.

We submitted our report, and the results were favorable, so NASA assigned us a larger team of seasoned engineers, namely the team that had almost finished the IUE (International Ultraviolet Explorer), led by Jerry

Longangecker. This was a time when the Space Shuttle was being considered by Congress, and NASA made a deal that would set its future for a long time: all new launches would be made with the Shuttle, and all the expendable rockets like the Deltas would be canceled. We argued but we had no success, and we had to redesign the COBE to go on the Space Shuttle. This wasn't so easy, since the COBE needed a polar orbit, achieved by a launch from California, at around 900 km altitude. Most of the Shuttles would be launched from Cape Canaveral (then called Cape Kennedy), so our requirement was a challenge in any case. By around 1979, NASA decided to build the COBE satellite in-house at Goddard, meaning that engineers and scientists at Goddard would work together very closely. This is an exception to the usual way that NASA obtains satellites, which is by writing contracts to major aerospace organizations and university laboratories. In our case, two of the three Principal Investigators were already at Goddard, and the third (George Smoot at Berkeley) was willing to have Goddard build that instrument too, so this new arrangement was very good for us. We had daily interactions with our engineering colleagues, we could walk into any laboratory to talk about any problem, and we made significant forward progress, and I really enjoyed that part. On the other hand, part of the deal was that our project was a training project for new engineers, and a reservoir for talent. When other projects got into trouble, our team was raided for top talent to go solve emergencies elsewhere, and of course there were many of those. I was very frustrated about this, but I had to admit, the Hubble Space Telescope really did have priority.

In 1980, I made a major decision, to marry Jane Hauser (no relation to Mike). I had met Jane in New York back in 1974 while I was taking a workshop in re-evaluation counseling, one of many personal growth experiences that I sought as a part of my emotional education. (My sister Janet became a teacher of this subject for many years, and so did one of my many bosses at Goddard.) Jane is a ballet teacher, but she was taking computer programming and math courses as she completed her undergraduate education, and I was very impressed. So on Nov. 22, 1980, a hundred scientists, engineers, and dancers threw us a potluck banquet after our wedding, and I have never seen so much good feeling and good food at one time and place, at least until I arrived in Stockholm. Jane has been my life partner, my best friend, my best editor, and my best advisor ever since. With her I have traveled to many amazing places, and become quite fascinated with understanding how ancient civilizations managed to accomplish their engineering feats. We've seen Tycho Brahe's observatory in Denmark, we've seen Ulugh Beg's observatory in Samarkand, and I think most amazing of all, we've seen Pompeii, with plumbing, faucets, running water, and so many signs of modern life that one can hardly imagine how that knowledge was lost. Sometimes I think it would be a lot of fun to write books about how great cities were built, but I seem to have something else to do right now.

From 1980 on through the rest of the COBE project, my professional life was almost entirely consumed with the COBE. For a while I was a Branch

Head at Goddard, in charge of the group that Mike Hauser had created. I went off for training courses in all kinds of personnel matters. It was an interesting education, and reinforced the idea that the manager is really working on behalf of the employees. It also emphasized what has become a recurring theme for me: life is a team sport, and it matters who's on the team, and which team(s) one chooses to be on. For the year after the COBE was launched, Werner Neupert acted on my behalf. Later on, Chuck (Charles L.) Bennett became the branch head as well as continuing as Deputy Principal Investigator for the Differential Microwave Radiometers on the COBE; he's also one of my other favorite advisors and great heroes.

I can't imagine telling anything like a complete biographical story about my work on the COBE. I made an attempt in the book *The Very First Light,* written with John Boslough, a professional science writer. Some people have told me that they were exhausted after reading this book, the story was so full of terrifying moments. Needless to say, the COBE team was exhausted too at various times. For more details, please see the Nobel Lecture accompanying this note, and the numerous technical publications from our team. But the main point I need to make is that the COBE mission was a team effort. Our team gave their complete concentration and support for a very long time, they dealt with having to redesign the mission after the Challenger explosion, they tested the observatory extensively, and they fixed the problems that they found. The analysis team found ways to compensate for "systematic errors" that were built into the designs, and in the end got measurements far beyond the formal requirements for sensitivity and accuracy.

After the COBE work was completed, I was wondering what do next. For years I had successfully repelled all challenges to my concentration on one overwhelming responsibility. Now, it was done, and I switched my attention to developing new mission concepts. My colleague Harvey Moseley was working on the IRAC (Infrared Array Camera) for the SIRTF mission (Space Infrared Telescope Facility, later named the Spitzer Space Telescope). He said the next telescope needed more angular resolution, because the IRAC was so sensitive that its long exposures would be confusion-limited, i.e. that the fuzzy images of distant galaxies would be so numerous that they would overlap. I started thinking about this question and thought we needed to build a small (2 meter) telescope that would be deployed after launch, so it could be squeezed into an inexpensive launch vehicle. I presented this idea at a colloquium one day and my colleagues laughed and said NASA would never fly such a radical departure from tradition, and anything with a mechanism was dead before starting. I also learned about the Edison mission concept, being developed by Tim Hawarden and Harley Thronson and an extensive international team. This mission was proposed to NASA but was summarily rejected, based on grossly inaccurate thermal calculations made by some reviewer. Curiously enough, the marriage of these two rejected ideas has become the concept for the James Webb Space Telescope (JWST), which is now my major passion.

My involvement with the JWST began in the fall of 1995, when I received a phone message from Ed Weiler at NASA Headquarters, asking that I submit a proposal the next day for a study of the Next Generation Space Telescope. I was completely astonished – I had no awareness of this topic, or of the fact that an entire conference had been held at the Space Telescope Science Institute to argue for such an observatory. However, I wasted no time and said yes immediately, and then called around to find out the background information. John Campbell, Project Manager for the Hubble Space Telescope, already had an idea, and there was a committee, chaired by Alan Dressler, preparing a report on "HST and Beyond". That report called for an infrared-optimized telescope to study topics from the early universe to the formation of stars and planets near home. It also called for an interferometer called the Terrestrial Planet Finder, to examine nearby stars for planets like our own Earth. So with this background, my creative juices were flowing, and so were those our colleagues. This initial phase of trying out wild ideas and hunting for ways to go far beyond anything ever done before is one of my biggest thrills. When Alan briefed our NASA Administrator, Dan Goldin, there was a real resonance, and Goldin told the January meeting of the American Astronomical Society that Alan's vision was much too small, and NASA would build a bigger telescope. Goldin got a standing ovation from the meeting. In the next few months, we started up two serious industrial/university partnerships to develop concepts that competed with the NASA concept, we had public briefings of the results, and we were well on our way. At the time, all the studies concluded that an 8 meter observatory could be built for the target price of $500 M in FY96 dollars, not counting civil service salaries, technology development, or the operations phase, but the designs were at the level of "viewgraph engineering" in the days when NASA was under the spell of "faster, better, cheaper," so it should not be surprising that later details have driven up the cost. We initiated technology developments for all the main inventions that were required for the mission, and those have all been successful. We negotiated a partnership with the European and Canadian space agencies, and when we finally chose a prime contractor (TRW, later purchased by Northrop Grumman), the observatory was named after James Webb, the second NASA administrator. Webb is a very appropriate honoree, as he is the person responsible for getting human beings to the Moon with the Apollo project, and he also insisted to President Kennedy that for the good of the Nation, there had to be a scientific research program at NASA and in universities. The National Academy's Decadal Survey ranked the JWST project as top priority in 2000, and thanks to this endorsement is one of the few large projects still continuing forward in NASA's science portfolio. I think the others will be revived as soon as budget can be found for them, since the need has not disappeared. At the moment, the JWST is in excellent technical shape. All the major technological developments have been completed to the required level, called TRL-6, which means they have been tested in the relevant environments. Also, the most difficult items to obtain, the mirrors, the detectors, and the microshutter arrays, are being fabricated with their final flight designs as I write this note. The JWST is now planned for launch in 2013. My role is called "senior project scientist," and I chair the science working group and ensure that the mission will meet the scientific requirements. Now, after 11 years of this project, it is quite mature, which means that huge teams of people are doing the serious work.

The JWST is not the only wild idea that I've been pushing forward. From conversations with Harvey Moseley came the concept for a far infrared interferometer to map the sky with the same image quality that we get with the Hubble Space Telescope; this mission is now called the SPECS, the Submillimeter Probe of the Evolution of Cosmic Structure as David Leisawitz named it. One of these days (but not very soon) it will fly. On another day, I talked with David Bennett at Notre Dame, and we created the idea of a satellite mission to find planets around other stars using the micro-lensing phenomenon. David took it seriously and has submitted several proposals for it, and I think it will fly one day too, because it does things no other planet-finding mission can do. Probably my wildest idea was to send a miniature telescope to the outer solar system to see the cosmic infrared background light directly, without interference from interplanetary dust. This idea was half-baked but it was fun to work on it, and I got a little money to have a young technician build a miniature radiative cooler. That part worked brilliantly but it wasn't enough for a mission. I still enjoy developing new mission concepts, and have recently been trying to persuade people to work on yet another way of hunting for planets around nearby stars.

Now, as I have passed the age of 60, and the Nobel Prize has recognized our COBE work, my life has changed again. I am giving many public lectures, to help the public understand the work we have done and hope to do in the future, and to inspire young people to be as excited about science as I am. I am also broadening my perspective one more time, trying to learn about the entire range of space science, and helping to guide NASA science towards the discoveries of the future. On April 2, 2007, I will take on the job of Chief Scientist of the Science Mission Directorate of NASA, so I will have the opportunity and responsibility to advise NASA on the proper balance of scientific programs from Earth science to cosmology. The panorama of amazing research programs is almost overwhelming, and I am looking forward to seeing it.

Any biographical statement would be empty without thanking the people who helped me through life. My parents, my sister, and my wife have all helped me immeasurably in finding my way through the challenges, and maintaining my faith in humanity despite all the disappointments that happen. My teachers in high school, college, graduate school, and my postdoctoral advisor Pat Thaddeus have led me to water and urged me to drink, and I have sometimes followed their advice. Their enthusiasm was contagious and I do my best to pass it on to my colleagues and to the public. My professional mentors at NASA have shown me how to work successfully and cheerfully with a giant organization full of talent. NASA's review panels have saved our projects over and over, though we often hate to hear their opinions, and I especially thank the people who told us when we were doing things wrong. It is so much better to know about it before we push the launch button! And I suppose it is obvious, but the technical infrastructure developed by our modern society, partly in response to the Soviet Union and its scientific and engineering accomplishments, has made all of this possible beyond any imagining in 1946 when I arrived on Earth.

FROM THE BIG BANG TO THE NOBEL PRIZE AND BEYOND

Nobel Lecture, December 8, 2006

by

JOHN C. MATHER

NASA's Goddard Space Flight Center, Greenbelt, MD 20771, USA.

ABSTRACT

NASA's COsmic Background Explorer satellite mission, the COBE, laid the foundations for modern cosmology by measuring the spectrum and anisotropy of the cosmic microwave background radiation, and discovering the cosmic infrared background radiation. I describe the history of the COBE project, its scientific context, the people who built it, and the scientific results. The COBE observed the universe on the largest scales possible, by mapping the cosmic microwave and infrared background radiation fields and determining their spectra. It produced conclusive evidence that the hot Big Bang theory of the early universe is correct, showed that the early universe was very uniform but not perfectly so, and that the total luminosity of post-Big Bang objects is twice as great as previously believed. The COBE concept was developed by a Mission Definition Study Team appointed by NASA in 1976, based on three competing proposals submitted in 1974. The COBE was built in-house by Goddard Space Flight Center, with a helium cryostat provided by Ball Aerospace, and was launched on a Delta rocket built by McDonnell Douglas. It is in a circular orbit 900 km above the Earth, in a plane inclined 99 degrees to the equator and roughly perpendicular to the line to the Sun. It carried three instruments, a Far Infrared Absolute Spectrophotometer (FIRAS), a Differential Microwave Radiometer with three channels (DMR), and a Diffuse Infrared Background Experiment (DIRBE). The helium cryostat cooled the FIRAS and DIRBE for 10 months until the helium was exhausted, but operations continued for a total of four years. Subsequent observations have confirmed the COBE results and led to measurements of the main cosmological parameters with a precision of a few percent.

I. SCIENTIFIC INTRODUCTION

A. CMBR Spectrum and the Big Bang

In the beginning was the Big Bang, so we now say with great certainty. The COsmic Background Explorer (COBE) satellite, proposed to NASA in 1974 and launched in 1989, provided very strong evidence for it: the spectrum of the cosmic microwave background radiation (CMBR) has the spectrum of an almost-perfect blackbody emitter at 2.725 ± 0.001 K, and the radiation is isotropic (the same in all directions) within 10 parts per million (rms) on angular scales of 7° and larger. This radiation is interpreted as the relic of an incredibly hot and dense early phase of the universe. In such a hot and dense phase, the creation, destruction, and energy equilibration of photons with one another and with all other forms of matter and energy would occur very rapidly compared with the expansion time scale of the universe. Such a state would immediately produce a blackbody radiation field. The expanding universe should preserve this blackbody spectrum, so measurement of any significant deviation from a perfect blackbody spectrum would either invalidate the whole idea of the Big Bang, or show that energy (e.g. from decay of primordial particles) was added to the CMBR after the rapid equilibration ended.

B. Isotropy

The fact that the radiation is isotropic to such a high degree is key evidence for its origin in the Big Bang. All the local systems that we know, from our Solar System to our Galaxy and the local cluster and supercluster of galaxies, are recognized because they are not uniform. Indeed, the isotropy itself was hard to explain, because it demands uniformity in the initial conditions of the great explosion of the entire observable universe. This was one of the great mysteries facing science in 1974.

C. Anisotropy

Moreover, according to the COBE, this radiation shows the imprint of the primordial density variations. On large angular scales the primordial radiation suffers a small gravitational redshift on exiting from denser regions (Sachs and Wolfe, 1967). These measured fluctuations have a nearly scale-invariant noise spectrum, in which the mean square variations are plotted against spatial (angular) frequency. The scale-invariant form of this spectrum was expected on the basis of general arguments (Peebles and Yu, 1970; Harrison, 1970; Zeldovich, 1972). When the theory of cosmic inflation was developed (Guth, 1981), it neatly explained the general isotropy of the radiation, since the currently observable piece of presumably infinite universe was at one time contained in a small region that existed for long enough to establish a uniform temperature.

D. Small angular scale anisotropy and primordial sound waves
On smaller angular scales, the measured fluctuation spectrum is dramatically modified, with a major peak at a typical angular size of a few degrees. According to the Big Bang theory, the universe became transparent when the temperature fell to about 3,000 K, around 380,000 years after the Big Bang. At that time, regions of the universe smaller than 380,000 light years would begin to exchange information and begin to erase or amplify the primordial fluctuations. We can describe these fluctuations as primordial sound waves in a multi-component fluid including photons (electromagnetic radiation), ordinary matter (baryons and leptons), dark matter (with mass but no electromagnetic interactions), and dark energy (causing the rate of expansion of the universe to increase).

E. Modern cosmology
Modern cosmology began in earnest with the recognition that galaxies are really far away and are made of billions of stars. This discovery depended on the continued production of ever larger telescopes, such as the four built by George Ellery Hale, each of which was the largest in the world for some years, and each making a wealthy donor proud. Einstein's theory of general relativity provided a way of computing the effects of gravity on large scales, and Georges Lemaître (1927, 1931) applied the theory to predict that the universe could not be static, but must be expanding or contracting. Einstein had introduced a constant of integration that could be adjusted to achieve a balance between the attractive forces of gravity and the "Λ constant," but the solutions were not stable. Edwin Hubble (1929) found that distant galaxies are receding from us and that the farther away they are, the faster they are going. This discovery, in the same year that the worldwide economy collapsed, changed cosmology from almost pure speculation to an observational subject. The apparent age of the universe was just the quotient of the distance divided by speed of recession, and the reciprocal of that number is called the Hubble Constant.

II. MY INTRODUCTION TO COSMOLOGY

A. Childhood
In the 1950's, when I was a child, the space age had not started yet, and there were famous debates between advocates of the Steady State theory and the Big Bang theory. In the 1940's, George Gamow was considering the consequences of the Big Bang idea, and with his graduate students Ralph Alpher and Robert Herman, was working on the question of the formation of the chemical elements. One key result was that only hydrogen and helium would come from the Big Bang, which means that the others had to be formed in nuclear reactions inside stars (Alpher, Bethe and Gamow, 1948). Even then it was already clear that we humans are made of recycled stellar material.

Their second key result was that the universe would be filled with the heat radiation left over from that great explosion. There were various estimates of its temperature, but in any case it would have been difficult to measure this radiation at the time.

By the 1950's, the modern age had begun in earnest. Transistors were invented, the H-bomb race began when the Soviet Union exploded a hydrogen bomb in 1953, and the US was worried. Then the Soviet Union launched the Sputnik in 1957. Suddenly, physicists and engineers were supposed to save the country from a really serious threat. Science fairs appeared in public schools, and government money flowed into civil defense and scientific education. I assembled a 5-tube shortwave radio kit, and I wanted to be a ham radio operator. Microwave technology grew to support the telephone network as well as radar, and in my rural neighborhood, a 4H club teaching children about electronics was sponsored by a small engineering firm. My father, one of the few PhD scientists in the county, had a Geiger counter. In high school, I went away to summer math and physics courses sponsored by the National Science Foundation. In a few short years, the US space program went from nothing to a declaration by President Kennedy that we would place a man on the moon within the decade.

B. College

By 1965, I was a freshman in college, and I had read one of George Gamow's (1947) popular books on the universe. I knew that the Big Bang might have produced a radiation field that filled the universe, so I wasn't surprised when it was discovered by Penzias and Wilson (Nobel Prize, 1978). Robert Dicke's group at Princeton was looking for the radiation at the same time. The story of the simultaneous publication of the results of the two groups has been widely told.

If the radiation is cosmic, it should have the spectrum of a blackbody and be equally bright in all directions (isotropic). Since the detections were made at relatively low frequencies, there was a contribution of emission from electrons in our galaxy that had to be measured and removed by some kind of modeling. At first, it was only possible to measure the Rayleigh-Jeans (long wavelength) part of the spectrum, and that part is a featureless power law, but at least the measurements were roughly consistent with a single cosmic temperature. There was also information from the measurements of the rotational temperatures of interstellar molecules, which could be accomplished by observing their absorption of starlight. It turned out that this measurement had already been made, by Adams and McKellar around 1941, but the significance was not recognized at the time. Then, in 1968, a rocket carried an instrument above the atmosphere, and found 50 times the brightness expected for the cosmic blackbody radiation (Shivanandan, Houck, and Harwit, 1968).

C. Graduate school

This was the situation when I started looking for a thesis project at the University of California at Berkeley in 1970. I met Michael Werner, then a new postdoctoral fellow working with Charles Townes (Nobel Prize 1964 for masers), and Paul L. Richards, a young faculty member with expertise in low temperature physics. They were starting up projects to measure the cosmic microwave background radiation in the wavelength range around 1 mm. The first was a ground-based measurement using a Fabry-Pérot interferometer to define wavelengths, and liquid helium-cooled far infrared detectors. We took this instrument to White Mountain in California and used it to measure the CMBR temperature at those wavelengths that were not too badly blocked by atmospheric emissions (Mather, Richards, and Werner, 1971). Following that project, Richards took a Miller Fellowship sabbatical to England and learned about the newly developed Martin-Puplett interferometer (Puplett and Martin, 1970). He developed a concept to fly such a device as a payload suspended below a high-altitude research balloon, above 99.5% of the atmosphere. This idea later developed into the spectrometer that flew on the COBE satellite. Richards returned to Berkeley and explained the idea to two of his graduate students, David Woody and me.

The instrument concept included the polarizing interferometer, a modern version of that invented by Michelson (Nobel Prize, 1907), immersed in liquid helium to keep it cold, a far IR detector (a bolometer), a cold reference blackbody, and a conical metal light collector that defined a 7 degree beamwidth on the sky and fed the light through a small hole into the instrument (Mather, Woody, and Richards, 1974). Compromises had to be made, due to the presence of a warm atmosphere in close proximity to the liquid helium. First, the conical antenna had to connect to a stainless steel reflector that made the transition from helium temperature (1.5 K) to atmospheric temperature. Second, a Mylar window kept the atmosphere out of the aperture. Third, a small warm calibrator body could be moved over the aperture to measure the sensitivity of the instrument. A NASA review panel visited the Space Science Laboratory at Berkeley in 1973, and we presented our story. The panel immediately told us that this instrument ought to be proposed as a space mission.

The instrument failed on its first flight but was retrieved in one piece. I wrote my thesis on the ground-based work and on the design of the balloon payload, and David Woody continued with the project. He built a test chamber, found the reasons for the initial failures, fixed them, and prepared the payload for re-flight. By now I had taken a postdoctoral position with Patrick Thaddeus in New York at the Goddard Institute for Space Studies (GISS). This time the flight was successful (Woody *et al.*, 1975).

III. ORIGINS AND DESIGN OF THE COBE

A. Initial Goddard concept

I joined Thaddeus's group to escape the perils of the CMBR field. However, a few months after I arrived at GISS at the end of January 1974, NASA issued an Announcement of Opportunity for new small satellites to be launched on Scout or Delta rockets. Thaddeus asked members of the lab for ideas. My thesis project had not worked right, but the main scientific difficulty was the atmosphere, and a space mission would be a lot better. Thaddeus was already very interested in the CMBR, and had measured its temperature using spectroscopy of interstellar molecules of CN, cyanogen (Thaddeus, 1972). He suggested that I call Rainer Weiss of MIT, Dave Wilkinson, one of the pioneers of the CMBR studies just a few miles away in Princeton, and Michael Hauser, who had just a few days before joined the Goddard Space Flight Center in Greenbelt, MD. We knew we would need a liquid helium cryostat, and we contacted Ball Aerospace in Boulder, CO to learn if it was possible to get one.

At that time, the Steady State theory still had strong advocates and ingenious defenders. The evidence for the age of the universe and for the age of the oldest stars seemed inconsistent. The evidence that the CMBR had a good blackbody spectrum was not very strong, and indeed most measurements were showing deviations from the prediction. There was not yet any serious theory for the fundamental anisotropy, but it was thought that the CMBR might not be the same brightness in every direction even if it were from the Big Bang.

We did not know we had competitors, but Luis Alvarez at Berkeley (Nobel Prize, 1968) had hired several people to work on this cosmic background radiation, including Richard A. Muller and George F. Smoot. Their team prepared a proposal for a single instrument to measure the anisotropy, much simpler than our Goddard proposal, and similar to concepts they were developing for receivers that would fly on U-2 aircraft and on balloons. Samuel Gulkis and Michael Janssen of the Jet Propulsion Laboratory in Pasadena also prepared a proposal.

NASA did not choose any of us immediately. There were about 150 proposals submitted altogether, covering a huge range of topics. NASA was very interested in doing the Infrared Astronomical Satellite (IRAS), in cooperation with the Netherlands and the UK. This mission would be a pioneering project, with a liquid helium cryostat in space, and new types of infrared detectors covering a huge wavelength range.

I started serious work on the flared horn concept. The balloon-borne thesis experiment used a design with an abrupt junction between the beam-defining cone and a stainless steel reflector to protect it from stray radiation from angles far off the line of sight. My new idea was to make a gradual transition, with a curved flare like those in musical instruments. The mathematics for this had been developed already, by Joseph Keller (1962), under the title of the "geometrical theory of diffraction." This approach said that light rays propagate in straight lines except at boundaries and obstacles, and that there were a variety of scattering and attenuation coefficients that could be computed for the boundaries and obstacles. The waves do diffract around curved surfaces, but they are attenuated exponentially as they go. This was just what I needed to protect the input of the instrument over a wide range of wavelengths (Mather, 1981; Mather, Toral, and Hemmati, 1986).

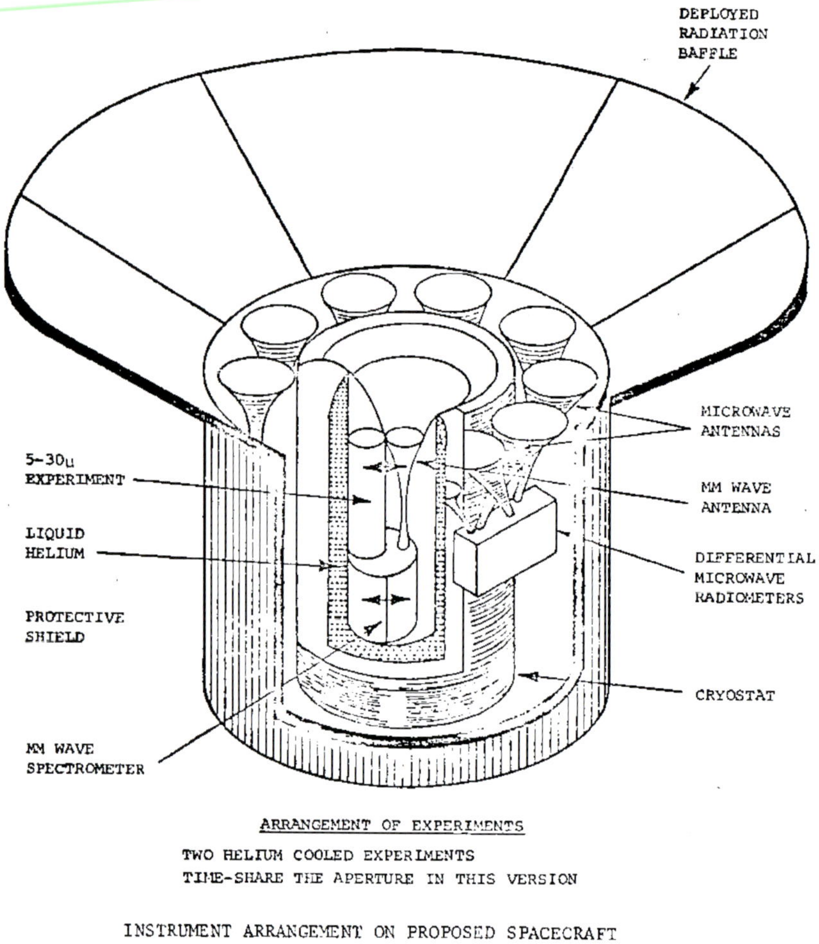

FIGURE 1

Figure 1. Original concept for COBE as proposed by Goddard group in 1974.

B. Building the COBE Team

In early 1976, Michael Hauser offered me a job at the Goddard Space Flight Center (GSFC) in Greenbelt, MD, and thought the COBE idea would go forward. In any case, it was worth a try.

In late 1976, NASA started a study of the COBE idea, but chose a new team composed of six members of the three competing groups. There were four members of my group, with me, Michael Hauser, David Wilkinson, and Rainer Weiss, plus George Smoot from the Berkeley group, and Sam Gulkis from the JPL group. We were to work with Nancy Boggess of NASA Headquarters, who was responsible for all of infrared astronomy at NASA, and a team of engineers at Goddard Space Flight Center, to define a new mission. We decided that one of the original four instruments on the Goddard concept would have to be omitted, and that the microwave radiometers should use corrugated horn antennas. We elected principal investigators and proposed to NASA Headquarters that these people be approved and given responsibilities for the individual instruments. These assignments were: G. Smoot, Differential Microwave Radiometers; M. Hauser, Diffuse Infrared Background Experiment; and myself for the Far Infrared Absolute Spectrophotometer. We chose Rainer Weiss as Chair of the Science Working Group, and NASA assigned me the job of Study Scientist, to coordinate the scientific requirements with the engineering teams.

We prepared our report with Martin Donohoe as engineering lead. Our study had to show that the mission could not be accomplished in any other way. We discussed whether a mission on board the newly approved Space Shuttle could possibly meet the scientific requirements, and we argued that all three instruments were really essential. We were soon directed to redesign our mission to use the Space Shuttle as a launch vehicle.

After this initial round of competition, we were told to continue and were given some funds to spend and people to work with. The Goddard engineering team assigned to us was led by Jerry Longanecker, project manager for the IUE, the International Ultraviolet Explorer. This engineering team was very experienced and quick to understand the challenges we brought them.

Around 1978, it was decided that Goddard would build the entire COBE mission in-house, which means that civil servants and neighborhood contractors would do most of the work. The advantage for Goddard was that this project was an excellent way to recruit bright young engineers and train them on a real space mission. The advantage for the COBE mission was that it would enable the scientists and engineers to work very closely together, without the impediments of contract management and physical distance. This was very important for the creative process and the give-and-take of solving problems. However, part of the plan was that the COBE had the lowest priority of all the Center's major projects.

Our team membership changed with time, but two photographs (Figs. 4 and 5) show the engineering and science teams as of 1988. The captions list the team members and their roles. Roger Mattson was our Project Manager and Dennis McCarthy was the Deputy for technical matters.

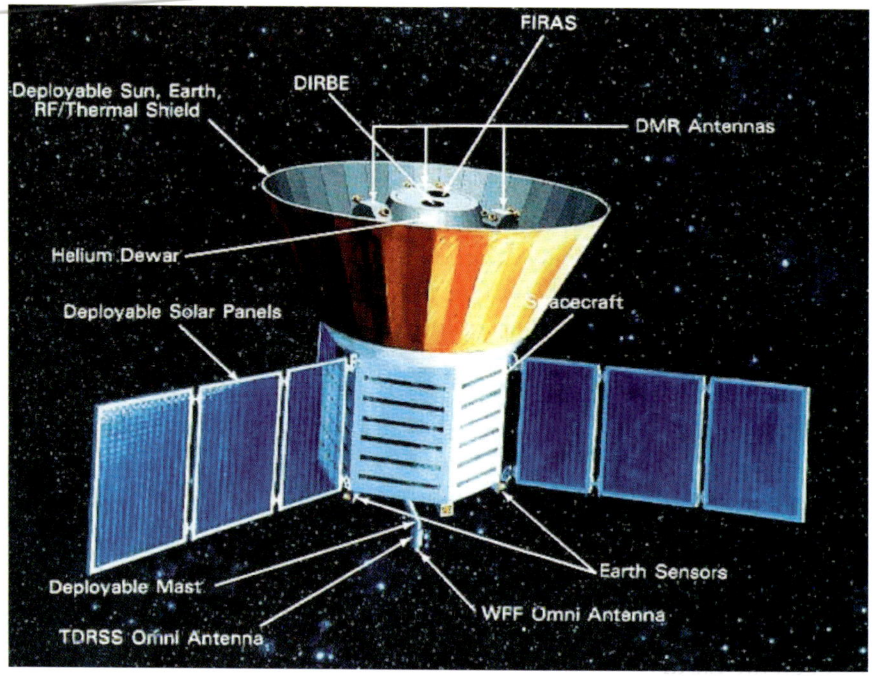

Figure 2. Artist's concept of COBE as flown. COBE was in orbit 900 km above the Earth, with the Sun to the side and the Earth below.

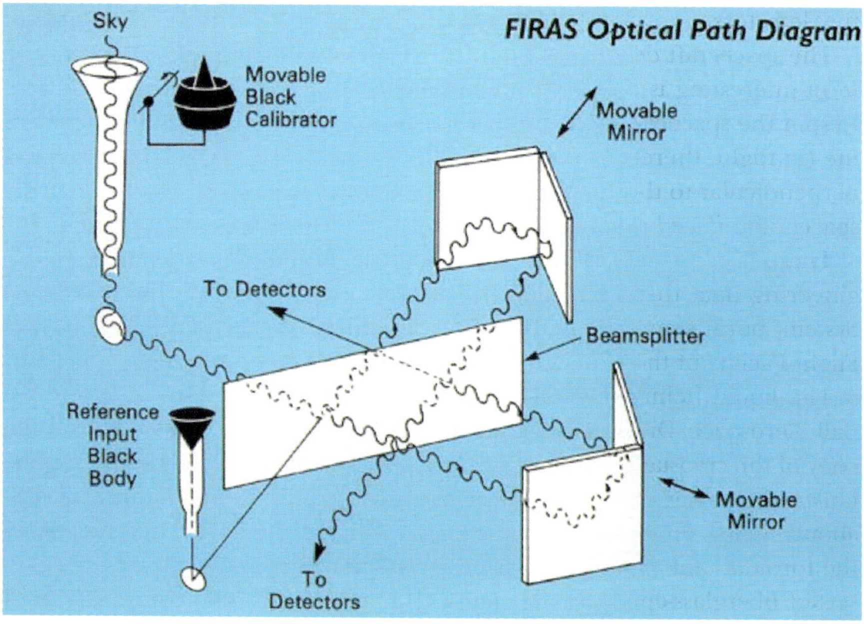

Figure 3. Concept for the Far Infrared Absolute Spectrophotometer. The movable black calibrator emits the same spectrum of radiation that is received from the sky.

C. Mission Concept and Design

The design included a protected region for the instruments, surrounded by a reflective shield. In the center is a liquid helium cryostat containing two instruments, a Far-Infrared Absolute Spectrophotometer (FIRAS), and a Diffuse Infrared Background Experiment (DIRBE). The Differential Microwave Radiometer (DMR) receivers were mounted in four (later three) boxes just outside the cryostat. Solar panels surrounded the spacecraft, and massive cylindrical pins projected from the sides to attach the COBE to the Space Shuttle. Inside, hidden from view, were tanks containing hydrazine fuel, to raise the orbit from the maximum height the Shuttle could reach. The orbit chosen for the COBE was circular, at 900 km altitude, with the orbital plane inclined 99° to the equator, and nearly perpendicular to the Sun line. The altitude and inclination are chosen in combination so that the Earth's equatorial bulge causes a torque on the orbit plane, just sufficient to make it precess a full revolution once per year and follow the Sun. With this choice of orbit, the COBE stays in full sun almost all the time, and the Earth's limb is only 60° from the nadir. However, for about 3 months each year, the combination of the orbital plane inclination (9° past perpendicular) with the tilt of the Earth's spin axis of 23.5° from the ecliptic means that the COBE flies through the Earth's shadow for up to 20 minutes per orbit, and flies between the Earth and the Sun for an equal period of time. Both events require attention. During shadow, the COBE must draw electric power from batteries, and some heat must be provided to those parts that would cool off too rapidly. During the opposite part of the orbit, the Earth shines over the top of the shield and illuminates the instrument apertures, causing stray light to reach the detectors.

The spacecraft design for the COBE had some very unusual features. The most interesting is perhaps the attitude control system, which was required to spin the spacecraft around its symmetry axis at about 1 rotation per minute (in flight, there was one spin per 72 sec), and keep the spin axis roughly perpendicular to the Sun line (actually 94° away), and roughly vertical, as the spacecraft orbited the Earth.

The spacecraft had a hybrid command and control system: it sent its engineering data through the TDRSS, the Tracking and Data Relay Satellite System, but it sent its scientific data directly to a ground station at Wallops Flight Facility of the Goddard Space Flight Center.

The liquid helium cryostat was a great technological achievement of the Ball Aerospace Division in Boulder, Colorado. Ours was a nearly-identical copy of the cryostat developed for the IRAS. The design has an outer vacuum tank with a cover that can be ejected after launch, a toroidal tank filled with about 500 liters of liquid helium at launch, and an instrument chamber inside the toroidal tank. The helium tank is suspended inside the vacuum shell by sets of fiberglass-epoxy straps in tension, and surrounded by concentric aluminum shells cooled by tubes carrying the escaping helium gas. Between the concentric shells are multiple layers of aluminized insulation blankets.

Figure 4. COBE Science Working Group. Eli Dwek is missing. Left to right, Ed Cheng, Dave Wilkinson, Rick Shafer, Tom Murdock, Steve Meyer, Chuck Bennett, Nancy Boggess, Mike Janssen, Bob Silverberg, Sam Gulkis, John Mather, Harvey Moseley, Phillip Lubin, Ned Wright, Mike Hauser, George Smoot, Rainer Weiss, and Tom Kelsall.

Figure 5. COBE engineering team leadership. Left to right, top photo, Don Crosby, Jeff Greenwell, Bill Hoggard, Roger Mattson, Ernie Doutrich, Herb Mittelman, Eileen Ferber, Bob Schools, Joe Turtil, Maureen Menton, Bob Sanford, Mike Roberto. Bottom photo, Pierce "Lee" Smith, Earle Young, Dennis McCarthy, Dave Gilman, Bob Maichle, Chuck Katz, Steve Leete, Bernie Klein, Loren Linstrom, Tony Fragomeni, John Wolfgang, and Jack Peddicord.

D. FIRAS

The purpose of the FIRAS was to determine whether the Big Bang radiation has the spectrum of a perfect blackbody. Even small deviations from perfection would signify that the universe is not simple, that there is some new phenomenon to be understood. Since the radiation comes almost equally from all directions, the instrument must be absolutely calibrated. Since there are processes in galaxies, and in our own Milky Way Galaxy, that produce radiation at wavelengths overlapping the CMBR, it is essential to map the radiation and see what part can be correlated with identifiable local sources. The main expected sources are dust grains, heated by starlight in the Milky Way, and interstellar molecules, atoms, and ions, heated by starlight and by collisions with other components of the interstellar medium. At longer wavelengths than those observed by the FIRAS, interstellar electrons, colliding with protons and spiraling in the magnetic field, are brighter than the CMBR, but all these local sources are easily recognized by their concentration in the plane of the Milky Way.

The FIRAS was an upgraded version of Paul Richards' original balloon-borne instrument. The main improvements were: 1. The whole instrument was in space, above the atmospheric emission and potential condensation on the optics. 2. The light gathering power (étendue) was increased, to collect more light, and extend the sensitivity to longer wavelengths. 3. The interferometer was symmetrized, providing access to both input ports and both output ports, so that it could be used as a differential comparator between the inputs, and so all the available input light could be detected. 4. The primary input from the sky could be entirely covered by an accurate external blackbody radiator, effectively a simulator of the Big Bang radiation, called the XCAL, the external calibrator. 5. The secondary input was fed by a smaller blackbody radiator, with an adjustable temperature to balance the input from the sky, called the ICAL, the internal calibrator. 6. The conical light concentrator of the balloon payload was improved to use a Winston cone, a non-imaging parabolic concentrator with a better beam profile, called the Sky Horn. 7. The stainless steel reflector that joined the balloon payload's input cone to the warm environment was replaced by a smoothly flared bell resembling that of a musical instrument. 8. The temperature of the Sky Horn and its flared section was controllable over a wide range of temperatures, as was the matching concentrator (the Reference Horn) on the secondary input. 9. The wavelength range was extended, and divided into two bands, Short and Long. 10. The detectors were improved. 11. The observing time was increased to 10 months instead of a few hours, and the entire sky was mapped with a single instrument.

The main sources of improved accuracy were the differential mode of operation, reducing the dynamic range (contrast) between the signal level and the detector noise, and the accurate external blackbody calibrator.

We have received many questions about the calibrator. According to Kirchhoff's laws, the emissivity of a body is the same as the absorptivity. The emissivity is the ratio of the emission from the actual body to that from an ideal blackbody at the same temperature. The absorptivity is the fraction of incident radiation that is absorbed. Hence, an ideal absorber is an ideal emitter. The chief aim of the calibrator design was to ensure that the only rays reaching the spectrometer input came from the calibrator, and not from other locations. We define the spectrometer input as the junction between the calibrator and the input concentrator. There are three possible sources of radiation crossing this boundary that do not come from the calibrator emission. These are: 1. Emission from the concentrator that strikes the calibrator and is reflected back into the acceptance angle of the Sky Horn. 2. Emission or reflection from the spectrometer that comes towards the calibrator and is reflected back towards the spectrometer by the XCAL. 3. Leakage of radiation from the sky or the calibrator support hardware or other objects above the XCAL, around the junction of the calibrator with the concentrator, or through the XCAL material. We analyzed all of these possible errors and as far as we know we avoided them at the level of a few parts per million. The details are given by Mather *et al.* (1998).

In order to achieve this level of performance we needed a highly absorptive material. We chose Eccosorb®, an epoxy filled with very fine iron powder. The epoxy has a refractive index of about 2, so that its surface reflections, about 10% at normal incidence, are not negligible (Hemmati, Mather, and Eichhorn, 1985). We therefore designed the XCAL as a re-entrant cone, like a trumpet mute, so that a light ray entering the cone would have to suffer at least 5 specular reflections before it could exit again.

Thermal gradients produce a second-order effect and we believe the effect was less than a few parts per million on the spectrum. However, for comparison with measurements by other instruments, we needed to know the absolute temperature very well. Our thermometers were germanium resistance thermometers traceable to the National Institute of Standards and Technology, but after launch there was doubt about the stability of their calibration, at the level of a few millikelvin. Fortunately, there were other methods to confirm the temperatures after launch.

To prevent leakage of radiation around the edge of the calibrator, where it meets the Sky Horn, the calibrator was provided with two rings of aluminized Kapton leaves, adjusted so that they barely made contact with the Sky Horn.

The detectors for the FIRAS were kept as cold as possible with copper cooling straps that reached directly from the detectors to the liquid helium cryostat attachment fittings. They were composite bolometers, built at Goddard. Each bolometer had a very thin diamond sheet, with a partially conductive coating optimized to absorb the incoming radiation, and suspended by thin Kevlar fibers. Attached to the diamond was a tiny cube of silicon, doped to become a temperature-sensitive resistor. Incoming radiation was absorbed by the coating on the diamond, converted to heat, and conducted to the silicon thermometer. A DC voltage was applied to the thermometer through a resistor, and the voltage on the thermometer was amplified by a JFET transistor nearby. The transistor would not work at the 1.5 K temperature of the rest of the instrument, so was suspended inside a small chamber on Kevlar threads, and heated electrically to around 70 K.

In developing concepts for the detectors, I pursued an idea I had from graduate school, to establish a convenient theory for the noise and ultimate sensitivity of bolometers. I worked on the manuscript while my future wife Jane was teaching ballet; I was driving her to work because we had broken her arm doing the samba. This work developed into a series of papers (e.g. Mather, 1984) which have ended up being my most cited publications. The reason was unexpected: our bolometers were good detectors for every kind of radiation, including cosmic rays. Harvey Moseley saw this as an opportunity, knowing that the detectors could be greatly improved. The improved detectors have been flown in space on the X-ray mission Suzaku, and they worked beautifully, able to measure the energy of a single 6 KeV X-ray photon with a precision of a few eV.

The interferometer mirrors had to move precisely, smoothly, without friction, for millions of strokes on a parallelogram linkage, with leaf springs at the joints, driven by a solenoid. This easily met the friction and lifetime requirement. The position of the mirrors was measured precisely by a scale. The scale was read out optically through fiber optics leading to light sources and detectors outside the cryostat. This led to a problem during flight, when cosmic rays hit the fiber optics and caused light flashes that confused the position measurement and led the control circuit to drive the motor hard against a physical stop. Fortunately, a timer circuit had been implemented to protect the equipment from such an event.

E. DMR

The DMR instrument was not inside the cryostat but shared the objective of measuring the CMBR. I trust that the story of this instrument will be more fully told by my co-recipient, George Smoot. Its purpose was to measure the anisotropy, the difference in brightness of the CMBR across the sky. When the COBE was first proposed, there was no serious theoretical prediction for the amount or pattern of such variations, but our view was that whatever it was, it had to be measured. Over the 17 years from proposal to our first data release, predicted amplitudes of the variations declined exponentially with time, as new equipment repeatedly failed to find anything but the dipole term due to the Earth's motion.

Not being guided by theory, we set the objective of measuring the anisotropy as well as our environment would allow. As with the FIRAS, there are local sources of radiation in the Milky Way Galaxy that are bright compared with the anisotropies we sought to measure. Our approach was to map the sky at three (originally four) frequencies, with identical receiver antenna patterns, and take advantage of the fact that only the cosmic microwave background radiation would have the spectrum of a blackbody. The electrons in our galaxy produce two kinds of radiation, by colliding with protons, and by spiraling in magnetic fields. Both of these types of radiation are strong at long wavelengths and drop rapidly as the wavelength gets shorter. The dust in our Galaxy has the opposite sort of spectrum and its emission spectrum is strongest at shorter wavelengths than the CMBR. In any case, measurements at multiple wavelengths are required to model and compensate for these emissions. We chose three frequencies: 31.4, 53, and 90 GHz.

The central idea of the instrument was the Dicke switch. This is a device that could switch the input of the microwave receivers rapidly between two sources, so that a difference in brightness seen from them could be recognized by a lock-in amplifier. For mapping the CMBR, it is possible to just compare two antennas pointed in different directions. Our special observing scheme swept the two antennas rapidly around the sky, measuring all possible pairs of directions separated by the fixed angle between the two antennas. Our spinning spacecraft, with its spin vector smoothly rotating around the Sun line on each orbit, and gradually moving around the Sun through the year, was an excellent solution. Over the course of the mission, the radiometer would observe hundreds of millions of differences in brightness between points on the sky 60° apart. Then, a least-squares fitting computer program would build a map from them that best represents all the data, including detailed modeling of the systematic errors.

The particular challenges for this instrument were sensitivity and immunity to systematic errors. The microwave receivers available in 1974 were almost all made with diode mixers, microwave circuits that combine the signal from the sky with a local oscillator signal to produce a new signal at an intermediate frequency. This intermediate frequency signal can then be amplified and its intensity measured. As the COBE design progressed, improvements in the mixers were made, and eventually we decided to sacrifice one of the original four frequency channels to gain the resources to use the new technology. The improvement was based on cooling the receivers. In our case there was plenty of cooling available, since the receivers were protected inside the sunshield and had to be heated to keep them warm.

Some systematic errors were obvious. First, the Dicke switches were made with ferrite beads activated by magnetic fields, and there was sensitivity to external magnetic fields from the Earth and from the magnetic torquer bars. We provided magnetic shielding, but we knew that we would have a residual problem. Second, the receiver antennas were susceptible to interference coming from other directions. We chose corrugated horn antennas, and learned how to make them so precisely that the stray radiation sensitivity was very small. We had to measure this effect after launch and compensate for it. The major source of stray radiation was the Earth, which is 10 million times as bright as the cosmic fluctuations that we eventually found. The Earth is hidden from the receivers most of the year, but some of its emission diffracts over the edge of the sunshield and reaches the antennas. Third, the receivers and switches were sensitive to temperature, and presumably to power supply voltages, and all sorts of other minor disturbances. Since we were looking for signals remaining after processing hundreds of millions of observations, we had to devise ways to find very small effects that could have ruined the data. Most of these were found and fixed before launch, but some had to be measured and compensated in software after launch. During the months when the Earth rose slightly above the plane of the sunshield for part of the orbit, and the satellite passed through the Earth's shadow on the opposite side of the orbit, the residual thermal, stray radiation, and other effects were sufficiently large that the data were not entirely trustworthy.

F. DIRBE

The DIRBE instrument was designed for a different purpose: to search for and measure the diffuse infrared background light at shorter wavelengths than the CMBR. The accumulated light of distant galaxies should produce a nearly-uniform glow in the sky. If the universe were infinite, stationary, and uniform in both time and space, then every line of sight would terminate on the surface of some star, and we would be bathed in light as though we were just inside some such star. The expanding universe gives only a finite observable volume and a finite time, and the most distant parts are highly redshifted. Nevertheless, the cosmic glow is one of the most important traces of the distant universe, the cosmic reservoir of lost photons. Measuring it would tell us about those faint, most distant early galaxies, even if no telescope could ever observe them.

The main obstacle to measuring the diffuse IR background light is our local astrophysical environment. There are several bright sources, beginning with the interplanetary dust in our Solar system. This grayish dust reflects some sunlight and absorbs the rest, coming to a temperature around 200 K, and re-emitting the light at mid-IR wavelengths. This dust is visible to the naked eye as the Zodiacal Light, and is much brighter at infrared wavelengths. Farther out, the interstellar dust behaves in similar ways. Its temperature depends on its distance from stars, and ranges from a few K to much higher values in shock waves or near stars. Very small dust grains are also heated momentarily to very high temperatures, up to 1,000 K or more, by absorption of individual visible or UV photons, or by cosmic ray impacts.

The observational strategy for understanding and compensating for these foreground sources was to measure at as many wavelengths as possible (10), over as wide a wavelength range as possible (from 1.2 to 240 μm), to measure the polarization at the three shortest wavelengths because the scattered sunlight is polarized, and to measure over as wide a range of angles from the Sun as possible because the zodiacal light is strongly concentrated towards the Sun and towards the ecliptic plane. Our range of angles was from about 64° to 124°, achieved by placing the DIRBE line of sight 30° from the spin axis of the COBE. The wide wavelength range required four different detector technologies: InSb photovoltaic detectors from 1 to 5 μm, Si:As photoconductors for 12 and 25 μm, Ge:Ga detectors for 60 and 100 μm, and bolometer detectors for 140 and 240 μm.

The instrument also had to be designed to be absolutely calibrated. To achieve this we needed the following features: 1. A dark interior of the instrument, representing zero signal. 2. A modulator, switching the instrument beam rapidly back and forth between the sky and the interior of the instrument, the infrared equivalent of the Dicke Switch. 3. A calibrator body that could intercept the instrument beam and replace it with zero emission. 4. A light source that could establish the gain of the instrument and verify its stability with time. 5. A precise measurement of the beam profile of the instrument, to compare calibration from standard stars with a diffuse background surface brightness. 6. Detailed understanding of the response to standard stars. 7. Suppression of stray light from bright objects (such as the Moon, Earth, Sun, Galactic Center, and Jupiter) outside the field of view.

The optical design for the DIRBE was optimized for these purposes. The aperture was only 20 cm, but the field of view was large (0.7°), to optimize sensitivity for the diffuse radiation. The DIRBE was enclosed in the cryostat, so its interior was truly cold and dark. The modulator was a specially-designed tuning fork "chopper" with vanes that opened and closed at 32 Hz. The telescope was a Gregorian design, so that field and pupil stops ahead of the chopper could limit stray radiation. A carefully designed baffle system in the telescope tube intercepted light that might be scattered into the instrument, a shiny cone at the entrance to the telescope tube intercepted and reflected away any light coming in from angles just grazing the top of the cryostat, and a tightly fitted cover protected the telescope from molecular or dust contamination before reaching orbit. All of the detectors had the same field of view, so that relative colors of the detected objects would always be correct. The field of view was square, so that a star passing through the field would always produce about the same response, regardless of exactly where it was in the field.

The DIRBE instrument was also used to provide accurate pointing information. As the COBE spins, the DIRBE beam sweeps rapidly around the sky, and the precise timing of signals from bright stars passing through the beam is available. With an algorithm developed by science team member Edward L. (Ned) Wright, these timing signals were combined with a model of the spin of the COBE and its other pointing sensors to give accurate pointing information.

IV. REBUILDING AND LAUNCHING THE COBE

The COBE project was officially approved in 1982, when it became clear that the IRAS was going to fly. The IRAS cryostat was extremely challenging, and the COBE depended completely on its success, since we had almost the same design. Also, the NASA budget was very tight in those days, with cost growth in the IRAS, the Hubble Space Telescope, and other projects. The IRAS was at last launched successfully on January 25th, 1983 and our way seemed clear. Michael Hauser was a member of the science working group for the IRAS and was responsible for the data analysis, so he was very familiar with the technological details and the scientific results of the IRAS. The IRAS produced some very remarkable surprises: many nearby stars were surrounded by dust clouds, like our own zodiacal light but far brighter; and distant galaxies were extremely bright at far infrared wavelengths, sometimes 100 or 1,000 times brighter than they are at visible wavelengths. Our project proceeded with many minor revisions and major review meetings. Our budget requirements also grew, but Headquarters did not have the extra funds to give, so we took many risky decisions and cut some corners. We were lucky about them.

By January of 1986, we were assembling flight hardware, and a full-scale plywood mockup of the spacecraft was built for setting up the electronic boxes and the electrical cables. Then came the fateful frozen day in Florida when the Space Shuttle Challenger and our 7 astronauts perished in flames. For us on the COBE team, and for most of NASA, the future looked bleak. Dennis McCarthy, the Deputy Project Manager, started looking around for alternate launch vehicles. NASA had been forced to abandon its trusty Delta launchers and all the other one-shot rockets in favor of the Shuttle, so we had nothing ready. Fortunately, there were still spare parts for the Delta rockets, and it was possible to assemble a complete set. However, the Delta could certainly not carry our payload as it was. The COBE was far too large and weighed too much. Our engineering team was not deterred. The COBE was proposed in 1974 for a Delta launch, and the Shuttle design we had was using 5000 lb of fuel and physical structure that would not be needed if a Delta rocket could take us all the way to the needed orbit. In the end, it was just barely possible to launch the COBE on the Delta.

One really new design was now needed though: the sunshield would have to be folded up and deployed after launch. The unfolding would be propelled by spring energy after a circular strap around the folded shield was cut by an explosive charge.

One minor change was then needed for the DMR instruments. The available space between the new folding sunshield and the cryostat shell was not quite enough to fit the microwave receiver boxes without modification. New box designs and some new waveguide hardware were needed, but this was also possible.

Now that we had a new mission concept, we presented it to NASA Headquarters. It was quickly recognized that the COBE could now be NASA's first scientific mission to fly after the Challenger explosion. We were given the green light in late 1986, and urged to be ready for launch as soon as possible. Suddenly our project went from the bottom of the priority totem pole almost to the top, behind the Hubble Space Telescope. We began round-the-clock work, as much as people could stand, to finish the redesign, rebuild the hardware, and fix all the known problems with the instrumentation.

Our payload was finally ready, and was loaded into a huge truck and driven around the Capital Beltway to Andrews Air Force Base. There, the truck was driven right into a waiting C5-A transport aircraft, and flown to the Vandenberg launch site in California. At the launch site, final assembly was completed, and final tests were made. On October 17th 1989, the Earth trembled with the San Francisco earthquake, but COBE was bolted down that day, because two of our engineers had gone off to be married.

On the evening before November 18, 1989, we were ready to launch. Ralph Alpher and Robert Herman had come to watch the launch of the mission that would test their theory. In the early morning of the launch day, the main question was about the winds. If the wind direction and speed changes too rapidly with altitude, the rocket control system will be unable to compensate and the rocket can break up. Weather balloons were launched and tracked to determine the wind shear, and the assembled crowds stood in a cold and dark field 3 km away from the launch pad. We watched the rocket ascend, dropping its spent booster rockets, and go out of sight, in just a few minutes. A few minutes later, the high altitude winds twisted the trail of rocket exhaust into contorted loops, making quite visible the hazard that the Delta had just traversed.

The COBE was not heard from again for about an hour, when it passed over the tracking station in Alaska, and it was alive and well. In the first few days it faced new hazards. First, the outside of the cryostat was cooling off rapidly. We worried that if it got too cold, the cover might not come off. Second, one of the gyros in the pointing control system failed a few days after launch. Fortunately our engineering team had designed the system to keep right on going. Then, we re-discovered Antarctica. The power system was designed to cope with a certain amount of sunshine, but we had not remembered how much sunlight would be reflected up to the COBE by the ice in Antarctica. Next, trapped electrons and protons in the Van Allen belts disrupted the functioning of the mirror control electronics for the FIRAS, but we commanded the FIRAS not to run there.

The liquid helium in the cryostat lasted about 10 months. The end was abrupt, and within a minute the temperature was rising rapidly. We continued to operate the DIRBE because its short wavelength detectors, made with InSb, worked just fine at the higher temperature. The interior of the cryostat ended up around 60 K. The DMR, which did not use the helium, was operated for a total of 4 years.

The fate of the satellite was to be used for communications practice. It is still in the same orbit, and still spinning in its proper orientation, as it needs no fuel to do so, and the guidance system is highly redundant and reliable. It was estimated that the orbit will decay in about 1,000 years.

V. DATA ANALYSIS AND INTERPRETATION

The command and control of the spacecraft and the analysis and interpretation of the data required up to 100 software engineers and scientists. All three instruments required special analysis to deal with instrument systematic errors and calibration, and then with understanding the local astrophysical environment.

A. FIRAS

The first instrument to produce scientific results was the FIRAS. We adjusted the temperature of the internal reference body to null the signal from the interferometer as well as possible. The depth of the null was a direct measure of the match between the internal body and the spectrum of the CMBR. Even before we had a precise calibration, we knew immediately that the match was excellent. With a rough calibration, we added the difference we observed to the Planck function for the matching temperature, and plotted the result. Not yet knowing the error bars, we assumed 1%. When I presented this spectrum at the January 1990 meeting of the American Astronomical Society in Crystal City (a suburb of Washington, D.C.), the audience of over 1,000 people rose in a standing ovation. None of us on the COBE team anticipated this response, since the spectrum was not a surprise to us.

In retrospect we now appreciate the reason for the enthusiasm. Not only was the spectrum beautiful to look at, but at one stroke it banished the doubts of almost everyone about the Big Bang theory. For so many decades, the intense combat between the Big Bang and Steady State advocates had continued, and for so many years, a series of small discrepancies between theory and measurement had been explained by ingenious people. Now it was over, although some Steady State proponents do not agree. The Big Bang theory was safe, and the universe was simple, simple enough for theorists to go on to the next problem.

Only a few weeks after the COBE was launched, Herb Gush and his team from the University of British Columbia launched a sounding rocket with their version of the FIRAS instrument (Gush, Halpern, and Wishnow, 1990). This payload had been flown several times, each without success, but this time it worked. Their results agreed with the FIRAS results, with the same temperature within the error bars, and showed no deviation from the blackbody spectrum either. With a little better luck they might have stolen some of our thunder years before the COBE was launched, but to achieve our final conclusions we really needed the all-sky maps that only COBE could provide.

Our spectrum plot now graces many astronomy textbooks, but people still ask one elementary question: why does the plot not look like the ones in other textbooks? The key point is that our plot gives the intensity versus frequency in reciprocal centimeters, the number of waves per centimeter. Textbooks usually plot the intensity versus frequency in Hz, or versus wavelength, usually in micrometers. The plots do not appear to match, not only because the scales are different, but because the differentials are different: the plots give the power per unit area per frequency interval or per wavelength interval, and those intervals also have to be converted. When these conversions are done, the plots agree.

The full calibration and analysis of the FIRAS data took many years. The systematic errors that we discovered in the data were not complete surprises, but developing accurate models for them required a least-squares fit with thousands of parameters, a few of which were critically important. Dale Fixsen was the main architect of this process. The main errors to be compensated were: 1. The cosmic rays that hit the detectors produced voltage impulses that had to be detected and removed. 2. Temperature variations from many causes caused the detector gain to vary. 3. There was a small amplitude vibration in the mirror mechanism, due to torsion in the parallelogram linkage, that was excited by the servo circuit. Although this was minimized in the servo by a notching filter at 57 Hz, there was a small residual effect on the shape of the interferograms. 4. The internal reference body was not a very good blackbody. 5. The thermometer calibrations apparently changed after they were done, years before launch, and the three thermometers on the external calibrator did not agree at the expected millikelvin tolerance. 6. There was a small optical effect in which radiation could make multiple passes through the interferometer, and appear to be modulated at twice or three times the proper frequency. Fortunately, the instrument had 4 detectors, two on each side and two in each frequency band, and it had two different stroke lengths, for a total of 8 different ways to observe. Comparison of the multiple detectors and stroke lengths gave many ways to understand and detect the errors. Observation of interstellar spectrum lines, particularly that of [C II] at 157.74 μm, gave ways to confirm the absolute frequency scale, and hence to confirm the temperature scale (Fixsen and Mather, 2002).

The understanding and compensation for the local astrophysical sources was also difficult. One result was that the dust in the Galaxy is almost all at about the same temperature and has about the same spectrum, but not quite. There are also some directions in the Galactic plane where there are clearly multiple dust clouds with different temperature in the same line of sight. Our analysts were amazed to see that some interferograms appeared to be "contaminated" with sine waves. These were just the expected response to bright interstellar spectrum lines. The fine structure line of ionized carbon [C II] at 157.74 μm is by far the brightest we saw, and carries about 0.3% of the total luminosity of the Milky Way. We observed the [N II] line at 205.178 μm for the first time; it was then observed in the laboratory. Other lines from CO, [C I], and H_2O were also seen (Bennett *et al.*, 1994). The 157.74 μm line is so bright that we were able to measure the differential Doppler shift of the line due to the rotation of the Galaxy, even though our spectral resolution was very modest.

The main measurement is that the spectrum of the CMBR matches a blackbody at 2.725 ± 0.001 K with an rms deviation of 50 parts per million of the peak brightness. The interpretation of that result is that less than 0.01% of the energy of the CMBR was added to it after the first year of the expanding universe. Energy added before that time would just change the temperature of the radiation (Wright *et al.*, 1994).

Energy added between redshifts of 10^5 to $3\text{x}10^6$, roughly the first 1,000 years, would give the radiation a modified spectrum with a chemical potential μ, as worked out by Zeldovich and Sunyaev (1970). In this case the photon mode occupation number is $\eta = 1/(e^{x+\mu}-1)$, where $x = h\nu/kT$, and h is Planck's constant, ν is the oscillation frequency, k is Boltzmann's constant, and T is the temperature. Our measured value (1994) was $\mu = (-1 \pm 4) \cdot 10^{-5}$, or $|\mu| < 9 \cdot 10^{-5}$, with 95% confidence.

Radiation added later would give a CMBR spectrum that is a mix of blackbodies at a range of temperatures, parameterized by y, as described by Zeldovich and Sunyaev (1969). Here $y = (1/m_e c^2) \int k(T_e - T_\gamma)d\tau_e$, where m_e is the electron mass, c is the speed of light, T_e is the temperature of the scattering electrons, T_γ is the temperature of the CMBR at the time, and $d\tau_e$ is the differential opacity of the scattering electrons. The distortion of the spectrum produced by this kind of mix of blackbodies is described by $dS_y/dy = T_0$ $[x \coth (x/2) - 4]\, dB_\nu/dT$, where S is the spectral brightness, B is the Planck function, and T_0 is the average temperature. We found $y = (-1 \pm 6) \cdot 10^{-6}$, a very small number.

There is a long list of hypothetical energy sources that had been used to explain prior measurements of deviations from the blackbody form, including turbulence, proton decay, other unstable particles, decaying massive neutrinos, late photoproduction of deuterium, explosive or normal galaxy formation, cosmic gravity waves, cosmic strings, black holes, active galactic nuclei, Population III stars, hot intergalactic medium, etc. Our results do not rule out small contributions from these sources but do show that they could not have been responsible for most of the universe that we see today.

The FIRAS instrument also measured: 1. The spectrum of the far IR cosmic background radiation, first detected by the DIRBE instrument (Fixsen *et al.*, 1998), 2. The spectrum of the far IR zodiacal light, showing that the responsible dust particles are large, ~ 30 μm in size, 3. The spectrum of the part of the CMBR due to the motion of the Earth through the cosmos, called the dipole (Fixsen *et al.*, 1994), and 4. Limits on spatial variation of the CMBR spectrum (Fixsen *et al.*, 1997).

The FIRAS also confirmed Planck's formula for the blackbody spectrum (Nobel Prize, 1918). If Planck's formula were incorrect, the calibration software would not have produced self-consistent results. The FIRAS calibration depends on temperatures through the form for the photon mode occupation number, $\eta = 1/(e^x-1)$, and this is the part of the Planck function that we tested.

A recent paper by Fixsen and Mather (2002) argued that modern detectors and instrument designs could produce a factor of 100 improvement in sensitivity and accuracy. In that case the astrophysical interference from dust and molecules would certainly limit the cosmological conclusions. However, if these foregrounds could be managed, it is not unlikely that the distortions of the CMBR spectrum from known forms of energy release (e.g. the reionization of the universe at a redshift of 10–20 as detected by WMAP) could be detected.

Several papers have been written about the possibility of detecting distortions of the CMBR spectrum from small effects during the recombination era. For instance, small opacities due to the molecule lithium hydride, LiH, might be seen, if the level populations of the molecule were slightly out of thermal equilibrium. A common question concerns the Lyman α photons remaining from the last recombination for each H atom. There would be approximately one photon per H atom, at a wavelength of $(1+z) \cdot 0.1216$ μm, where $z \sim 1{,}089$, and a fractional line width of a few percent. H atoms are very much less numerous than CMBR photons, by a factor of more than 10^9, and the expected wavelength is in a region of the spectrum that is filled with Galactic dust, atomic, and molecular emission, and zodiacal dust emission, so these few photons are very unlikely to be observable.

Improved measurements of the CMBR spectrum at longer wavelengths are now in progress. The ARCADE project (Kogut *et al.*, 2004) is a balloon-borne microwave radiometer with a full-beam external blackbody calibrator. Operating without a protective window, it depends on high-speed flow of helium gas to keep the residual atmosphere at balloon altitudes from falling into the instrument and condensing on the antennas. Preliminary results show that the measured temperature is consistent with the FIRAS number. Eventually, this approach will provide improved measurements of the μ parameter for early energy release, since this distortion is greatest at long wavelengths.

B. DMR

The DMR instrument was the second to produce cosmological results. The data analysis team worked diligently to understand the needed corrections for known systematic errors, particularly the magnetic and thermal sensitivities of the Dicke switches, and the stray light from bright objects like the Earth. The first hint that we had detected a cosmic anisotropy was shown to the Science Working Group at a special team meeting at Nancy Boggess's home in October 1991 by Ned Wright, who had written his own analysis program for the first year of DMR data. The immediate response of the science team was that this was very important, too important to release quickly. We were well aware of recent junk science results on polywater, cold fusion, and other topics, and we were determined to get the answer right.

There were three main issues. First, were all the known instrument errors properly modeled and compensated? Second, had we properly understood and removed the effects of the Galactic foreground electrons and dust? And third, were the elaborate computer programs reliable?

To tackle the first issue, we held team meetings devoted to brainstorming about everything that could possibly affect the accuracy of the data, and devising strategies to measure and analyze each effect. For each one, we needed two different people and computer programs to agree, and we needed a science team review of the results. Alan Kogut and Ned Wright were the key analysts, and David Wilkinson was our most determined skeptic. This process took many months.

The second issue was analyzed at Goddard by Charles Bennett and Gary Hinshaw, and by the rest of the team, and described in Bennett *et al.*, 1992. Their strategy was to represent the two kinds of foreground emission by Galactic electrons and the dust emission by models with adjustable coefficients, and then to determine those coefficients by comparing the maps made by the DMR and other equipment at different wavelengths. The result was that a linear combination of the three DMR maps, weighted with particular coefficients, would eliminate almost all the Galactic emission for directions outside the Galactic plane.

The third issue, that of software reliability, was managed by thorough tests of each computer code, and comparison of results of the personal code written by Ned Wright with the official code written by the DMR team at Goddard and the code developed independently by George Smoot and his team at Berkeley.

An additional verification came from the data from balloon-borne instruments. We had agreed to delete the long-wavelength channel (at 23 GHz) from the DMR, and to fly a balloon-borne maser instrument to map part of the sky at that wavelength. The maps from this instrument contained strong signals from the Milky Way galaxy, but they were consistent with the DMR data (Fixsen, Cheng, and Wilkinson, 1983). Also, the MIT-Princeton team of Stephan Meyer, Edward Cheng, Ken Ganga, and Lyman Page, including two COBE team members, flew a balloon-borne instrument with bolometer detectors at shorter wavelengths, and achieved enough sensitivity to see the cosmic fluctuations (Ganga *et al.*, 1993). Their data were processed just in time to show that they were consistent with the cosmic fluctuations seen by DMR, before the DMR data were made public. Indeed, if the DMR had not been built, it is possible that the balloon data would have eventually been accepted as the first detection of cosmic structure. For our purpose, it was enough that we knew the DMR data, covering the whole sky, were sound.

Finally, the results were ready and prepared for publication. The announcement was made at the meeting of the American Physical Society in April 1992 in Washington, D.C. There had been enough advance publicity, and some leaks, that the press conference was filled with TV cameras and reporters. Within the day, the results were reported worldwide, and George Smoot's famous remark about seeing the face of God made the news everywhere. Steven Hawking is quoted as saying something like this was "the most important discovery of the century, if not of all time." The next day, interpretive papers had been submitted and distributed on the worldwide web by leading cosmologists, who had everything written in advance except the conclusions. Within the year, there were thousands more papers citing our results.

The results we showed were in the form of maps and fluctuation spectra. The maps were nicely adjusted to show pink and blue blobs to represent hot and cold parts. As Sachs and Wolfe (1967) had pointed out, the colder regions represent higher density, because of the gravitational redshift of the photons leaving potential wells. The spectra were statistical characterizations of the spatial fluctuations, mathematically precise descriptions of the typical sizes of the blobs. To a first approximation, we found that there is no typical size – blobs of all sizes are equally likely and equally bright. This is called "scale-invariant", as predicted by Harrison and Zeldovich and by simple forms of the theory of cosmic inflation. On closer examination, the fluctuations are a little too weak on large angular scales (90 degrees and larger), and they increase a little on the smallest angular scales we could see (7 degrees), as they should according to theory. The first point has still not been explained, and may not be statistically significant. The second is very important, and is due to the motions of matter at the time of the cosmic decoupling.

So what had the DMR measured? We had indeed discovered and mapped the primordial density fluctuations of the universe. If these had not been found, theorists would have been extremely disappointed, because by 1992, there was a nearly complete theory of the origin of the large scale structure of the universe, built on the idea of cosmic inflation to set initial conditions and guarantee large scale uniformity of the universe. The theory holds that very small amplitude primordial density fluctuations are the seeds for large scale structure, and that ordinary matter falls into the regions with greatest initial density, leaving empty regions (cosmic voids) where the initial density is least. The only needed force is gravitation. A remarkable result of the theory is that the density fluctuations grow linearly with time, and not exponentially like so many other natural phenomena. The reason is that the gravitational attractions of distant parts of the universe diminish with the expansion, slowing down the exponential growth to linear growth. Therefore, measurements of the large scale structure of the universe as represented by galaxies and galaxy clusters should also represent the initial conditions, the primordial fluctuations. By 1992, we had measurements of the large scale fluctuations traced by galaxies, and so we thought we knew what to expect.

However, there was trouble with this picture. Ordinary matter alone gives the wrong patterns, and it is not free to move relative to the rest of the universe until it becomes a neutral gas at the cosmic decoupling era. Fortunately, it was already recognized that some kind of additional matter might fill the universe, called cosmic dark matter. By hypothesis, it is invisible, and has no interactions with light except through gravitation. On the other hand, since it is not tied to the CMBR radiation field, it is free to start moving before the cosmic decoupling, and can fall into the primordial gravitational wells and make them grow deeper. Also, nothing is known of this dark matter except what astronomers claim to know: there is no agreed theory of it, no measurement of any of its particles in laboratories, and no knowledge of the masses of the particles, their stability, or anything else. Only one kind of dark matter has so far been observed in the laboratory, the neutrinos, with their three flavors and their antiparticles, and it seems that their masses are not enough to explain the cosmic dark matter.

The other thing at issue in 1992 was whether the universe is spatially flat or not. Theorists felt that a zero curvature universe was simple and pleasing and somehow ought to be true. To make such a universe, we would have to have a cosmic acceleration term in the equations, like the Λ constant that Einstein (Nobel Prize 1921, but not for this work) proposed and later rejected. Perhaps just the right amount of such a term, which produces negative spatial curvature, could balance the positive curvature produced by ordinary matter and dark matter. That would be an amazing coincidence, unless there is some unknown law of nature that requires this to be true. This acceleration term is now called the cosmic dark energy, which points out that it might not be just a constant of mathematical integration as Einstein saw it, but could perhaps be a new kind of force or matter with its own peculiar equation of state. So the interpretation of the measured cosmic fluctuations has become a major scientific industry.

The DMR was operated for a total of four years, and the additional data were statistically consistent with the first year's data. The new data gave much better ways to hunt for and correct systematic errors of all sorts, so the final results improved much more than the factor of two in random noise levels.

The DMR data were analyzed to hunt for many interesting things, none of which have been found with much statistical significance. For instance, there was a suggestion that cosmic strings might exist, stretching across vast regions of space. At one time it was thought that they could produce the cosmic density fluctuations all by themselves. If they were strong enough, they would produce discontinuities in the temperature maps, and should be visible. Since none have been found, they apparently were not responsible for the CMBR anisotropy, but perhaps they are just very rare. Searches were also made for non-Gaussian fluctuations; perhaps there were specific localized objects, with either positive or negative temperatures, that could be found by close examination. None of these have been found either, above the levels that should appear from ordinary Gaussian fluctuations, and from known point sources like Jupiter. The DMR angular resolution is too coarse for it to respond much to fainter objects. Another possibility is that the universe has a peculiar topology: suppose that we found the same object or pattern in two different places on the map. In that case, the universe could have the topology of a sphere or a torus, in which the same object could be viewed in two different directions. None of these searches have returned positive results either.

Following the DMR announcement, there have been many more instruments built, and hundreds of scientists worldwide have continued to measure and analyze. Ground-based and balloon-borne instruments have measured with improved angular resolution. COBE team members Charles Bennett and David Wilkinson conceived, proposed, and built the MAP, the Microwave Anisotropy Probe, launched in 2001 and still operating. It was renamed the Wilkinson Microwave Anisotropy Probe in 2002, following Wilkinson's death on September 5th, 2002. The WMAP has extended the DMR all-sky maps to much higher sensitivity and angular resolution, and has confirmed that the DMR data are accurate. With the WMAP data, we now know many of the cosmic parameters (matter, dark matter, dark energy density, age of the universe, etc.) to precisions of a percent or two. It has also detected a wonderful surprise: evidence of the effects of the reionization of the universe, at a redshift around 13. This is recognized through the polarization of the CMBR, produced as a result of the quadrupole anisotropy of the CMBR seen by the electrons when they scatter the radiation towards us for the last time. The angular scale of the polarization pattern we observe measures the redshift at which the scattering occurs, and the amplitude measures the optical depth of the scattering.

There are several remaining challenges for future CMBR anisotropy measurements. The one receiving the most attention now is the hunt for the polarization induced by gravitational waves in the Big Bang itself. A Task Force on Cosmic Microwave Background Research, chaired by Rainer Weiss, prepared a beautiful summary report about the benefits, challenges, and strategy for measuring this polarization. The polarization map is a vector field on a sphere, and can be decomposed into two parts: the divergence of a scalar field, called E-mode, and curl of a vector field, called B-mode.

According to theory, the primordial gravity waves should produce a polarization pattern with a curl, and no other subsequent process should be able to do so. The polarization pattern would be much fainter than the temperature anisotropy map, and the curl component would be much fainter than the divergence component. Hence this is very difficult to measure, but it may already be possible with current generations of detectors. The generic name for this space mission in the US is "Inflation Probe" and three design studies have been supported by NASA. Technically, such a mission could be flying in a decade, but competition for scarce resources may delay it. In any event, the measurement of the B-mode polarization is the most direct method available to learn about the forces prevailing in the Big Bang itself, could help us come to the long-sought Theory of Everything, and presumably would be a Nobel-winning discovery. It is no surprise that elementary particle physicists, both theorists and experimenters, have been turning to CMBR studies as their next exciting opportunity.

C. DIRBE

The DIRBE instrument was the last to produce cosmological results, largely because the local foregrounds at DIRBE wavelengths are very bright and complex. The definitive DIRBE results are given in a series of papers (Hauser *et al.*, Kelsall *et al.*, Arendt *et al.*, and Dwek *et al.*, 1998). The surprise found in the DIRBE data is that the universe is twice as bright as previously believed from measurements of individual galaxies. There is a general glow called the cosmic infrared background or CIB, composed of two parts, at near IR wavelengths of a few μm, and far IR wavelengths of a few hundred μm. The near IR background is not yet understood (see the review by Hauser and Dwek, 2001), but the far IR background is apparently produced by a previously unknown population of very bright dusty galaxies at redshifts of a few (2–3).

To get to this result, the DIRBE team had to go through a much more complex process than was required for the other two instruments, because the main foreground faced by DIRBE is variable in time and space. It is produced by the interplanetary dust, which is smoothly distributed in a thick disk orbiting the Sun. This disk is not so simple though. It has several sources, in collisions among asteroids, in the disintegration of comets, and in the migration of small particles from the outer solar system. From the IRAS data, we know that there are at least three rings of dust orbiting as though they are collision debris from certain families of asteroids. Also, these particles move under the influence of gravity, radiation pressure and the Poynting-Robertson drag, and electromagnetic forces for those particles with electric charges. The drag force makes the particles spiral in towards the sun over time scales of thousands to millions of years, depending on size, and the particles may experience close encounters with the planets, and repeated gravitational impulses when their orbital periods are commensurate with the planets. Some become locked in orbital resonances for long periods of time, as the gravitational forces overcome the Poynting-Robertson drag. Some even become locked in orbital resonance with the Earth, producing leading and trailing blobs that were seen in the DIRBE data. This phenomenon, an annoyance for those seeking the cosmic IR background radiation, is of great interest for those hunting for planets around other stars. Indeed, it has already been seen in the dust clouds of bright stars such as Fomalhaut, where a large planet is presumed to be organizing the dust into a ring (Kalas, Graham, and Clampin, 2005).

In addition, the Earth, and the COBE with it, move through the dust cloud. The major effects are that the plane of the Earth's orbit is not the symmetry plane of the dust cloud, which feels greater forces from Jupiter, and that the Earth's orbit is not circular, so the Earth moves in and out as well as up and down in the cloud. Moreover, the dust cloud is not centered on the Sun, since it feels the strong pull of Jupiter.

The model for this dust was built in a heuristic way, and contains many free parameters, adjusted to fit the time and spatial variations of the sky observed by the DIRBE. Full documentation of the model is given by Kelsall *et al.*, 1998. The parameters include the dust temperature and its power-law variation with distance from the Sun, the dust number density and its power-law variation with distance, the emissivity and reflectivity at each wavelength, and the phase function of the scattering. There is a model of the particle resonances with the Earth, to explain the leading and trailing blobs. There are parameters to describe the thickness of the dust disk, and the shape of its density distribution near the ecliptic plane, and parameters for the tilt and forced eccentricity of the dust disk presumed to be due to Jupiter. Even the symmetry plane of the dust distribution is warped, presumably by the competing gravitational perturbations of nearby planets. Even with all this complexity, there are significant residuals from the model, and Kelsall believes there are real time variations at the level of a percent that are not explained.

Understanding the foreground emission from the Galaxy was also difficult and done in a heuristic fashion. The majority of the sky at most wavelengths has detectable stars and dust clouds, the brightest of which can be masked out and ignored. The fainter ones are not resolved by the large DIRBE beam and can be modeled based on a priori Galactic models with no free parameters (Arendt *et al.*, 1998).

The result of all this subtraction is that there are a few favored directions where the Galactic foregrounds are least, possibly due to supernovae which have cleared out the dust and gas for large distances. As any true cosmic background must be roughly isotropic, it was important that observations in these favored directions must give the same answers.

The DIRBE results were essential to both the FIRAS and DMR interpretations. First, they showed that the local universe was well understood, so that the FIRAS and DMR observations really do represent the distant Big Bang. This was important, for instance, in arguing that it is only a coincidence that the cosmic dipole due to the Earth's motion lies in the Ecliptic Plane. Also, the DIRBE maps of Galactic dust agreed with the FIRAS maps, showing that there were no new and strange effects. Conversely, after the DIRBE team determined that there is a far IR cosmic background field, the FIRAS data were used to confirm it and measure its spectrum (Fixsen *et al.*, 1998).

VI. SUMMARY: COBE'S PLACE IN HISTORY AND WHERE ARE WE NOW?

The COBE mission, started in an era when slide rules were common and aerospace designers used pencils and large sheets of paper, led to a revolution in our understanding of the universe. It confirmed the Big Bang theory, and discovered the primordial density fluctuations that formed the large-scale structure of the universe. It found that the universe is twice as bright as previously thought. It led to a series of ever more powerful instruments to measure the CMBR, one of which (the WMAP) has already tested the idea that the earliest times in the universe included an exponential expansion called inflation. We now have precise values (especially from WMAP) of many cosmic parameters, and CMBR studies now in preparation could reveal the nature of the forces in the Big Bang by discovering the primordial gravity waves. The Planck mission, a project of the European Space Agency with NASA participation, is due for launch in 2008 and will extend the WMAP results to smaller angular scales by using shorter wavelengths. It will also have greater sensitivity, using bolometric detectors, and has a chance to measure the cosmic polarization signals even better than WMAP.

Our team members have continued on with many other projects. Rainer Weiss is one of the leaders of the LIGO project, the Laser Interferometer Gravitational-Wave Observatory. George Smoot is a full professor at the University of California at Berkeley. Charles Bennett, now at Johns Hopkins University, and David Wilkinson (who died in 2002), led the WMAP, with many of the engineers and scientists who built the COBE. Edward Wright is Principal Investigator for the WISE mission, the Wide-field Infrared Survey Explorer, which will survey the entire sky with 1,000 times the sensitivity of the IRAS. Michael Hauser is Deputy Director of the Space Telescope Science Institute, which operates the Hubble Space Telescope and will operate the James Webb Space Telescope. Edward Cheng started a small company, Analytical Concepts, and Stephan Meyer is Deputy Director of the Enrico Fermi Institute and Associate Director of the Kavli Institute for Cosmological Physics at the University of Chicago. Most of the senior project managers and engineers have retired.

The COBE observations also lead on to new missions to observe the first stars and galaxies, such as the James Webb Space Telescope (JWST). This is my current project, for which I serve as the senior project scientist. The JWST is a deployable 6.5 m infrared telescope, to be launched in 2013 to an orbit around the Sun-Earth Lagrange point L2. With its protected environment and the latest in modern infrared detectors and instruments, it could also produce stunning discoveries. Theory confidently predicts that star formation began at very high redshifts (> 20), and that some of the first proto-galaxies and supernovae may be observable at redshifts of 15 or more.

ACKNOWLEDGEMENTS

The COBE mission was first proposed in 1974, and our team owes tremendous thanks to those who read our thin booklets and recognized their importance. Dr. Nancy Boggess, Program Scientist for infrared astronomy at NASA Headquarters, was one of those people. She advocated the COBE even when many astronomers were not interested, and she backed other infrared astronomy missions like the IRAS, the Kuiper Airborne Observatory, the SOFIA (the Stratospheric Observatory for Infrared Astronomy) and the Spitzer Space Telescope. After the COBE was started, we benefited from strong and steady support from a series of managers at NASA Headquarters and Goddard Space Flight Center, who had patience and faith in our work. Our team was large, and most of the team members (not quite all) are listed in the books by Mather and Boslough (1996) and by Smoot and Davidson (1993). We also thank the taxpayers, who, whether they knew it or not, were funding the COBE mission. Benjamin Franklin, one of the greatest scientists of his day, and Thomas Jefferson, the first US president to support a major scientific research project, would be proud to see that the country they helped to found has carried on with its support of science for the public good.

We also thank our families, who were participants in the COBE mission whether they knew it or not, and supported or at least tolerated the long hours and weekends that we worked to make the mission a success. For myself I thank especially my wife Jane, who has known me since before the COBE was imagined in 1974, and who has taken a keen interest all along. My parents sent me to school, led by example (my father was a geneticist working with dairy cattle, and my mother was a teacher), and paid astronomical sums for tuition. My teachers showed me the way and encouraged my curiosity. I especially thank my thesis advisor Paul L. Richards, who started our balloon payload, and my fellow graduate student David Woody, who made the balloon payload work after all. For sponsoring and participating in the 1974 COBE proposal, my postdoctoral advisor Patrick Thaddeus gets the credit – without his influence, there would not have been a COBE, or at least not this one. Within the COBE team I owe special thanks to our project manager Roger Mattson and his deputy Dennis McCarthy, and to the instrument manager for the FIRAS, Robert Maichle, the FIRAS instrument engineer, Michael Roberto, and to Richard A. Shafer, deputy principal investigator. Dale Fixsen was the architect of the FIRAS calibration, and Richard Isaacman and Shirley Read led the two software teams for the FIRAS. I also thank my co-author John Boslough, without whom our book "The Very First Light" would not have been written.

With the funds from the Nobel Prize and the Peter Gruber Foundation prize in cosmology, and the concurrence of my wife Jane, I have started the John and Jane Mather Foundation for Science and the Arts. George Smoot has also set up a foundation with similar purposes. This is one way of giving thanks to the many people who contributed to the COBE project.

REFERENCES

The COBE project archive is now located at http://lambda.gsfc.nasa.gov/ and includes project information, images, data files, and documentation. Ned Wright's online Cosmology Tutorial, http://www.astro.ucla.edu/~wright/cosmolog.htm includes a history of the COBE project, with online images of early versions of the COBE. The Wikipedia entry on COBE is well-written (http://en.wikipedia.org/wiki/COBE) and Richard A. Muller's summary of the history of the Berkeley anisotropy measurements is at http://muller.lbl.gov/COBE-early_history/preCOBEhistory.html. My favorite textbook on cosmology is by Peebles (1993). An excellent recent summary of cosmology is given by Bennett (2006).

1. Alpher, R. A., Bethe, H. and Gamow, G. 1948, Phys. Rev. **73**, 803.
2. Arendt, R. G. *et al.*, 1998, Ap. J. **508**, 74.
3. Bennett, C. L., 2006, Nature **440**, 1126-1131.
4. Bennett, C. L. *et al.*, 1992, Ap. J. Letters **396**, L7-L12.
5. Bennett, C. L. *et al.*, 1994, Ap. J. **434**, 587-598.
6. Dwek, E. *et al.*, 1998, Ap. J. **508**, 106.
7. Fixsen, D. J., Cheng, E. S., and Wilkinson, D. T., 1983, Phys. Rev. Lett. **50,** 620-622.
8. Fixsen, D. J., Dwek, E. , Mather, J.C., Bennett, C. L., and Shafer, R. A. 1998, Ap. J. **508**, 123.
9. Fixsen, D. J., and Mather, J. C. 2002, Ap. J. **581**, 817-822.
10. Fixsen, D. J. *et al.*, 1994, Ap. J. **420**, 445-449.
11. Fixsen, D. J., Hinshaw, G., Bennett, C. L., and Mather, J. C. 1997, Ap. J. **486**, 623.
12. Fixsen, D. J., and Mather, J. C. 2002, Ap. J. **581**, 817-822.
13. Gamow, G., 1947, "One, Two, Three… Infinity", Viking Press.
14. Ganga, K., Cheng, E., Meyer, S., and Page, L. 1993, Ap. J. Lett. **410**, L57.
15. Gush, H. P., Halpern, M., and Wishnow, E. 1990, Phys. Rev. Lett. **65**, 537.
16. Guth, A. H., 1981, Phys. Rev. **347**, D23.
17. Harrison, E. R., 1970, Phys. Rev. **D1**, 2726.
18. Hauser, M. G., and Dwek, E. 2001, Ann. Rev. Astron. Astrophys. **39**, 249.
19. Hauser, M. G. *et al.*, 1998, Ap. J. **508**, 25.
20. Hemmati, H., Mather, J. C., and Eichhorn, W. L. 1985, Appl. Opt. **24**, 4489.
21. Hubble, E., 1929, Proc. Natl. Acad. Sci. U.S.A. **15**, 168–173.
22. Kalas, P., Graham, J., and Clampin, M., 2005, Nature **435**, 1067.
23. Keller, J., 1962, J. Optical Soc. Am. **52**, 116-130.
24. Kelsall, T. *et al.*, 1998, Ap. J. **508**, 44K.
25. Kogut, A. *et al.*, 2004, Astrophys. J. Suppl. Ser. **154**, 493–499.
26. Lemaître, G., 1927, Annals of the Scientific Society of Brussels **47A:** 41. Translated in: (1931) MNRAS **91:** 483–490. Primeval atom: Lemaître, G., 1931, Nature **128** Suppl.: 704.
27. Mather, J. C., 1984, Appl. Opt. **23**, 584-88.
28. Mather, J. C., 1981, IEEE Trans. Antennas Propag. **AP-29**, 967.
29. Mather, J. C., and Boslough, J. 1996, "The Very First Light," Basic Books, New York.
30. Mather, J. C., Toral, M., and Hemmati, H., 1986, Appl. Opt. **25**, 2826-30.
31. Mather, J. C., Richards, P. L., and Werner, M. W. 1971, Astrophys. J. Lett. **170**, L59.
32. Mather, J. C., Fixsen, D.J., Shafer, R.A., Mosier, C., and Wilkinson, D.T. 1998, Ap. J. **512**, 511-520.
33. Mather, J. C., Woody, D. P., and Richards, P. L. 1974, IEEE Trans. Microwave Theory Tech. **22**, 1046.
34. Peebles, P. J. E., and Yu, J. T. 1970, Ap. J. **162**, 815.
35. Puplett, E., and Martin, D. H. 1970, Infrared Phys. **10**, 105-109.
36. Sachs, R. K., and Wolfe, A. M. 1967, Ap. J. **147**, 73.
37. Shivanandan, K., Houck, J. R., and Harwit, M. O., 1968, Phys. Rev. Lett. **21**, 1460–1462.
38. Smoot, G., and Davidson, K. "Wrinkles in Time," Avon Books, 1993.
39. Thaddeus, P., 1972, Ann. Rev. Astron. Ap. **10**, 305.
40. Wright, E. L. *et al.*, 1994, Ap. J. **420**, 450-456.

41. Woody, D. P., Mather, J. C., Nishioka, N. S. and Richards, P. L. 1975, Phys. Rev. Lett. **34**, 1036.
42. Zeldovich, Ya. B., 1972, MNRAS, **160**, 1.
43. Zeldovich, Ya. B. and Sunyaev, R. A., 1969, Astrophysics and Space Science, **4**, 301.
44. Sunyaev, R. A., and Zeldovich, Ya. B., 1970. Astrophysics and Space Science **7**: 3.

APPENDIX – COBE TEAM MEMBERS

SCIENCE TEAM

Charles L. Bennett, Deputy Principal Investigator, DMR
Nancy W. Boggess, Deputy Project Scientist for Data; formerly Program Scientist for Infrared Astronomy, NASA Headquarters
Edward S. Cheng
Eli Dwek
Samuel Gulkis
Michael G. Hauser, Principal Investigator, DIRBE
Michael A. Janssen
Thomas Kelsall, Deputy Principal Investigator, DIRBE
Philip M. Lubin
John C. Mather, Project Scientist, and Principal Investigator for FIRAS
Stephan S. Meyer
S. Harvey Moseley, Jr.
Thomas L. Murdock
Richard Arrick Shafer, Deputy Principal Investigator, FIRAS
Robert F. Silverberg
George F. Smoot, Principal Investigator, DMR
Rainer Weiss, Chairman, Science Working Group
David T. Wilkinson
Edward L. Wright, Data Team Leader

ENGINEERING AND MANAGEMENT TEAM

Donald F. Crosby, Instrument Engineer, DIRBE
Ernest C. Doutrich, Flight Assurance Manager
Irene K. Ferber, Project Secretary
Anthony D. Fragomeni, Observatory Manager
Thomas J. Greenwell, Integration and Test Manager
David Gilman, Program Manager, NASA Headquarters
William D. Hoggard, Delta Liaison/Launch Operations Manager
Charles Katz, Systems Engineer, Instruments
Bernard J. Klein, Instrument Engineer, DMR
Loren R. Linstrom, Systems Engineer, DIRBE
Robert J. Maichle, Instrument Engineer, FIRAS
Roger A. Mattson, Project Manager
Dennis K. McCarthy, Deputy Project Manager
Maureen J. Menton, Secretary
Herbert J. Mittelman, Resources Officer
Stephen Servin-Leete, Systems Engineer, DMR
Earnestine Smart, Project Support Specialist
Pierce L. Smith, Ground Data Processing Systems Manager
Jack W. Peddicord, Deputy Project Manager (Resources)
Michael Roberto, Systems Engineer, FIRAS
Robert G. Sanford, Mission Operations Manager
Robert T. Schools, Project Support Manager
Joseph F. Turtil, Systems Engineer
John L. Wolfgang, Software Systems Manager
Earle W. Young, Instruments Manager

Portrait photo of Dr. Mather by photographer Pasco Izzo.

George F Smoot

GEORGE F. SMOOT III

Winning a Nobel Prize is a life-changing event. It is a turning point. In addition to many people asking about how you arrived at this point, the winner is asked to produce his biography. All this and the events related to the Nobel Prize have resulted in each one of us going back and reviewing his life and influences that led to the present.

There were many key people, events and circumstances that had a substantial affect upon my development. However, it is clear that family has by far had the largest effect.

Both of my grandfathers were judges, along with my uncle and my cousin. One grandfather, Tal Crawford, served as Chief Justice during the Nuremberg War Crimes trial. Although I knew him only as a child, he had an influence upon me directly from his great kindness and careful discussions, and also through my mother Talicia. My father even started college with the intent to major in law, but World War II and interest in engineering and science diverted him.

Another key person was my Ph.D. thesis advisor at MIT, Professor David Henry Frisch. I was recently at a meeting with his son, University of Chicago Physics Professor Henry Frisch, and recalled how Professor Frisch related from time to time how he had descended from thirteen generations of rabbis, and that now there were two generations of physicists. (Henry is sure that the next generation will change the trend more quickly.) My family had a tradition in law, specifically as judges. We thought that all three of these professions had a number of traits in common: in particular, being able to reason, remember, and listen to others (perhaps less obvious for physicists but still true in collaborations).

The first pivotal event I remember in my development as an inquiring scientist began with a visit to my cousins, as we took a night drive across the state of Alabama. At that time, this was a very long distance to me. Excited from the visit with my cousins, I stayed awake, looking out the window rather than taking a nap. I noticed that the moon was following us mile after mile tagging along like my dog but with greater speed and persistence. I asked my parents, "How does the moon know to follow us?" They told me that the moon followed all the cars, not just ours. I was intrigued and wanted to know how the moon did this. Was it immensely superior to a dog? My parents patiently explained that the moon was very big and very far away and thus the angle did not change noticeably as we drove for miles. They gave examples of near and far objects that we could see along the way, so I could realize this for myself. I was impressed by how big and far away the moon must be but even more impressed that one could understand what they saw in the world and that it was so beautifully simple and clear when visualized in the proper way. It was a startling revelation that the world could be understood by simple rational evaluation.

Not long after that, I got another glimpse at the importance of education, study, and reason. My father (George Smoot II) and mother both had resumed their college educations after the war was over, and life settled down following the birth of myself and my sister Sharon. They would drive from our hometown to their new college, Auburn University (closer than their original Tulane and Sophie Newcomb alma maters). At that time, I attended Auburn day care. That semester, my mother had a biology/zoology class that ran late so I would sometimes attend the class with my mother, sitting on her lap, reading the text, and listening to the lectures. It was clear how my parents, the professors and students valued education, and that they were discovering and understanding things about the outdoors that I played in everyday. I liked the respect that the other students and professors gave my mother for being a good student, and that they tolerated me being in the class and sometimes explained things to me. The seriousness of learning was tempered with fun, so it was a wonderful place to be and opened up a whole new way of understanding the world as a child. I still spent a lot of time outside exploring nature but now tried to tie that to reason and a network of knowledge.

Figure 1. Young George Smoot with sister Sharon and mother Talicia at the age of discovery.

This thoughtful, educational life was very different from the other difficulties that my family was facing. With two kids, the budget was tight. Part of my father's education was covered by the G.I. bill, but my mother's had to be raised separately along with paying for other expenses. My father had started a lumber business after the war to provide additional income. The business was successful, providing just enough for us to eke by. Then, a problem came up. In one of the stands of trees that my father purchased, a large, productive still was producing bootleg alcohol for this dry portion of the state. As soon as the tree cutting began, the group operating the still approached my father. They offered him cash not to cut the trees and expose their substantial operation. My father, at some personal risk, informed the police, and the trees were promptly impounded (no longer allowed to be cut for the saw mill) by the judge handling the case. He was apparently on the payroll of the illegal distillery operation as some police must have been. This caused an immediate financial crisis for us, as much of the saw mills/lumber company funds were tied up in trees, which were to be cut down and then milled to lumber and sold. Suddenly there was a shortage of trees to mill. Needless to say, our lumber company was soon in bankruptcy. We quickly moved out of our house and in with my grandmother, while my parents struggled to finish their college courses and graduate.

Meanwhile, my father had migrated from law into engineering. I later wondered if it was due to the contrast between the spirit and the letter of the law as practiced in this lumber case or whether his natural talent and interest in engineering and mathematics emerged. Shortly after graduation, my father landed a job with the U.S. Geological Survey in hydrology (water resources). We were on the road to getting out of debt with no asset but our family education.

Soon, an opportunity opened up in the form of a USGS job in Alaska, a location where my father had served as a fighter pilot during World War II. He jumped at the chance to work in the field and its pioneering efforts. Our family moved to Alaska, piling all our belongings into a homemade trailer and making the trip up the Alaska-Canada highway. We felt like the early settlers going west in covered wagons to a new life and land. There were mishaps – failures of the trailer – and other adventures on the long trip north to the Matanuska Valley (in sight of Mt. McKinley). We were still very poor and did not have one of the new-fangled TV sets that could get reception from the few stations in Anchorage. Instead, I spent my time exploring the outdoors, viewing the night sky, and studying. By now (1954), I needed glasses due to the long hours reading and studying I did during the winters. My mother also worked, starting her career as an elementary school teacher. I was in the Palmer, Alaska unified school district from grades 2–6 where I received a good education. Many of my classmates were Eskimo and Indian orphans as the state orphanage was near by presumably a big step up from being left on an ice floe. The experience seemed like a great opportunity to meet people from all over the Valley and learn things. We only occasionally had to do the duck and cover drills in case of nuclear or air attack. In this isolated area, we felt fairly safe except for the occasional moose wandering onto the playground. Because the ground froze hard during the winter we could also ice skate at the low end. There I got my first classic lessons in conservation of angular momentum for spinning ice skaters (arms in and out) as well as learning not to stick my tongue on the cold metal figures of the playground equipment.

During this period my father showed me how to be an intrepid explorer as he regularly procured data for hydrology and hunting. We did not buy the meat we ate during our four years in Alaska but acquired our own from hunting and fishing. My father and his friend Frank Trainer would go into the wild and bring back whatever was in season: moose, mountain sheep and goats, caribou, and bears. We also had a large vegetable garden. Though the growing season was short, the light days were long. The Matanuska Valley is famous for its large vegetables. For example, one year that we lived there, a 40-lb (18 kg) cabbage won at the state fair. During the winter, moose would come down into our now frozen vegetable garden and dig out some of the remaining edibles.

Figure 2. My father as naval aviator in World War II.

My father would also reconnoiter on field trips to measure the properties and flow of rivers, often having to chop through many feet of ice to get to the water in question. He invented a number of devices and techniques, which we were able to see. Eventually, word spread of his efficiency, and he was called to the research station in Columbus Ohio, and then later to headquarters in Washington, D.C. As a leader in the field, my father traveled worldwide to measure rivers and other water sources. Our family still has clippings of the time that my father went to Brazil to work with the local agency to make the first measurement of the water flow rate of the Amazon River. Shortly after that, he went to measure the Mekong River and its tributaries something new and fairly dangerous. He traveled to many places, and later worked for the World Bank on issues related to irrigation projects and water resources. The role model he provided and message he sent was: go anywhere and do anything necessary to get the data or the meat for meals.

My mother also was a tremendous role model. Though soft spoken, she continued to develop her career as a teacher. When we moved to Columbus Ohio, she entered the Masters Degree program at Ohio State and received credentials to teach science. I remember very well how she performed her teaching job, took care of my sister and I, and also went to night courses studying at Ohio State; this while my father was traveling. Some nights my sister and I accompanied my mother and studied in the library. Again, the importance that my parents attached to education, both by word and deed was apparent. While traveling, my father found a bargain used bookstore and over a number of visits, bought for me the complete Modern Library of what were considered the great books. At first I was encouraged to read these but soon eagerly awaited the arrival of the next batch.

Figure 3. Serious student preparing to attend MIT.

While I was in high school, Sputnik appeared in the October sky (1957), and the nation and world became interested in science and space. My mother tutored me in additional science and history, while my father drilled me in the basement, force-feeding trigonometry and introductory calculus courses that were not given at my high school. I also put together a crystal radio set and later, a transistor radio kit.

During this time, I worked to save money for college – as a paperboy for the Columbus Dispatch, mowing lawns, and working as a caddy at the Scioto Country Club (all while studying and reading). After three years of this work,

I had barely enough money to pay for little more than my first semester at MIT. During my undergraduate years at MIT, I worked various jobs to help pay for my education, in addition to support from my parents and student loans. The hardest part was taking the rigorous courses and a heavy course load, but I wanted to take advantage of a good education. It was clear that MIT was definitely pushing us with as much information as we could soak up, and then some.

At the end of the four years, I received dual degrees in Mathematics and Physics along with the clarity that I really was going to be a physicist. I had come into college with a thirst for knowledge and an interest in many areas. I considered premed and other things but eventually gravitated to math and physics. At the time, MIT made every entering freshman take courses in math, chemistry, physics, and the humanities plus a choice of elective. Soon, it was clear to me that I enjoyed and did my best in mathematics and physics, and so focused in those areas. By the time four years had gone by, my interest in physics pulled ahead of mathematics. While I had a number of good and influential teachers in elementary and high school, when I got to MIT, I found professors who were also excited by the research at the forefront of knowledge and understanding. They had an attitude of anticipation and excitement about what the next day would bring and what challenges they could overcome.

My fellow students were much the same way and we studied hard. However, we had to let off steam in an MIT traditional way of hacks that is generally high-tech pranks and feats of technical derring-do. Some were simple, like welding rails in a 3-D cross in someone's dorm room or filling a refrigerator completely full of Jell-O (you put it on its back and then right it carefully once the Jell-O gels). Some were more elaborate, like a steam catapult for hard-boiled eggs or water-balloon crossbow. One received peer kudos for technical difficulty.

We carried forward the occasional prank into graduate school. My fellow graduate student and research colleague, Orrin Fackler and I were working shifts together, putting together an experiment at Brookhaven National Laboratory's AGS accelerator. We had determined that the most effective target for turning a positive kaon beam into a neutral kaon beam would be made of osmium. Now that was rare, more so than gold or platinum, and thus we did not think we could get it so were thinking of alternatives. Our advisor, Professor David Frisch said, "Your calculations look correct. I will go and personally sign out the osmium from the strategic metal and you can use it for your theses experiments." This was an inspiring show of confidence in us and our willingness to do what was necessary for the experiment. Frisch was personally on the hook for a small cylinder of metal worth twice what his new car was.

No good deed goes unrewarded. While working a half-day shift 7 PM to 7 AM, we were preparing the target box to hold the new target that Frisch had brought down to Brookhaven after his Friday class. It turns out that it did not fit in the carefully designed target box because it had warped a bit during the welding process. Orrin and I discussed this and realized that we would have to spend the entire night hack sawing and filing down the inside portion of the target box. By early morning, we had a significant pile of silver colored aluminum chips and dust similar color to osmium. We knew that Professor Frisch would soon (typically 6:30AM) come bounding in to see how much we had accomplished during the night (not much). I convinced Orrin that we should play a trick on Professor Frisch. When he came in enthusiastically greeting us, I wiped my brow as I finished screwing back together the target box and complained how hard osmium was to work with. I told him how it did not fit and how we had to saw and file on the osmium all night. Dave looked stricken with concern, but he knew I played pranks so he looked over at the always serious and more mature Orrin who added, "Yes, there was a lot to do but we worked hard to shave it down to have it ready for you today." Dave now looked back with his hand over his heart. I responded, "What is the matter? Is it toxic?" Dave gasped, "No, but it is precious." Orrin and I then confessed the prank, and after he recovered, Dave had a good laugh about it.

Though I was interested in many fields in physics, particle physics offered funds and opportunities for me to do a senior thesis. I was able to build equipment, take data, and learn things that no one had previously known. Soon I was hooked, and while accepted to a number of graduate programs, I chose to continue at MIT doing particle physics experiments. Working with researchers, including Professors Frisch and Osborne; other people in their groups; Kendall and Friedman; and later Professor Sam Ting, was an important experience in the rigor and attitudes of seeking the best science.

I was working on my senior thesis in the 1965–1966 academic year when I heard of the Penzias and Wilson discovery of the 3K Big Bang relic radiation. I was already interested in astrophysics, and paid some attention to the discovery, but I was busy with my schoolwork and research so I filed it away as something I had learned.

A few years later in 1967, Joe Silk was writing his paper on primordial fluctuations in the CMB (which was very influential for me) and Steven Weinberg began teaching his course at MIT that eventually became the book *Gravitation and Cosmology*. This informed me that there was something exciting and fundamental happening in this area. However, I was newlywed, finishing my thesis, and my wife Maxine was applying for graduate school in English Literature. I would soon need a postdoctoral scholar position, so I could not act right away. In 1970, I finished my Ph.D. at MIT and then went out seeking a new position. I was twenty-five years old. I interviewed at a number of places including a few positions at Berkeley, where I was persuaded by 1968 Physics Nobel Prize winner, Professor Luis W. Alvarez to choose Berkeley, and his group in particular, because they were willing to venture into completely new areas. This was also a position that would make good use of my newly acquired skills and knowledge.

A major reason that I came to Berkeley (University of California and the Lawrence Berkeley National Laboratory) was the I-can-do attitude along with the emphasis on first-rate science. There were many excellent researchers in Berkeley including a number of Nobel Laureates. Luis Alvarez was among the notables and it was his recruitment that attracted me so directly. However, there were many others whom I knew and interacted with besides Luis Alvarez including Emilio Segrè, Owen Chamberlain, Edwin McMillan, Glenn Seaborg, Don Glaser, Melvin Calvin, Yuan Lee, and more recently Steven Chu. There were many more outstanding scientists, engineers, and technicians that made Berkeley an excellent environment conducive to outstanding research. Interacting with such individuals provided many role models, along with the knowledge that these were real people who organized and applied themselves to do exceptional work.

I left MIT in 1970 and came to the University of California at Berkeley to work with Luis Alvarez on the NASA-funded High-Altitude Particle Physics Experiment (HAPPE, pronounced happy) at Lawrence Berkeley National Laboratory. The goal of HAPPE was to probe for interactions at higher energies than accelerators could produce at the time. The unfortunate Pacific Ocean crash of the HAPPE instrument, followed by the approval to construct a new high-energy accelerator (at what is now FermiLab) resulted in the reevaluation and new direction of the HAPPE program. Our goal was to design an experiment to find evidence of the Big Bang, which had become scientists favored explanation for the formation of the universe. The Belgian priest and astronomer Georges Lemaître first proposed the Big Bang theory in the late 1920s, after Edwin Hubble discovered that distant galaxies were rapidly receding from us; hence the universe was expanding. Its many critics, particularly Fred Hoyle, scornfully called the theory the "Big Bang" but the concept gained wide acceptance after 1964 when Arno Penzias and Robert Wilson of Bell Telephone Laboratories fortuitously discovered the cosmic background radiation - microwave remnants from the creation of universe. The existence of the cosmic background radiation was initially predicted by George Gamow in 1948, and eventually supported by other Big Bang theorists.

Figure 4. Wild-eyed Berkeley researcher at work in the lab wearing the mandatory safety glasses.

According to the basic Big Bang theory, 14 billion years ago our observable universe began with a rapid expansion of an extremely small but extraordinarily dense region in which energy and matter were one, and it has been expanding ever since. The initial subatomic particles and antiparticles arose from the primal energy, and eventually their daughters formed protons, neutrons, and electrons. After only 3 minutes, light nuclei formed. For the first 380,000 years, the temperature was too hot for the electrons to attach to the protons and nuclei to form atoms. The radiation that filled the universe could not travel freely without being scattered by the free electrons; therefore no light could escape and the universe was opaque. After 380,000 years of darkness, the universe cooled sufficiently for electrons to affix to nuclei, freeing radiation to escape (a process known as decoupling) and allowing light to become visible. The radiation that filled the universe during the period of decoupling continues (filling the universe) in the form of cosmic background radiation, but it has cooled from about 3,000 Kelvin at decoupling to less than 3 Kelvin. Kelvin is a scale of temperature in which 0 K, or -273.15° Celsius, is the temperature that molecules have no random thermal kinetic energy left. Following decoupling, the atoms joined to form molecules, and these aggregated under the force of gravity to create the universe known today.

Figure 5. On Lawrence Berkeley Laboratory Bldg 50 roof. Left to right, people are Giovanni de Amici, Charles Lineweaver (kneeling), Luis Tenorio, George Smoot (bending), John Gibson (in back), and Jon Aymon. They are looking at a 90 GHz differential microwave radiometer. This radiometer was used on the roof, at White Mountain, and at the South Pole to measure the emission from the Cosmic Microwave Background and also from the atmosphere.

Our first balloon-borne detectors were designed to see if a portion of the cosmic rays were antimatter. We knew from particle experiments that whenever matter is created from energy, an equal amount of antimatter also forms. Simple extrapolation implies that the early Universe had essentially equal amounts of matter and antimatter present. So the primary question was: Where is the antimatter? We designed a balloon-borne superconducting magnetic spectrometer to search for antimatter in the incoming cosmic rays, the only sample of material we had from beyond the solar system.

Beginning in 1971, I served as team field leader in such remote areas as Palestine, Texas and Aberdeen, South Dakota on several balloon launches designed to study the sky from a high altitude, away from misleading signals on the ground and produced in the intervening atmosphere. Our search for antimatter found no evidence of it, changing the question to: Why was there this slight part-per-billion excess of matter over antimatter in the early universe? Our limits for anti-carbon and anti-oxygen reached the one part per 10,000 level, and the observation was recognized in 1973 by the American Institute of Physics as one of the world's twelve outstanding physics experiments of the year.

In 1973, I turned my attention back to the cosmic background radiation, which was thought to be relic from the Big Bang and it could be used to indicate whether the universe was rotating, as some had theorized, or simply expanding without rotation. It could test if the expansion was uniform in all directions and also should have residual traces from the primordial perturbations that eventually made clusters, galaxies, and eventually stars and planets. I chose to work on measuring cosmic background radiation partly because I knew this: whatever we learned would be fundamental. Regardless of what we found, our observations would tell us about the early universe. James (P.J.E.) Peebles's 1971 book *Physical Cosmology* helped convince my colleagues that this was a worthwhile area of research.

With the backing of Luis Alvarez and the help of NASA, we developed and used a differential microwave radiometer (DMR) mounted in a high-flying U-2 spy plane to study the cosmic background radiation in 1976. The DMR measured differences in temperature as small as one-thousandth of a degree in the microwave radiation between two points. It consisted of two rotating measuring horns at sixty-degree angles pointed through the upper hatch of the plane. The observations showed that the universe is not rotating and that it is apparently expanding with uniform speed in all directions. Those findings were important, and the only signal we found was a dipole anisotropy. The Doppler effect makes the radiation appear warmer in the direction towards which we are moving, varying smoothly to cooler in the opposite direction – forming a dipole pattern. This observation determines the direction and speed of movement. The most startling discovery was that it was in the opposite direction of our galaxy's rotation. That dipole indicated that the Milky Way is moving at over a million miles per hour relative to the rest of the universe. This means that an object of enormous size (such as a supercluster of galaxies) must be exerting gravitational pull on our galaxy. Billions of years of acceleration brought it to this rapid speed.

Announced in 1977, astronomers regarded the discovery suspiciously, because most of them believed that objects were spread evenly throughout the universe. We won over skeptics by repeating the experiments in the Southern Hemisphere, eliminating the possibility that we had been misled by a location affect.

In a homogeneous universe, objects large enough to produce that much gravity could not have formed. The data implied that some regions of the universe are virtually devoid of galaxies while others are filled with billions of galaxies – forming superclusters. That revelation demanded a reassessment of the origin of the universe. In the era of decoupling, it was assumed that matter was evenly distributed throughout the universe, hence the cosmic background radiation (CMB). The CMB is an image of that time, and should show a uniform temperature (as tests had indicated). With the new knowledge of the uneven distribution of matter, scientists came to believe that for the universe to develop as it has there must have been some areas denser than others at decoupling, with the consequence that there should be variations in the temperature of the cosmic background radiation visible today.

Since 1974, I had worked on a proposal to NASA for a satellite to measure and map cosmic background radiation, in addition to the balloon-borne and U-2 experiments. Out of the 120 proposals submitted to NASA, my group and two others interested in cosmic background radiation were selected. In 1976, NASA told us to join forces. The collaboration was named the Cosmic Background Explorer, or COBE (pronounced COH-bee), which carried three instruments to examine the beginnings of the universe: (1) three DMRs similar to the U-2 (but more sensitive) to map the cosmic background radiation; (2) a far infrared absolute spectrophotometer (FIRAS) to measure the spectral curve of cosmic background radiation, which would tell if the CMB was truly from the Big Bang; and (3) the diffuse infrared background experiment (DIRBE) to look for cosmic infrared background – the glow from the earliest galaxies. It took six years to convince NASA that we knew what we wanted to do and knew how to do it. During that time, as head of the DMR team, I began commuting between Berkeley (where I was working on various ground-based, balloon-borne and the U-2 experiments) and NASA's Goddard Space Flight Center, in Greenbelt, Maryland, where COBE was based.

NASA finally accepted the proposed mission and scheduled it for a late 1988 launch on the space shuttle instead of its original rocket concept, but after the January 28, 1986 Challenger shuttle tragedy, the COBE team was told that they would not have a place on future shuttles. Two other rocket mishaps in 1986 resulted in an indefinite postponement of all launches, leading the team to consider using a French rocket. Confronted by the possible embarrassment of having a major American science project launched on a French rocket, NASA gave the go ahead to launch COBE in 1989 on the last small Delta rocket, which required the size and weight of COBE to be greatly reduced. Having less than three years to redesign the satellite forced a hectic schedule. Scientists from the three instrument teams had to work closely with engineers to bring COBE to fruition. According to the Goddard Engineering

Newsletter, "the transformation of COBE from [the space shuttle] to a Delta launch is probably one of the greatest engineering challenges ever undertaken by the [Goddard Space Flight Center]." On November 18, 1989, the $160 million COBE satellite was launched from Vandenberg Air Force Base in California. Controlled from Goddard, COBE traveled about 16,800 miles per hour in a polar orbit about 560 miles above Earth and completed a survey of the sky every six months.

The first important discovery showing that the cosmic background radiation resulted from the Big Bang came from the FIRAS instrument, led by Dr. John Mather. The radiation measured by FIRAS perfectly matched the blackbody curve, meaning that the CMB had a thermal origin in the early universe.

In early 1991, anisotropies at many angular scales became evident in the data from the DMRs. By fall of that year, evidence of smaller ripples, or wrinkles of temperature fluctuations, had been detected by the DMR team. In light of the many invalidated theories and discoveries, I was obsessed with detecting possible causes of error. I demanded secrecy until we were sure that we were seeing cosmic fluctuations instead of signals from the sun, moon, galaxy, the instruments themselves, or anything else. To inspire diligence in seeking errors, I offered anyone on the team two free plane tickets to anywhere in the world if they could prove a mistake had been made.

The scientific community grew restless waiting for the results, and there were papers stating that the Big Bang was in trouble since the temperature variations had not been found. There were heated arguments among the COBE team about whether the findings were ready for publication. In late 1991, we had become convinced that the analysis was correct, and had a program of tests and studies to ensure this. I wanted to check for one more possible source of error: radio interference from our galaxy that was greater and different from the meager amount we anticipated. There were already maps of our galaxy's microwave emissions, but if they were wrong, then the DMRs' data could be as well. Therefore, in November I led a team to Antarctica for a month to make our own celestial map and to check the brightness of the sky in a new area. Antarctica was chosen because in the cold, dry air, it has less interference from earthly signals. Even though it was austral summer and the sun never set, the working conditions were harsh. Oxygen was thin and the temperature could easily reach seventy-five degrees Fahrenheit below zero. Withstanding the elements, we confirmed previous celestial maps. After my return, I donned a tuxedo to signify and lead the formal portion of the review that certified the COBE DMR results were a success in terms of data quality, collection, and analysis. The review included NASA officials and COBE and DMR team members on the testing, checking, and progress of COBE DMR results. However, we held back on announcing the finding until the fluctuations could be extensively examined and characterized, and publications were ready with interpretation and back up.

This great care was necessary because of the technical difficulties of finding such a small signal in a large background. In *Wrinkles in Time*, I compared looking for the cosmic background radiation to "listening for a whisper during a noisy beach party while radios blare, waves crash, people yell, dogs bark, and dune buggies roar." The key was removing unwanted signals from the picture to identify temperature fluctuations of a mere 1/100,000 of the CMB intensity. We examined the hundreds of millions of measurements in a myriad ways. The team confirmed the previously discovered dipole with relative ease, but identifying the smaller wrinkles proved difficult. The observed quadrupole was lower than anticipated compared to the higher order fluctuations (having smaller angular scale). Finally, removing the quadrupole effect from the data, I found that the observed fluctuations fit predictions for a scale invariant spectrum. To confirm this finding, I suggested to graduate student Charles Lineweaver to look at the data with the quadrupole removed, without telling him what he would find. He used a different version of the analysis program that he had written rather than the one I wrote. When Lineweaver left the office in the early morning hours, he slipped his results under my office door with the message: "Here are the plots you asked for. Eureka?" They agreed precisely with my own calculations. Everything had fallen into place.

After three more months of intensive work, and fine-tuning the details and text, the team submitted manuscripts for publication in the *Astrophysical Journal* and announced the results at a meeting of the American Physical Society in Washington, D.C., on April 23, 1992. I had the lead talk at the meeting and was followed by six other COBE members. The announcement that COBE had not only detected a quadrupole effect, the first evidence of structure in the early universe, but also smaller ripples in the temperature of the cosmic background radiation that are consistent with Big Bang theory, stunned the audience of scientists and touched off a media frenzy rarely seen at scientific conferences. Much of the attention and excitement outside of the scientific world focused on my comment at the press conference that "If you're religious, it's like seeing God." Michael Turner, an astrophysicist at the University of Chicago, declared, "They have found the Holy Grail of cosmology," and Stephen Maran, the editor of the *Astronomy and Astrophysics Encyclopedia*, said "It's like Genesis." The commotion was capped off by a *Newsweek* headline reading "The Handwriting of God." These creation analogies brought widespread notice and discussion of our findings.

The detected ripples in space-time, which measure about a part in 100,000 or 30 millionths of a degree differences in temperature, started as small lumps (quantum fluctuations) but over time expanded to the immense size that COBE observed. The denser areas of the early universe, which appear slightly cooler in the cosmic background radiation, would have had sufficient gravity to attract more matter. This resulted in a snowball effect, in which more matter causes more gravity that attracts more matter, and the lumpy areas and voids become more and more defined.

The major importance of the COBE team's findings is that it gives quantitative measures for what had been merely speculation, providing firmer ground on which to base further study. The discovery of wrinkles in space-time neither confirms nor refutes many of the abundant variations of the

Big Bang theory. Indeed, while weeding out some theories, the findings have invigorated many theories and new concepts, with scientists arguing over whether the wrinkles offer proof or refutation. This richness has resulted in a blossoming of the field and attracted many new young people to the discipline of cosmology. Successive experiments, including the satellite missions WMAP and Planck, have exploited these fluctuations as a tool to understand the properties and evolution of the universe with unprecedented clarity and precision.

In the years following this discovery and especially since receiving the Nobel Prize in 2006, I have seen that it is now my turn to be a mentor and role model as well as a spokesperson supporting basic science research in general. I want to encourage the next generation and hand over a science enterprise to my successors, which is as good or better than the excellent one I came into as a young scientist.

Figure 6. George F. Smoot announcing the COBE DMR finding at the April 23, 1992 press conference following the scientific talks.

COSMIC MICROWAVE BACKGROUND RADIATION ANISOTROPIES: THEIR DISCOVERY AND UTILIZATION

Nobel Lecture, December 8, 2006

by

GEORGE F. SMOOT III

Lawrence Berkeley National Laboratory, Space Sciences Laboratory, Department of Physics, University of California, Berkeley, CA 94720, USA.

1 THE COSMIC BACKGROUND RADIATION

Observations of the Cosmic Microwave Background (CMB) temperature anisotropies have revolutionized and continue to revolutionize our understanding of the Universe. The observation of the CMB anisotropies angular power spectrum with its plateau, acoustic peaks, and high frequency damping tail have established a standard cosmological model consisting of a flat (critical density) geometry, with contents being mainly dark energy and dark matter and a small amount of ordinary matter. In this successful model the dark and ordinary matter formed its structure through gravitational instability acting on the quantum fluctuations generated during the very early Inflationary epoch. Current and future observations will test this model and determine its key cosmological parameters with spectacular precision and confidence.

1.1 Introduction

In the Big Bang theory the CMB Radiation is the relic radiation from the hot primeval fireball that began our observable universe about 13.7 billion years ago. As such the CMB can be used as a powerful tool that allows us to measure the dynamics and geometry of the universe. The CMB was first discovered by Penzias and Wilson at Bell Laboratory in 1964 [1]. They found a persistent radiation from every direction which had a thermodynamic temperature of about 3.2K. At that time, physicists at Princeton (Dicke, Peebles, Wilkinson and Roll) [2] were developing an experiment to measure the relic radiation from the Big Bang theory. Penzias and Wilson's serendipitous discovery of the CMB opened up the new era of cosmology, beginning the process of transforming it from myth and speculation into a real scientific exploration. According to Big Bang theory, our universe began in a nearly perfect thermal equilibrium state with very high temperature. The universe is dynamic and has been ever expanding and cooling since its birth. When the temperature of the universe dropped to 3,000 K there were insufficient energetic CMB photons to keep hydrogen or helium atoms ionized. Thus, the primeval plasma of charged nuclei, electrons and photons changed into neutral atoms plus background radiation. The background radiation could then propagate through space freely, though being stretched by the continuing expansion of the universe, while baryonic matter (mostly hydrogen and helium atoms) could cluster by gravitational attraction to form stars, galaxies and even larger structures. For these structures to form there must have been primordial perturbations in the early matter and energy distributions. The primordial fluctuations of matter density that will later form large scale structures leave imprints in the form of temperature anisotropies in the CMB.

1.2 Cosmic Background Radiation Rules

As a young undergraduate I heard of Penzias and Wilson's [1] discovery of the 3°K background radiation and its interpretation by Dicke, Peebles, Roll and Wilkinson [2], but not until two or three years later did I begin to understand the implications and opportunity it afforded. I was a first year graduate student at MIT working on a high-energy physics experiment when Joe Silk, then a graduate student at nearby Harvard, published a paper [8] entitled "Fluctuations in the Primordial Fireball" with the abstract *"One of the overwhelming difficulties of realistic cosmological models is the inadequacy of Einstein's gravitational theory to explain the process of galaxy formation*[16]*. A means of evading this problem has been to postulate an initial spectrum of primordial fluctuations*[7]*. The interpretation of the recently discovered 3°K microwave background as being of cosmological origin*[8,9] *implies that fluctuations may not condense out of the expanding universe until an epoch when matter and radiation have decoupled*[4]*, at a temperature T_D of the order of 4,000°K. The question may then be posed: would fluctuations in the primordial fireball survive to an epoch when galaxy formation is possible?"*

My physics colleagues dismissed this work as speculation and not a real scientific enquiry. It seemed to me a field ripe for observations that would be important no matter how they came out. Obviously, there were galaxies. Determining if the radiation was cosmic was critical. If the 3°K microwave background was cosmic, it must contain imprints of fluctuations from a very early epoch when energies were very high. Silk's work also made me realize the enormously important role of the cosmic background radiation in the early universe. Going back to earlier times when the universe was smaller, one would reach the epoch when the radiation was as bright as the Sun. At this epoch the universe was roughly a thousand times smaller than present. This is impressively small but one could readily and reasonably extrapolate back another thousand in size and then the radiation would be a thousand times hotter than the Sun.*

* *If the radiance of a thousand suns*
were to burst into the sky,
that would be like
the splendor of the Mighty One
I am become Death, the shatterer of Worlds.

reported J. Robert Oppenheimer quote of the Gita at the first atomic bomb test 16 July 1945. Since the Gita's first translation into English in 1785, most experts have translated not "Death" but instead "Time". The atomic fireball when first visible would roughly be a thousand times the temperature of the Sun.

But in truth, if this was the relic radiation, then the pioneering calculations of Gamow *et al.* [3] tell us we can comfortably and reliably look back to the point where the universe was a billion (10^9) times smaller. This is the epoch of primordial nucleosynthesis when the first nuclei form and their calculations correctly predicted the ratio of hydrogen to helium and the abundance of a few light elements. At that epoch the temperature of the radiation was a million times greater (and 10^{24} times brighter) than that of the Sun. Any object placed in that radiation bath would be nearly instantly vaporized and homogenized. Even atoms were stripped apart. At such early times the nuclei of atoms would be blown apart. The very early universe had to exist in a very simple state completely dominated by the cosmic background radiation which would tear every thing into its simplest constituents and spread it uniformly about.

Also in 1967 Dennis Sciama published a paper [11] pointing out that if this were relic radiation from the Big Bang, one could test Mach's Principle and measure the rotation of the universe by the effect that rotation would have on the cosmic microwave background. It could rule out Godel's model of a rotating universe and its implied time travel supporting Mach's Principle and keeping us safe from time tourists. Here was another fundamental physics and potentially exciting observation that one could make, if the CMB were cosmological in origin.

Not long after (submitted October 1967, published April 1968) Stephen W. Hawking and George F. R. Ellis published a paper "The Cosmic Black-Body Radiation and the Existence of Singularities in Our Universe" [10] which used the early singularity theorems of Penrose, Hawking, and Geroch to show that if the CMB was the relic radiation of the Big Bang, and if it were observed to be isotropic to high degree, e.g. a part in 100, that one could not avoid having a singularity in the early universe. The rough argument goes that, if the CMB is cosmological and uniform to high level, say one part in X, then one could extrapolate the universe backwards to a time when it was 1/X smaller. If X is sufficiently large, then the energy density in the CBR (microwaves now much more intense and hotter) would be sufficient to close the universe and cause it to extrapolate right back to the singularity. The only premises in the argument were: (1) the CMB was cosmological, (2) it would be found to be uniform to about a part in 10,000 (X = 100 in their original optimistic argument but actually 10,000 in present understanding), (3) General Relativity or a geometric theory of gravity are the correct description, and (4) the Energy Condition that there is no substance which has negative energy densities or large negative pressures. Hawking and Ellis provided strongly plausible arguments against violation of the Energy Condition. This observation would certainly be a death blow to the numerous popular oscillating universe models and other attempts to make models without a primordial singularity. Once again we see theorists providing arguments for the cosmic implications that could be drawn from observations of the CMB – if it were truly cosmological.

One needed to be of two minds about the CMB: (1) be skeptical and test carefully to see that it was *not* the relic radiation of the Big Bang and (2) assume that it was the relic radiation and had the properties expected and then look for the small deviations and thus information that it could reveal about the universe. Early on one had to make a lot of assumptions about the CMB in order to use it as a tool to probe the early universe, but as more and more observations have been made and care taken, these assumptions have been tested and probed more and more precisely and fully. The history of the observations and theoretical developments is rife with this approach. The discovery of the CMB by Penzias and Wilson was serendipitous. They came upon it without having set out to find it or even to explore for some new thing. In retrospect the discovery, though serendipitous, was not in a vacuum. There were ideas back to the time of Gamow [3], Doroshkevich and Novikov [4], reinvented by Dicke and Peebles [5] that there should be a relic radiation. There were plenty of observations that in retrospect pointed that there was something there, e.g. McKellar's 1941 observations of the anomalous temperature of CN molecules in cold clouds, followed by a string of others having noticed something unusual. However, Penzias and Wilson made the definitive observations in the sense that they observed a signal, checked for potential errors, added calibrations, and otherwise made their case air-tight so that the world took notice.

This tremendously important observation was rapidly interpreted and then a number of theorists began to work out the possibilities and potential implications and make these known to possible observers. Observers, and often their funding sources, who have to invest a significant amount of effort, time, and resources like to have some assurance that the observations are likely to prove worthwhile.

I immediately understood that what we can actually observe of the relic radiation is its electric field, $\vec{E}\ (v, \theta, \phi, t)$ (or magnetic field $\vec{B}$), so I made a table in text book fashion of the various things one could measure about the radiation based on observing the electric field here and now. My idea was to check each of these in a systematic way to establish clearly that the CMB was or was not the relic radiation from the Big Bang and then find out what it could tell us about the early universe. First is the frequency v spectrum of the radiation. If the 3°K radiation were truly the relic radiation from the early hot universe in thermal equilibrium, then it would have the famous blackbody spectrum whose careful formulation by Max Planck in 1900 initiated quantum theory:

$$\bar{n} = \frac{1}{e^{\frac{hv}{kT}} - 1} \qquad B(v) = \frac{8\pi h v^3}{c^2} \frac{dv}{e^{\frac{hv}{kT}} - 1} \qquad B(\lambda) = \frac{8\pi h c^2}{\lambda^5} \frac{d\lambda}{e^{\frac{hc}{\lambda kT}} - 1} \tag{1}$$

where $\bar{n}$ is the mean photon occupation number per quantum state and $B(v)$ (and $B(\lambda)$) is the brightness in units of energy per unit area per second per unit bandwidth (per unit wavelength). This spectrum has the property that it is precisely well-prescribed by only one parameter, its temperature T_{CBR}. The demonstration that this was likely to be true took years of effort with many misleading results along the way. The theory of potential slight distortions from the blackbody shape and what that might reveal also took time to develop and be absorbed by observers.

Likewise, one could map the incoming radiation as a function of position on the sky designated by the angles θ and Φ. In the simplest possible Big Bang model, the relic radiation would be isotropic, that is, independent of the angles θ and Φ on the sky. To first order, as Penzias and Wilson had shown, the 3°K radiation was isotropic, but as Joe Silk [8], Sachs and Wolfe [7], and others pointed out, there must be some residual perturbations to give rise to galaxies and clusters of galaxies and they give rise to temperature fluctuations across the sky. In these earliest days the fluctuations were anticipated to be fairly large (slightly below the 10% level limit by Penzias and Wilson) but after careful study they were predicted to be at the one part in a thousand level ($\Delta T/T \sim 10^{-3}$). Later the theoretical predictions were to get much smaller.

The vector direction plane of the oscillating electric field $\vec{E}$ is expected to be completely random from purely thermal radiation of a universe in complete thermal equilibrium and high opacity. However, in 1968 Martin Rees [13] pointed out that the small temperature fluctuations and Thompson scattering at the last scattering surface would give rise to a very slight linear polarization of the CMB.

The time dependence t of the electric field $\vec{E}\,(t)$ shows up in two ways. The first way is that thermal radiation has not only a well-defined distribution but also a well-defined statistical fluctuation spectrum. Specifically, the variance of the number of photons per unit mode n due to the thermal statistical fluctuations should be of the form

$$< n^2 - \bar{n}^2 > = \bar{n}^2 + \bar{n} \tag{2}$$

where the first term is called wave noise and the second term is called the shot noise of the individual photons. At low frequencies $h\nu << k_B T$ (Rayleigh-Jeans regime) then the wave noise dominates and the rms fluctuations are simply $\bar{n} = 1/(e^{h\nu/k_B T} - 1) \simeq k_B T/h\nu$. The rms fluctuations are proportional to the temperature T. At high frequencies $h\nu >> k_B T$ (the Wien tail), the shot noise dominates. This is a phenomenon that my group tested at low frequencies using correlation radiometers in the 1970s. Likewise, the early bolometer experiments tested the other regime indirectly and this is an assumption that continues forward in present observations, particularly those near the CMB peak where both effects are significant.

There is another second order effect in the correlations of the photons first made manifest in the Hanbury-Brown and Twiss interferometer and, though tested, it is not so central to CMB observations.

The second time dependence is that as one were to observe the radiation in the distant past, its temperature should increase in direct inverse to the scale size of the universe: $a(then)T_{then} = a(now)T_{now}$ or $T_{then} = (1+z)\,T_{now}$ where $1+z = a(now)/a\,(then)$ with a being the scale size of the universe at the epochs of interest. This is simply the stretching of wavelengths with the scale change of the universe combined with the Planck law. A number of groups have done experiments to check this dependence and found reasonable but so far limited evidence that supports this dependence. There is abundant evidence that the CMB is not a local phenomenon in that otherwise cold dense molecular clouds in our galaxy and nearby galaxies show additional excitation which just matches the energy input from the CMB. However, as we and others have found out, it is more difficult to make these observations in very distant galaxies.

z	T (K)	Molecule	Quasar	Reference
1.776	<16@2σ	C I	QSO 1331+170	Meyer *et al.* 1986, *Ap.J.*, 308, L37
1.776	7.4 ± 0.8	C I	QSO 1331+170	Songaila *et al.* 1994b, *Nature*, 371, 43
1.9731	7.9 ± 1.0	C I	QSO 0013-004	Ge *et al. Ap.J.*, 1997, 472 astro-ph/9607145
2.309	<45K@2σ	C II	PHL 957	Bahcall *et al.* 1973, *Ap.J.*, 182, L95
2.909	<13.5K@2σ	C II	QSO 0636+680	Songaila *et al.* 1994, *Nature*, 368, 599
4.3829	<19.6K@3σ	C II	QSO 1202-07	Lu, Sargent, Womble, Barlow 1995, Preprint

Table 1. The Temperature of the Cosmic Background Radiation for a few redshifts *z*. Values of the CMB temperature from the observation of the fine-structure transition of the C I and C II.

In taking into account the real universe with real galaxies and clusters of galaxies, there was another probe of the fact that the CMB fills the universe and another eventual cosmological probe with it. In 1970 and more explicitly in 1972, [14] Rashid Sunyaev and Yacob B. Zel'dovich predicted that the hot ionized medium in galactic clusters provided sufficient free electrons to scatter a small percentage of the CBR photons passing through the cluster. On average, since the electrons were hotter than the CBR photons, they would scatter the photons preferentially to higher frequencies causing a diminution of photons at low frequencies and a surplus at high frequencies. This meant that a cluster of galaxies would cast a faint shadow at low frequencies and glow at higher frequencies, since the CBR photons would come from the greatest possible distances. It was also clear that this was a spectral effect and would be independent of red shift and could be used to observe galaxy clusters across the full observable universe. In 1974, to look for the SZ effect, Rich Muller and I went to use the Goldstone radiotelescope and its new maser receiver, a key part of NASA's Deep Space Net to observe the Coma cluster (match of beam size and low frequency of observation). Unfortunately, the observations were not quite sufficient to make the detection. However, Mark Birkinshaw [15] and others continued to pioneer these observations over the next two decades, improving the approach and level of detection. A significant breakthrough came with the use of the Hat Creek Observatory by Carlstrom, Holzapfel, *et al.* [80] with clean high signal-to-noise observations of galaxy clusters showing the expected effect and correlation with X-ray observations. These established without a doubt that the CMB fills the universe and comes from far beyond the most remote galactic clusters observed. We are soon to see a substantial step forward in the utilization of the SZ effect beginning in 2007 with the observations from new instruments such as APEX-SZ and the South Pole Telescope (SPT) [81].

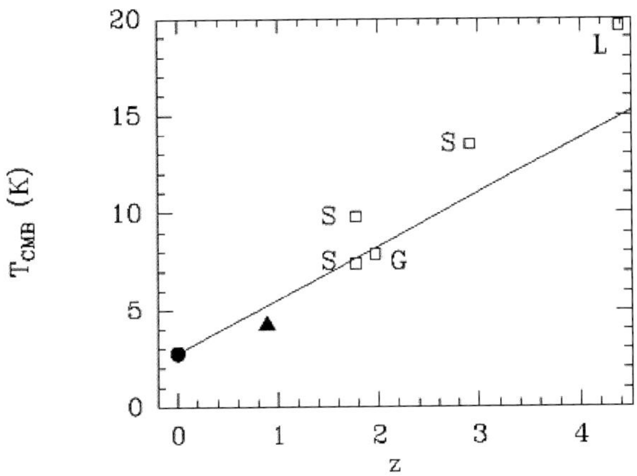

Figure 1. Summary of CMB temperature measurements as a function of redshift. The filled dot is from COBE (Mather *et al.* 1994). The squares are upper limits obtained on the CI or CII from Songaila *et al.* (1994ab, S), Lu *et al.* (1996, L) and Ge *et al.* (1997, G). Combes *et al.* is the filled triangle. The line is the (1+z) expected variation. Figure from Combes *et al.* 1999.

1.3 Transition to Cosmology

Though intrigued and highly motivated by the fledgling science of cosmology in the early 1970's, I first focused on finishing my graduate research to get my Ph.D. I did continue to pay attention to cosmology. One important factor was that Professor Steven Weinberg was at MIT at the time giving a cosmology course whose notes eventually turned into his very good book, *Gravitation and Cosmology*. I was not able to attend all the lectures, but did get a lot of the notes and later the book. Weinberg's clear interest and seriousness added credibility to cosmology among my colleagues. This provided a foundation and piqued my interest in the field while I was spending most of my graduate student time doing particle physics.

My Ph.D. research involved testing a rule of weak force decays that the change in charge of a kaon in the decay was equal to the change in strangeness. For this research, four graduate students: Orrin Fackler, Jim Martin, Lauren Sompayrac, and myself, under the direction of my advisor Professor David Frisch (MIT Physics Department), used a special beam of K^+ into a compact platinum target in the front of a magnetic spectrometer to produce K^0's and observe their decays in particle detectors inside the magnetic field. This was a highly technical and exacting experiment. We [17] found that the $\Delta S = \Delta Q$ rule (change in strangeness is matched by the change in charge of the particle decaying) was followed in the weak force decays. This rule is now understood as an automatic consequence of the quark model. This effort illustrates the temporal progress of science and how new young students are trained to do science. In our case, Professor Frisch gave us great and challenging tasks and responsibilities. He had us work independently much of the time, but there were people we could easily ask for advice and training. Now I was ready to move on and begin life as a newly minted postdoctoral scholar and find such a position. I investigated and interviewed for a number of jobs. Most of these were in particle physics which matched my training and my advisor's contacts. However, one interview was with Professor Luis Alvarez's group at Berkeley and, in particular, with a section that had been involved in trying to use energetic cosmic rays to push the frontiers of particle physics. They had met with a ballooning disaster in the program, were slowed in the original goals, and were looking to move in a new area. They were interested in flying a superconducting magnetic spectrometer to investigate the cosmic rays. Alvarez, like nearly all particle physicists at that time, knew that in every high energy interaction the conversion of energy into matter involved the production of equal amount of antimatter. Berkeley, in particular the Lawrence Berkeley National Laboratory, had been the scene for the discovery of antiproton and antineutron which established in everyone's mind that for every particle there was a matching antiparticle. Classically trained particle physicists thought at that time that in the Big Bang Model there would be equal amounts of matter and antimatter. The question was then, "Where was the antimatter?" We had a good idea that there was none on Earth and probably not in the solar system or we would be witnessing annihilation of matter and antimatter. Hannes Alfvén, an acquaintance of Alvarez, had a cosmological model in which there was an annihilation Leiden frost barrier that kept most of the matter and antimatter regions separate on a moderately large scale. Alfvén encouraged Alvarez and the group to search for some leakage between the regions in the most likely sample of material from great distances, the cosmic rays. A cosmic ray magnetic spectrometer was an ideal instrument for this antimatter search. The skills and techniques I had learned as a graduate student matched well with those needed for this research and Luis Alvarez and his colleagues, specifically Larry Smith, Mike Wahlig, and Andrew Buffington, recruited and encouraged me to join them in this effort.

We, including a number of very able technicians and engineers, designed, built, and flew, a number of times, superconducting magnetic spectrometers observing a sample of cosmic rays. As our search progressed our limits got progressively lower down to one in a thousand or less and then one in ten thousand or less. The first limit gets one out past the near neighborhood of stars. The second takes one to our whole Galaxy, and perhaps beyond, with evidence that there was little or any antimatter compared to the matter on that scale. To me the question changed from "Where is the antimatter?" to "Why is there an excess of matter over antimatter in our universe?" This currently remains one of the major questions of cosmology. Bear in mind that we have strong reasons to believe that there was an equal amount of matter and antimatter in the very early universe. At early times the Cosmic Background Radiation photons had enough energy to produce particle-antiparticle pairs and a simple thermal equilibrium would have essentially the same number of each species of particle and corresponding antiparticle as photons in the very early universe. Currently there are more than a billion CMB photons for every proton and neutron (and thus every electron). In the very early universe there would have been essentially the same number (per degree of freedom weighting) of every particle and antiparticle and all would have been relativistic behaving very much like photons or neutrinos all in strong thermal equilibrium. As the universe expanded and cooled, eventually the particles and antiparticles annihilated into lighter things including the CBR photons which by then were too cool to drive the reaction back the other way. Without some imbalance developing, there would both be much less matter around in the present and there would still be equal amounts of matter and antimatter separated in their sparseness.

In 1964, updated and clarified through 1986 as the need grew, Andrei D. Sakharov put forth the necessary conditions for what he called the baryon-asymmetry (matter over antimatter excess) to exist: (1) Baryon number violation, (2) CP violation, and (3) Non-equilibrium. Since that time theorists have been trying to find the correct theory and experimentalists evidence for these conditions.

During the later phases of these antimatter-search observations, I began to consider what to do next. Should we make an improved version of the experiment and probe deeper or should I strike off on something new? Alvarez offered the advice that one should periodically review what new developments had taken place. I distilled and codified his advice and other experience into: When you reach a natural pause, check as to see what new avenues are open because of (1) new scientific knowledge and ideas, (2) new instrumentation and techniques that open new areas for research, and (3) new facilities, infrastructure, or other support. An important ingredient was: What things could be brought together to enable significant research progress? A lot of judgment is necessary in this process.

In 1973 I dug out the 1967 paper by Dennis Sciama pointing out that one could test Mach's Principle and measure the rotation of the universe by the effect that rotation would have on the cosmic microwave background. It was not that specific on what the anisotropy pattern should be. There was also a 1969 paper by Stephen Hawking [16] which did provide cases for many Bianchi models, but was difficult slogging for the non-expert. Unfortunately, there was no clear idea of how fast the universe should be rotating except by analogy with the rotation of every thing in the universe from electrons to galaxies. This was insufficient to convince my colleagues or others that this was a measurement worth pursuing.

In 1971 Jim Peebles published his book, *Physical Cosmology* which was much more astrophysically and observationally oriented than Steven Weinberg's *Gravitation and Cosmology*. In *Physical Cosmology* Peebles had a section called

"Applications of the Primeval Fireball". In this section Peebles had a well developed discussion of the topic of what were the implications of the "possibly discovered primeval fireball", i.e. the cosmic microwave background. Peebles' writing was clear and easy to understand by nonspecialists. One application that Peebles laid out was entitled "The Aether Drift Experiment" in which one could use the CMB (zero net momentum of the radiation frame) as a reference to measure one's motion relative to the natural frame to describe the Big Bang expansion of the universe. The predicted temperature variation with angle θ to direction of motion due to the Doppler effect produced by the observer's motion is

$$T(\theta) = T_0\gamma/(1-\vec{\beta}\cdot\hat{n}) \simeq T_0(1+\beta\cos\theta) \qquad (3)$$

where $\vec{\beta}=\vec{v}/c$ and $\hat{n}$ is the direction of observation. Here was a well-defined project with an easy to calculate minimum signal. Astronomers knew that the solar system was moving as it orbited along with the rotation of our Galaxy. The orbital speed is known to be about 200 km/s or about $v/c = \beta = 0.7 \times 10^{-3}$. This gives an expected signal of about 2 mK (0.002 K). Astronomers who thought about it also thought our galaxy and Andromeda were co-orbiting each other so that there was an additional component of motion. But very, very few even thought about it at the time. There were a couple of papers with predictions.

The first was Dennis W. Sciama's 1967 paper, "Peculiar Velocity of the Sun and the Cosmic Microwave Background" which predicted: "*The sun's peculiar velocity with respect to distant galaxies is roughly estimated from the red-shift data for nearby galaxies to be* $\sim$*400 km/sec toward* $l^{II} \sim 335°$, $b^{II} \sim 7°$. *Future observations on the angular distribution of the cosmic microwave background should be able to test this estimate, if the background has a cosmological origin. If the test is successful it would imply that a "local" inertial frame is nonrotating with respect to distant matter to an accuracy of* 10^3 *sec of arc per century, which would represent a 5000-fold increase of accuracy.*" The second paper was a follow up of the first by J. M. Steward and D.W. Sciama [19] entitled "Peculiar Velocity of the Sun and its Relation to the Cosmic Microwave Background". Its abstract summarized: "*If the microwave blackbody radiation is both cosmological and isotropic, it will only be isotropic to an observer who is at rest in the rest frame of distant matter which last scattered the radiation. In this article an estimate is made of the velocity of the Sun relative to distant matter, from which a prediction can be made of the anisotropy to be expected in the microwave radiation. It will soon be possible to compare this prediction with experimental results.*"

1.4 Why not seek the Seeds of Galaxy Formation First?

Why not seek the seeds of galaxy formation which at the time was predicted to be at the same level? The angular scales of the anticipated signals were very different. One of the largest (angular-size) clusters on the sky was the Coma cluster which is approximately half a degree on the sky. Most clusters are in the arcminute range and galaxies are in the arcsecond range. With the receiver technology of the time, observations would have to be made at long wavelengths and that would require very large radiotelescopes dedicated for long periods of time. The radiotelescopes were not designed for this type of observation and thus prone to a number of potential systematic effects including significant ground signal pickup. One could readily estimate the expected angular scales for what was then thought to be a universe full mostly of isolated galaxies in some Poisson distributed fashion. One could estimate the causal horizon to be of order 2 degrees and primordial galaxy seeds as one hundredth that angular size (roughly an arcminute or so). In this old picture one would expect a sky speckled with tiny arcminute spots at the mK level, while the Doppler Effect from the Aether Drift, non-uniform Hubble expanson, or the rotation of the Universe promised signals that were large features and coherent on the sky that might unveil new physics.

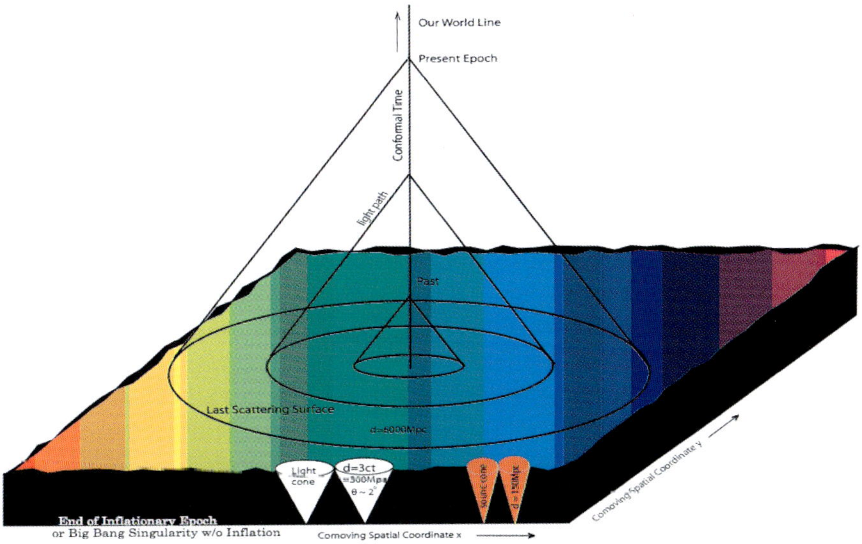

Figure 2. A Space-Time diagram in units of conformal time (vertical) $\eta=\int cdt/a(t)$, where $a(t)$ is the scale factor of the universe, and comoving coordinates which are converted to physical distance by multiplying by the scale factor $a(t)$. In these coordinates light travels on a 45° angle. The line in the center is our matter's path through time shown with no peculiar motion (very small in practice). The universe is shown opaque until the last scattering surface from either the end of the Inflationary epoch or the Big Bang singularity. The thickness of this is shown exaggerated relative to the subsequent elapsed time until the present (now) so as to show the causal horizon (distance that could be covered by the speed of light $d=3ct$ in physical units) and the sound horizon (distance that would be covered by speed of sound in the early universe). These two horizons have special imprint upon the physical structures in the universe.

1.5 *Beginning the New Aether Drift Experiment*

So now here was a project that had a guaranteed signal of well-defined angular dependence, and amplitude. This made it a good candidate to propose to colleagues, funding agencies, etc. One problem to overcome was the strong prejudice of good scientists who learned the lesson of the Michelson and Morley experiment and Special Relativity that there were no preferred frames of reference. There was an education job to convince them that this did not violate Special Relativity but did find a frame in which the expansion of the universe looked particularly simple. More modern efforts to find violations of Special Relativity look to this reference frame as the natural frame that would be special so that perhaps the suspicions were not fully unfounded. We had to change the name to "The New Aether Drift Experiment" and present careful arguments as the title "Aether Drift Experiment" was too reminiscent of the Michelson and Morley Ether Drift Experiment.

With that behind us my colleagues Rich Muller and Terry Mast were interested enough to learn more and begin outlining the experiment and encouraging and winning over others. Eventually, enough colleagues were convinced that some of the skilled technical staff in the group could be used to help develop the experiment. Key technical people were Jon Aymon – software, Hal Dougherty – mechanical, John Gibson – electronics, Robbie Smits – rotation system, John Yamada – technical assembly. Seed funding for components and shops was first procured and then proposals to NASA and so forth as the experiment began to form. A key step was recruiting graduate student Marc V. Gorenstein to work on the project. I had known Marc as an undergraduate at MIT before he came to Berkeley Physics graduate school and this connection helped just as Mike Wahlig and Andy Buffington had been at MIT in Frisch's group while I was an undergraduate and then new graduate student before they had come to Berkeley. There was a chain of contacts, familiarity, and confidence that helped make connections.

Now we had a nucleus of a team and a well-defined objective – build an instrument with sufficient sensitivity and precision to measure a CMB anisotropy at the 2 mK (10^{-3}) level on large angular scales. We began work and, with previous unfortunate experience with scientific ballooning, were considering using a differential microwave radiometer (DMR) on a U2 aircraft. Terry Mast peeled away to work on the nascent 10-m Telescope project. We were luckily joined by Tony Tyson taking a sabbatical to Berkeley from Bell Labs where he was working on developing gravity wave detectors. Tony had become expert on low noise detectors and vibration isolation, two key technologies we would need in this endeavor and which were important in the development of the DMR.

1.6 Context

1973 was when I began moving into work on CMB but it was a field that was already active on the east coast in significant part due to the activities of Jim Peebles and Bob Dicke leading to pioneering work by David Wilkinson, first with P. Roll in 1965 and then a succession of graduate students, e.g. beginning with Bruce Partridge [12] in 1967. There was spin-off from Princeton to MIT of Rainer Weiss who worked with Dirk Muehlner there. Both of these groups began with observations of the CMB spectrum and branched to anisotropy measurements. I chose to begin with anisotropy and move to the spectrum and other aspects later.

In 1970 Joe Silk came to Berkeley and began the theoretical cosmology effort creating a west coast effort and began to influence his colleagues to consider cosmological observations. Soon afterwards Prof. Paul Richards began a program taking on graduate students John Mather and then Dave Woody. Richards' program develops bolometers and Michelson Interferometer for spectrum observations and these are the precursor for COBE FIRAS. Significantly later these bolometers descendants become a key detector for CMB anisotropy observations. See the proceedings by my co-recipient John Mather.

Joe Silk and I developed a symbiotic student training program. Those that he wanted to get involved in analysis and understanding of observations would apprentice with me for a semester or a year. These students then helped with defining possible observations or working out some theory needed. Some of the students involved in this over the years were Mike Wilson, John Negroponte, and Eric Gawiser.

1.7 Why did we need such a strong team and effort?

The anticipated signal was at the level of one thousandth of the CMB (~3K) which in turn was one hundredth of the ambient temperature ~300K. The equivalent radio-signal receiver noises were in the same range. Thus the anisotropy was anticipated to be at a part in one hundred thousand (10^{-5}) of the noisy backgrounds. To have a significant measurement we would need to probe down to one tenth that level or a part in a million (10^{-6}). Thus we needed sensitivity to low signal levels which meant relatively long observation and stability of the instrument.

What were the techniques we could use? First we could use a technique championed by Bob Dicke in the 1940s that rapidly switches the receiver input between two sources and looks at the difference. The more the comparison was done with signals at the same level and the more quickly the inputs were switched, the less important would be the inevitable instrumental drifts due to the intrinsic $1/f$ electronic device noise and the roughly $1/f^2$ thermal environmental fluctuations that would prevent direct measurements of the CMB to the part in a million level. For measurements of the CMB one needed a reference at or near its 3K temperature. For spectrum measurement one would use a reference load cooled with liquid helium to achieve this. Our approach for the anisotropy experiment was to use two identical antennas pointing at different portions of the sky and switch rapidly between them. This configuration we called a Differential Microwave Radiometer (DMR).

It was then necessary to exclude, reject, average out other signals and sources of noise. We had to choose an observation frequency in which the CMB fluctuations would be larger than (or at least distinguishable from) those from other sources, particularly our own Galaxy. This led us to choose a roughly 1 cm wavelength and chose where we looked in the sky. Except near the galactic plane the CMB anisotropy should dominate and 1 cm was a wavelength that was relatively minimal atmospheric emission and so had been chosen by microwave pioneers as K-band. When it was realized that

K-Band had a water-line in it, the band had been readjusted by microwave engineers to be K_A band. Thus there were standard microwave components that were optimized for this wavelength range.

The electrical noise of the receiver produced background fluctuations that were of order

$$\delta T_{rms} = \frac{2T_{system}}{\sqrt{B\tau}} + \frac{\Delta G}{G} T_{diff} \sim \frac{27mK}{\sqrt{\tau/\sec}} + 100mK \frac{\Delta G}{G} \sim \frac{0.5mK}{\sqrt{\tau/hour}} \quad (4)$$

where T_{system} ~$300K$ was the effective receiver noise temperature of ambient temperature receivers of that epoch, B was the bandwidth on the order of 500 MHz and τ was the observation time, and ΔG was the change in receiver power gain G in the time period of the observations preferably set by the switching between inputs of receiver whose effective temperature difference was T_{diff}. (Note that these two effects should in general be added in quadrature as they would be uncorrelated.) The first is simply due to the variance of $\bar{n}^2 + \bar{n}$ in number of photons observed due to thermal fluctuations of blackbody radiation and the second to receiver gain drift. By making the temperature difference T_{diff} small - $|T_{diff}| < 0.1K$ was possible to achieve, one could hope not to increase the rms noise significantly as long as the receiver gain variation was kept significantly less than 0.5% for the switching time for the required sensitivity of about 0.3 mK. To achieve this level of sensitivity we would need to observe each patch of the sky for about two hours.

Thus our plan was to average down the random noise in two hour chunks but we also had to exclude signals that were not random. A key issue was the rejection of signals coming from off the main beam axis. A fundamental property of optics is that diffraction will cause the beam to have off-axis response. The usual antenna technology of the time with the lowest sidelobes (off-axis response) was the 'standard gain horn' which has the optimum gain for a simple pyramidal horn configuration. This horn basically is a smoothly expanding waveguide and has in one plane (E-plane which is in the same plane as the electric field vector) a uniform illumination to the edge of the horn. The illumination in the orthogonal direction (H-plane) varies as a sine wave with zero amplitude at the waveguide (horn) edges and peaking in the middle. This field configuration is simply the lowest and best supported mode of the waveguide and the one for which all the other components are designed to utilize.

The far field (equivalent to the beam response) is simply the Fourier transform of the aperture electric field. The Fourier transform of an electric field that is zero outside the horn and uniform inside the horn is the familiar to physicists sin(x)/x pattern. For a reasonable horn size the beam is fairly broad, but also, more importantly, the sidelobes are typically only down by a factor of ten thousand at 90° to the beam axis. Since the ground is non-uniform and a million times greater than the anticipated anisotropy signal level, we needed a better solution. I decided that I had to learn antenna theory and find what could be done to get to lower sidelobes. The H-plane pattern with its sine wave illumination, specifically the tapering of the field to zero at the edges of the horn aperture, has quite low sidelobes and points the path towards the solution. One would want electric field illumination that tapered smoothly to zero at the aperture edges. Ideally one would like the field and its derivative to be zero at the edge, even though that meant that for the given aperture diameter the forward gain was lower since it was under-illuminated relative to uniform illumination (hence the optimum standard gain horn design). The fact that the H-plane beam pattern was quite low meant that one could achieve the necessary low off-axis response as long as the electric field tapered reasonably to zero.

Another way to look at the issue is that the energy in the wave is stored in the electric field and considering a wave going in the time-reversed direction, the issue was how to take the field tightly coupled to the waveguide and send it out the antenna and have it separate from the antenna and match into propagating freely in space. If the electric field is not zero at the metal on the end of the horn near the aperture, then the electric field generates currents in the metal to make the field close to zero in the conducting metal. These currents then cause field to propagate out at other directions. So again by the end of the antenna we need the electric field decoupled and zero at the metal surface. There are two approaches to this that eventually were used in the two CMB instruments on the COBE satellite. The first approach is to flare the ends of the horns very much like the bell on a trumpet or a trombone which as musical instruments have a similar issue of emitting sound waves from tightly coupled at the mouth piece but freely propagating once they leave the horn. Hence the pictures I showed of the Princeton (Wilkinson group) anisotropy experiment with the musical instrument bells on the end of their horns – but for receiving EM-waves, not transmitting them. The electromagnetic wave prefers to propagate along a straight path and effectively peels away successively along the curve. This approach has the benefit of working for a large range (bandwidth) of frequencies and the disadvantage of extending the size of the aperture substantially – the more one needs off-axis rejection the larger the flare must be. This was the approach used in the COBE FIRAS instrument where there was a single large external horn antenna that had to work well over an extended wavelength range.

The second approach which I eventually pursued was to separate the electric field from the antenna very early and use the rest of the antenna to keep defining and shaping the beam and then to put in quarter wavelength deep grooves at the ends of the aperture. Quarter wavelength deep grooves would then force currents exactly out of phase with the electric field (1/4 wavelength down and 1/4 wavelength back meant 1/2 wavelength or 180° out of phase). This chokes off the surface currents in the horn aperture and does not allow them to go out and around the horn to make far and back lobes. The issue at the horn throat is to excite a second mode that has the property that at the center of the beam its field is in phase with the standard mode but at the E-plane edges its field is out of phase and just cancels the electric field from the standard first mode giving a field pattern that is very similar to the H-plane and has very low sidelobes. I studied the literature, consulted with engineers from TRG Alpha in Boston Massachusetts, and got from JPL a copy of their software JPLHORN for calculating beam patterns which I modified and used. Soon it was clear that one could do this quite well with what is called a corrugated-horn antenna, especially in the case of a conical horn. The first groove needed to be a half wavelength deep so as to not develop too much reflection and then one could either tune and go directly to quarter wavelength (in the cone) deep grooves which was easier to fabricate or, as we did on later horns, taper the groove depth from half-wavelength to quarter wavelength depth in a few (5 to 10) grooves and have the remaining thirty or so grooves at quarter wavelength depth. This configuration produced very low far sidelobes, in a very compact configuration, and had very low losses in the antenna since the electric field did not produce significant currents in the antenna wall. It had the draw back that it was relatively expensive to make since it required a very good machinist working on fairly large forged aluminum blocks to cut in all those grooves precisely. This development was sufficiently successful that eventually we had to develop new techniques to observe sidelobes this low (necessary for the COBE DMR) and used an antenna range at JPL sited on the edge of a mesa [25]. This work was repeated for the COBE DMR antennas at GSFC in a specially developed range [26]. This was a key development since to measure the CMB anisotropy precisely, one must achieve off-axis rejection to a part in a billion level or better, and for the DMRs the antennas needed to be sufficiently compact to fit within the available space.

The design called for two identical horns whose output was rapidly and alternately switched into the receiver. As long as the horns were identical and they looked out through identical atmosphere the measurement should be sufficient. However, things are not perfectly identical and so we had to have a back up which was that we must rotate the receiver and interchange the position of the antennas on the sky so we could separate out any intrinsic signal from the instrument from that coming from the sky. This was a generic issue which over the years my students referred to as Smoot's Switch Rule: As soon as one introduced a switching (or technique) to cancel out or correct for an effect, one had two new effects to be concerned with (1) did the device produce a signal itself and (2) did the process (e.g. rotating the instrument) produce a signal. These effects always occur at some level so one has to make sure that they are small and compensated for in the design. For example, the switching of the receiver input from one antenna output to another always introduces a spike, step or some form of extra signal during the process, so one does not include that as part of the signal stream that continues on. Likewise the switch has a slight offset when connected to one antenna compared to the other. So one measures and adjusts this as well as possible and then makes sure to rotate the apparatus so that the sky signal is interchanged (and thus of opposite sign) as to which horn antenna it enters. Then one must check that the process of rotation does not change the state or performance of the DMR, e.g. from the Earth's magnetic field, or other effects. In general, since we are measuring such a small signal, one had to be concerned to roughly third order in things as well as conduct many tests and analyses.

1.8 The DMR and U-2 Observations

Finally we had developed the instrumentation and approach to observe the CMB and to detect the first order anisotropy due to the motion of the instrument relative to the last scattering surface and thus the zero momentum frame for the CMB. The instrument used in the experiment was a DMR, eventually described in detail in a paper [22]. The vehicle of choice was the high-flying and very stable U2 jet aircraft famed for making flights from Turkey to Scandinavia as well as over other hot spots where high resolution (thus stable platform) photographs taken from high altitude were of use. The U2 had been converted to performing environmental and earth resources observations in a program run by the NASA Ames Center from Moffett Field, California. After a series of flights and subsequent data processing and analysis we detected [20] this first order, dipole anisotropy, as well as showing that it was the dominant large scale signal on the sky [21]. Later we were to put together a program to make these observations from Peru so that we could see that the pattern held over the southern sky as well [23]. The U2 experiment revealed that the temperature varied smoothly from -3.5mK in a direction near the constellation Aquarius to +3.5mK in the direction near the constellation Leo (see Figure 3). Surprisingly, this meant that our solar system was moving at 350km/s, nearly opposite to the direction that was expected from the rotation around the Galactic center. This result forced us to conclude that the Milky Way was moving at a speed of about 600km/s in the direction near the constellation Leo. This motion was not expected in the model where galaxies are simply following undisturbed world lines given by simple Hubble expansion of the Universe, which was the idealized version that most cosmic astronomers held to at the time. It also implied that the Andromeda Galaxy as well as the many smaller members of the local group were also moving along with similar velocities. This group motion implied that there is a gravitational center (later named the "Great Attractor") of a huge clump of matter relatively far away so that its pull was uniform enough not to disrupt the weakly bound local group.

Even though the dipole was not of direct cosmological origin, it carried a significant meaning on how matter is organized in the universe and therefore on what conditions must exist at the onset of the Big Bang. Once we could convince astronomers that this was correct and the Great Attractor or equivalent could be found, then we would also achieve Dennis Sciama's test of Mach's Principle and our relative rotation with respect to the distant matter in the universe. It took some time before astronomers took this seriously (some encouragement from UCLA astronomer George Abell helped) and work began that eventually led to understanding that the existence of clusters and superclusters of galaxies and our motion as well as other bulk motions were a natural consequence of the large scale organization of matter. At the same time searches on larger and larger scales (shells at greater radius and redshift) began to converge on the 'right' answer given by the CMB. The current best observed dipole, (3.358 ± 0.017 mK), indicates that the Solar System is moving at 368 ± 2 km/sec relative to the observable Universe in the direction galactic longitude l = 263.86° and latitude b = 48.25° with an uncertainty slightly smaller than 0.1° [82]. This is quite far from the Galactic rotation direction (nominally 250 km/s toward l = 90° and b = 0°).

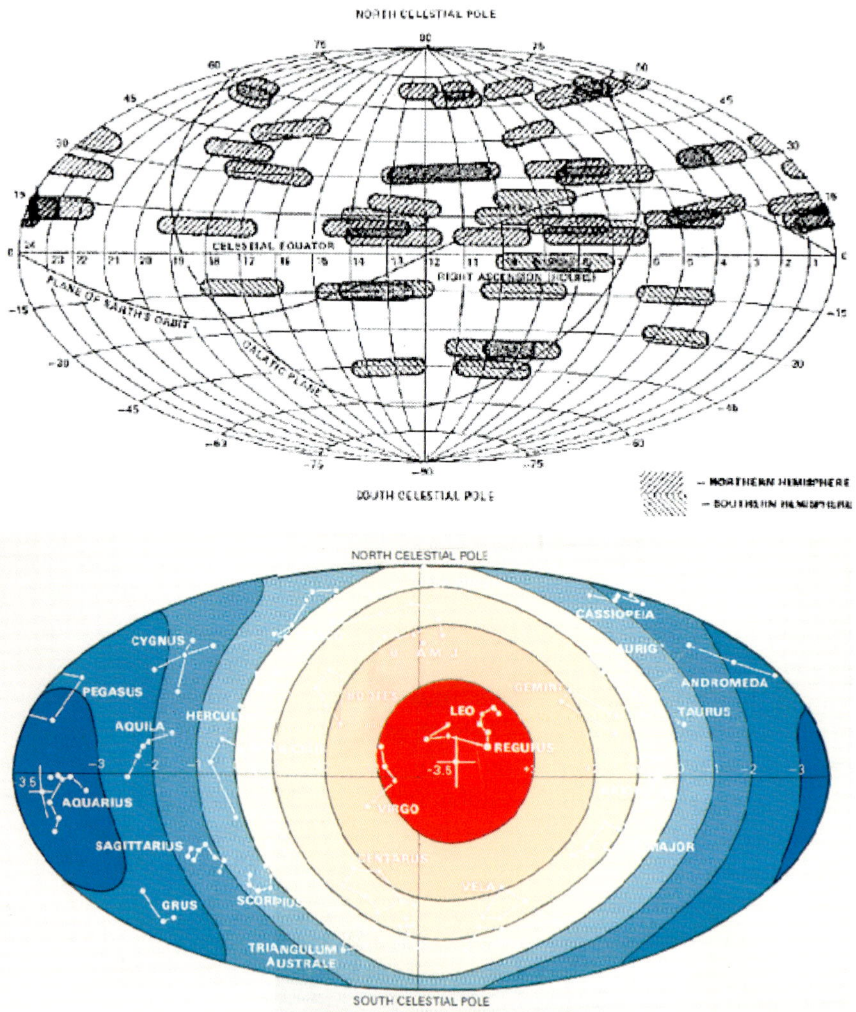

Figure 3. Dipole anisotropy measured by U-2 flight experiment (1976). Top panel: Sky covered by U-2 flight experiments in northern and southern hemispheres. Bottom panel: The dipole map made by U-2 experiments. The red spot centered near the constellation Leo indicates +3.5mK from the median background temperature and the blue spot centered near the Aquarius is -3.5mK region [20, 22].

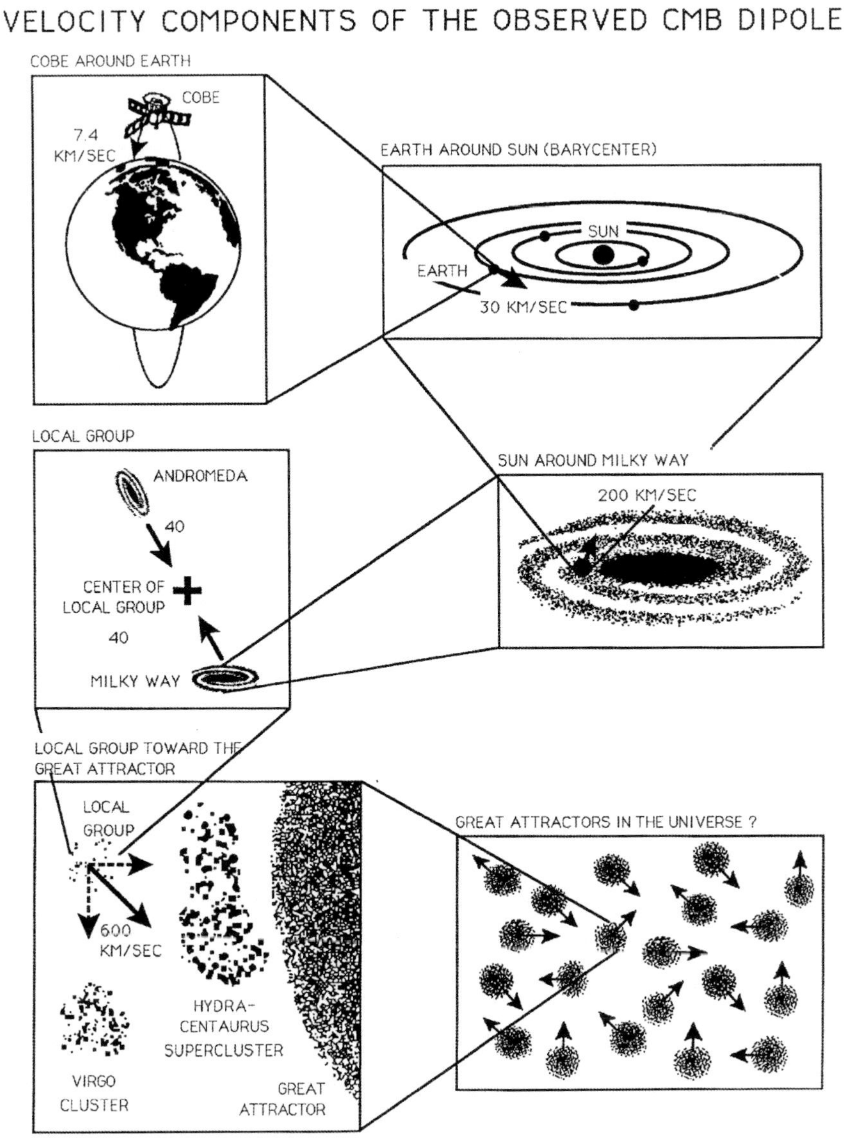

Figure 4. Dipole anisotropy results from the sum of many components of velocity due to gravitational attraction of various mass concentrations.

Surprisingly, it is closer to the original prediction in his 1967 papers by Dennis Sciama of "~400 km/sec toward $l^{\mathrm{II}} \sim 335°$, $b^{\mathrm{II}} \sim 7°$" which was based upon very sketchy observations of the time. The meaning we could take from this was that stepping back and taking a skeptical look with a clear mind did allow one to realize that there was large scale structure in the universe and some expected variation from simple Hubble flow due to the small acceleration of distant gravitational attraction operating over billions of years. If large voids and superclusters could form, then there must be some bulk flows. The motion of our Galaxy is just a bit above the norm; hence the need for the "Great Attractor" which is one of many in the universe.

Detection of the intrinsic CMB anisotropy was a great technical challenge on its own right, for it required an accuracy of one part in 100,000. Galactic and extragalactic emissions, foreground emissions and noise from the instruments themselves added noise to the measurement which were much larger than the CMB signal. Even U-2 flight experiments, which were carried out at altitudes above 65,000 feet (20 km), and which were above 95% of atmosphere (balloon experiments are also performed at similar condition), were not enough to catch such a subtle primordial whisper. Space-based observations would provide far better results, so the next phase of the CMB measurements moved on to satellite-borne experiments.

1.9 Polarization of the CMB

Penzias and Wilson set the first limit on the polarization of the CMB at less than 10%. Then Martin Rees came forward with his prediction [13] that the CMB should be linearly polarized at what should be a few per cent of any intrinsic anisotropy. In the 1970s then graduate student George Nanos in Wilkinson's Princeton Group [27] and our group [28, 29] began observations to improve the limits on the polarization of the CMB and test Rees's prediction. At the time we were still thinking in terms of classical large scale anisotropy. The simplest version would be the case if the universe was expanding at different rates in the three orthogonal directions. The optimal case would be a quadrupolar axisymmetric expansion which would couple maximally to linear polarization.

Nanos completed his thesis observations and work and published in 1979. Nanos's abstract reads *"An attempt is made to detect linear polarization in the 2.7 background at a wavelength of 3.2 cm, using a Faraday-switched polarimeter pointed to the zenith where the earth's rotation carries through a circle of constant (40.35 deg N) declination. A two-step calibration process was employed. First, the change in dc voltage at the second detector was measured with a 300 K absorber to determine the gain of the microwave front end. Then, by inserting a known ac voltage at the lock-in frequency at the same point, the rest of the receiver was calibrated."* The null result was interpreted using Rees' axially symmetric model as an upper limit on asymmetry in the Hubble expansion. Nanos then went onto a career in the Navy becoming Vice Admiral and later director of Los Alamos National Laboratory.

Once the U2 experiment was progressing well and we had developed techniques and instrumentation, I felt it was time to begin polarization observations. I recruited graduate student Phil Lubin and convinced him to do, for his Ph.D. thesis, a polarization experiment at 1-cm wavelength. We calculated that observations could be made from the ground using the basic instrumentation that we developed for the U2 DMR but with a single antenna that accepted both linear polarizations and a switch to chop the receiver input quickly between the two polarizations. After study we determined that the atmosphere was sufficiently unpolarized that we could make our observations from the ground. We needed only point the antenna up and the rotation of the Earth would sweep out a strip on the sky. Once that worked we could tip the radiometer slightly north or south and obtain strips on the sky. Hal Dougherty produced the mechanical system construction and John Gibson produced the electronics and power supplies. We had to design the system to rotate about its axis to separate instrumental effects from sky signals. Phil then tuned the system up and began observations and ultimately found it worked well and we gathered data mostly from Berkeley but also some from the southern hemisphere (Lima, Peru).

Later, with improved design and construction by our outstanding mechanical tech, Hal Dougherty, we extended to 3-cm wavelength for Philip Melese d'Hospital's project and then on to 0.3-cm wavelength. Our data were continuing measurements of the linear polarization of the cosmic background radiation as well as provided the first measurement of the circular polarization. We surveyed eleven declinations for linear polarization and one declination for circular polarization, all at 9 mm wavelength [28, 29]. We found no evidence for either a significant linear or circular component with statistical errors on the linear component of 20–60 μK for various models. Our linear polarization, a 95 percent confidence level limit of 0.1 mK (0.00003) for an axisymmetric anisotropic model was achieved, while for spherical harmonics through third order, a corresponding limit of 0.2 mK was achieved. For a declination of 37°, a limit of 12 mK was placed on the time-varying component and 20 mK on the dc component of the circular polarization at the 95 percent confidence level. At 37° declination, the sensitivity per beam patch (7°) was 0.2 mK.

After these observations interest in the CMB polarization died down significantly until the COBE DMR discovery of intrinsic CMB anisotropies and then investigations of the CMB polarization has become a major interest of the field.

1.10 Balloon-borne Anisotropy at 3-mm Wavelength

After the detection of the dipole anisotropy and first estimated map of the CMB anisotropy, it was time to develop instruments to make a real map. Phil Lubin, now a new Ph.D., and I conceived of a new detector with much greater sensitivity than the U2 receiver so as to be able to make a map with observations of reasonable time duration. We knew that to improve the sensitivity of the receiver it would need to be cooled to cryogenic temperatures. We obtained a liquid-helium dewar and Hal modified it to hold the antenna and the receiver front end and so forth. We needed to move to higher frequency to keep the system compact and then, as a result, needed to be higher in the atmosphere to avoid atmospheric fluctuations. This led us to a cryogenic, balloon-borne system design. It was then up to us to produce the front end of the receiver while Hal and John produced the mechanical systems and the electronics. A critical piece of the effort was developing a large mechanical chopper to switch quickly where the beam intersected the sky and a reliable pop-up calibrator. I also recruited a new graduate student, Gerald Epstein, for whom this was his Ph.D. research project. After two flights in the northern hemisphere, Gerald Epstein had enough results for his thesis but we wanted to continue map making to cover the southern hemisphere. We then recruited Brazilian graduate student, Thyrso Villela, which was natural for balloon flights from Brazil. We were able to produce a map covering about 70% of the sky which showed the dipole anisotropy well and showed clearly that the higher order (quadrupole moment) and above were significantly lower. Equally important, the project showed that we could put together a compact cryogenic system that was capable of more sensitive observations of the CMB. This provided the confidence and heritage needed to improve the COBE DMR receivers, in mid-COBE DMR development. This was a key piece of evidence which was presented to the GSFC engineering management and ultimately to NASA Headquarters to convince them to allow the change to passively (radiatively) cooled COBE DMRs as the suborbital experiments and theory began to indicate that we were going to need every bit of sensitivity we could muster.

This balloon-borne project later morphed into the MAX, MAXIMA, BOOMERANG, and MAXIPOL experiments when a collaboration with Paul Richards' group and the developing bolometers technology made bolometer arrays a real possibility. Phil Lubin moved to be a Professor at UCSB and continue collaborating. Gerald Epstein moved to work in science policy beginning at OTA. Thyrso Villela took a position as a researcher and professor at INPE (Brazilian Space Agency) in San Jose dos Campos.

1.11 Spectrum of the CMB

Observations of the spectrum of the CMB started with the discovery observation by Penzias and Wilson combined with the original speculation that this was the relic radiation from the Big Bang and would to first order be a blackbody (Planckian) spectrum. Later theoretical studies confirmed that to high order one expected that the relic radiation should have blackbody spectrum because of the high level of thermal equilibrium expected in the early Universe.

A non-interacting Planckian distribution of temperature T_i at redshift z_i transforms with the universal expansion to another Planckian distribution at redshift z_r with temperature $T_r / (1 + z_r) = T_i / (1 + z_i)$. Hence thermal equilibrium, once established (e.g. at the nucleosynthesis epoch), is preserved by the expansion, even during and after the photons decoupled from matter at early times $z \sim 1089$. Because there are about 10^9 photons per nucleon, the transition from the ionized primordial plasma to neutral atoms at $z \sim 1089$ does not significantly alter the CBR spectrum [33].

Shortly after the Penzias and Wilson discovery and initial estimate of the CMB temperature, there were a number of observations and determinations/estimations of the temperature at various wavelengths which were the beginnings of the effort to establish that the CMB spectrum was blackbody. Dave Wilkinson and Peter Roll [34] were pioneers in radiometric observations beginning on the roof of Jadwin Hall at Princeton. Wilkinson and colleagues, first Stokes and Partridge [35], continued with a set of long wavelength ob servations from the White Mountain Research Station in California. This is a high altitude site operated by the University of California and one that is a good site for CMB observations because of its high altitude (12,000 feet), dryness, and reasonable accessibility (a road that is open for about half the year). As mentioned above, by 1974 Prof. Paul Richards began a program taking on graduate students, John Mather and Dave Woody. Richards' program develops bolometers and Michelson Interferometer for spectrum observations and these are the precursor for COBE FIRAS. The FIRAS instrument design came directly from the original White Mountain instrument, which was then morphed to be the Woody and Richards balloon-borne instrument. The FIRAS team, led by John Mather, studied the results and performance of the Woody and Richards instrument and experiment and designed FIRAS to be both as symmetric as possible and to operate at the same temperature as that from the sky input. Another key feature was the sky simulating blackbody which was carefully designed, crafted and tested to be a very good blackbody at a well-defined temperature.

ν (GHz)	λ (cm)	T_{CMB}^{th} (K)	Reference
0.408	73.5	3.7±1.2	Howell & Shakeshaft 1967, *Nature*, 216, 753
0.6	50	3.0±1.2	Sironi *et al.* 1990, *Ap.J.*, 357, 301
0.610	49.1	3.7±1.2	Howell and Shakeshaft 1967, *Nature*, 216, 7
0.635	47.2	3.0±0.5	Stankevich *et al.* 1970, *Australian J. Phys*, 23, 529
0.820	36.6	2.7±1.6	Sironi *et al.* 1991, *Ap.J.*, 378, 550
1	30	2.5±0.3	Pelyushenko & Stankevich 1969, *Sov. Astr.*, 13, 223
1.4	21.3	2.11±0.38	Levin *et al.* 1988, *Ap.J.*, 334, 14
1.42	21.2	3.2±1.0	Penzias & Wilson, 1965, *Ap.J.*, 142, 419
1.43	21	$2.65^{+0.33}_{-0.30}$	Staggs *et al.* 1996, *Ap.J.*, 458, 407
1.44	20.9	2.5±0.3	Pelyushenko & Stankevich 1969
1.45	20.7	2.8±0.6	Howell & Shakeshaft 1966, *Nature*, 210, 1318
1.47	20.4	2.27±0.19	Bensadoun *et al.* 1993, *Ap.J.*, 409, 1
2	15	2.5±0.3	Pelyushenko & Stankevich 1969, *Sov. Astr.*, 13, 223
2	15	2.55±0.14	Bersanelli *et al.* 1994, *Ap.J.*, 424, 517
2.3	13.1	2.66±0.7	Otoshi & Stelzreid 1975, *IEEE Trans on Inst & Meas*, 24
2.5	12	2.71±0.21	Sironi *et al.* 1991, *Ap.J.*, 378, 550
3.8	7.9	2.64±0.06	De Amici *et al.* 1991, *Ap.J.*, 381, 341
4.08	7.35	3.5±1.0	Penzias & Wilson, 1965, *Ap.J.*, 142, 419
4.75	6.3	2.70±0.07	Mandolesi *et al.* 1986, *Ap.J.*, 310, 561
7.5	4.0	2.60±0.07	Kogut *et al.* 1990, *Ap.J.*, 355, 102
7.5	4.0	2.64±0.06	Levin *et al.* 1992, *Ap.J.*, 396, 3
9.4	3.2	3.0±0.5	Roll & Wilkinson 1966, *PRL*, 16, 405
9.4	3.2	$2.69^{+0.16}_{-0.21}$	Stokes *et al.* 1967, *PRL*, 19, 1199
10	3.0	2.62±0.058	Kogut *et al.* 1988, *Ap.J.*, 325, 1
10	3.0	2.721±0.010	Fixsen *et al.* 2004, *Ap.J.*, 612, 86
10.7	2.8	2.730±0.014	Staggs *et al.* 1996b, *Ap.J.*, 473, L1

Table 2. Values of the CMB temperature at $v = 11$ GHz.

ν (GHz)	λ (cm)	T^{th}_{CMB} (K)	Reference
19.0	1.58	$2.78^{+0.12}_{-0.17}$	Stokes *et al.* 1967, *Phys . Rev. Lett.*, 19, 1199
20	1.5	2.0±0.4	Welch *et al.* 1967, *Phys . Rev. Lett.*, 18, 1068
24.8	1.2	2.783±0.089	Johnson & Wilkinson 1987, *Ap.J. Let*, 313, L1
30	1.0	2.694±0.032	Fixsen *et al.* 2004, *Ap.J.*, 612, 86
31.5	0.95	2.83±0.07	Kogut *et al.* 1996, *Ap.J.*, 470, 653
32.5	0.924	3.16±0.26	Ewing *et al.* 1967, *Phys . Rev. Lett.*, 19, 1251
33.0	0.909	2.81±0.12	De Amici *et al.* 1985, *Ap.J.*, 298, 710
35.0	0.856	$2.56^{+0.17}_{-0.22}$	Wilkinson, 1967, *Phys . Rev. Lett.*, 19, 1195
37	0.82	2.9±0.7	Puzanov *et al.* 1968, *Sov. Astr.*, 11, 905
53	0.57	2.71±0.03	Kogut *et al.* 1996, *Ap.J.*, 470, 653
83.8	0.358	2.4±0.7	Kislyakov *et al.* 1971, *Sov. Ast.*, 15, 29
90	0.33	$2.46^{+0.40}_{-0.44}$	Boyton *et al.* 1968, *Phys . Rev. Lett.*, 21, 462
90	0.33	2.61±0.25	Millea *et al.* 1971, *Phys . Rev. Lett.*, 26, 919
90	0.33	2.48±0.54	Boynton & Stokes 1974, *Nature*, 247, 528
90	0.33	2.60±0.09	Bersanelli *et al.* 1989, *Ap.J.*, 339, 632
90	0.33	2.712±0.020	Schuster *et al.* UC Berkeley PhD Thesis
90.3	0.332	<2.97	Bernstein *et al.* 1990, *Ap.J.*, 362, 107
90	0.33	2.72±0.04	Kogut *et al.* 1996, *Ap.J.*, 470, 653
154.8	0.194	<3.02	Bernstein *et al.* 1990, *Ap.J.*, 362, 107
195.0	0.154	<2.91	Bernstein *et al.* 1990, *Ap.J.*, 362, 107
266.4	0.113	<2.88	Bernstein *et al.* 1990, *Ap.J.*, 362, 107

Table 3. Values of the CMB temperature at frequencies v = 11 GHz. The measures from FIRAS, COBRA and CN molecules are excluded.

In the early 1980s, during the Woody and Richards experimental effort, my group started an international collaboration to measure the low frequency portion of the spectrum complementary to Woody and Richards high frequency observations. This collaboration included the University of Milano group headed by Giorgio Sironi, the Bologna group of Nazareno Mandolesi, University of Padua theorists Luigi Danese and Gianfranco DeZotti, and the Haverford group led by Bruce Partridge. We carefully developed special radiometers to measure the spectrum at wavelengths of 12, 6, 3, 1 and 0.3 cm (frequencies of 2.5, 5, 10, 30, and 90 GHz). We developed very large (0.75 m diameter) liquid-helium-cooled (3.8 K) reference loads so as to have a blackbody source near to the temperature of the sky and CMB. We had to develop techniques, model of galactic emission and where it was low, and most importantly models of the residual high altitude atmospheric signal. We not only modeled the atmosphere but also conducted zenith scans and the Haverford group operated a full time atmospheric monitor. We conducted a series of campaigns in successive summers at White Mountain and then for two successive years at the South Pole which is also a high and very dry site with a very stable and relatively small atmospheric signal. We published a series of papers on the theory, observations, interpretation leading to a much improved set of measurements at long wavelengths [36–41, 45]. The few additional observations that have tried to improve on these measurements have been generally balloon-borne versions with the same basic concept but above much more of the atmospheric signal. Over time a large number of people were involved including G.F. Smoot, LBL/UCB, G. De Amici, UCB, S.D. Friedman, UCB, C. Witebsky, UCB, N. Mandolesi, Bologna, R.B. Partridge, Haverford, G. Sironi, Milano, L. Danese, Padua, G. De Zotti, Padua, Marco Bersanelli, Milano, Alan Kogut, UCB, Steve Levin, UCB, Marc Bensadoun, UCB, S. Cortiglioni, Bologna, G. Morigi, Bologna, G. Bonelli, Milano, J.B. Costales, UCB, Michel Limon, UCB, Yoel Rephaeli, Tel Aviv.

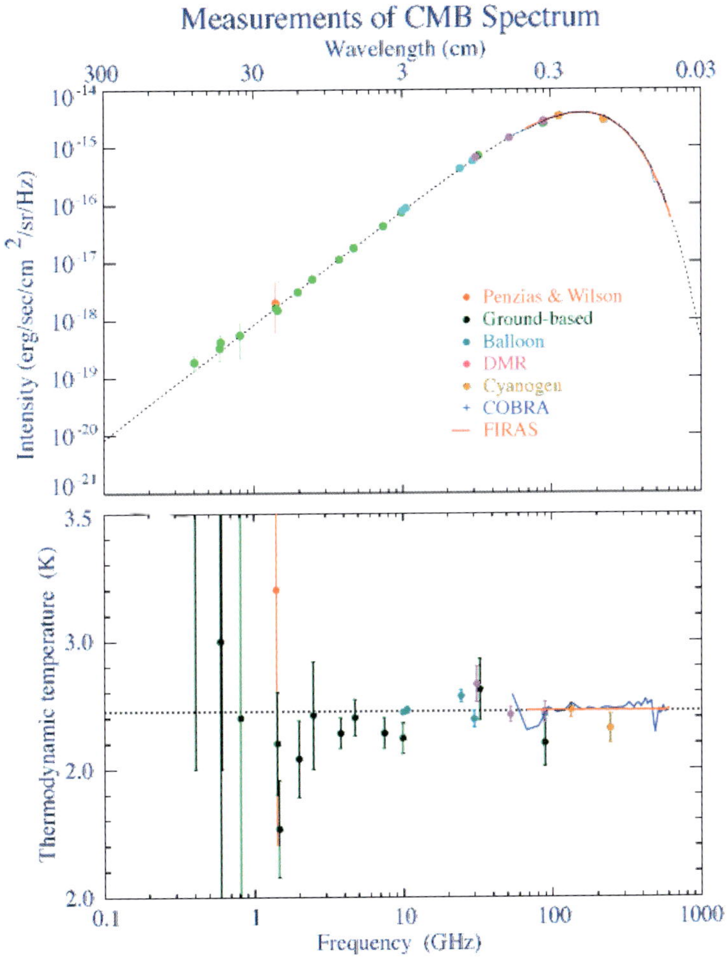

Figure 5. Selected observations of the CMB frequency spectrum.

1.12 The Cosmic Background Explorer (COBE) Mission

In 1976 NASA formed the COBE Science Study Group consisting of Sam Gulkis from JPL, Michael Hauser (P.I. for DIRBE Instrument) and John Mather (P.I. for FIRAS Instrument) from GSFC, George Smoot (P.I. for DMR Instrument) from SSL/LBL/UC Berkeley, Rainer Weiss, the chair, from MIT and Dave Wilkinson from Princeton.

After many years additional scientists were added to form the COBE Science working group in the 1980's. These included Chuck Bennett, GSFC, Nancy Boggess, NASA/GSFC, Ed Cheng, GSFC, Eli Dwek, GSFC, Mike Janssen, JPL, Phil Lubin, UCSB, Stephan Meyer, U. Chicago, Harvey Moseley, GSFC, Tom Murdock, General Research Corp, Rick Shafer, GSFC, Bob Silverberg, GSFC, Tom Kelsall, GSFC, and Ned Wright, UCLA.

ν (GHz)	λ (cm)	T^{th}_{CMB} (K)	Observed star	Reference
113.6	0.264	2.70±0.04	z Per	Meyer & Jura 1985, *Ap.J.*, 297, 119
113.6	0.264	2.74±0.05	z Oph	Crane *et al.* 1986, *Ap.J.*, 309, 822
113.6	0.264	2.76±0.07	HD21483	Meyer *et al.* 1989, *Ap.J. Lett.*, 343, L1
113.6	0.264	$2.796^{+0.014}_{-0.039}$	ζ Oph	Crane *et al.* 1986, *Ap.J.*, 136
113.6	0.264	2.75±0.04	ζ Per	Kaiser & Wright 1990, *Ap.J. Lett.*, 356, L1
113.6	0.264	2.834±0.085	HD154368	Palazzi *et al.* 1990, *Ap.J.*, 357, 14
113.6	0.264	2.807±0.025	16 stars	Palazzi *et al.* 1992, Ap.J., 398, 53
113.6	0.264	$2.729^{+0.023}_{-0.031}$	5 stars	Roth *et al.* 1993, *Ap.J.*, 413, L67
227.3	0.132	2.656±0.057	5 stars	Roth *et al.* 1993, *Ap.J.*, 413, L67
227.3	0.132	2.76±0.20	ζ Per	Meyer & Jura 1985, *Ap.J.*, 297, 119
227.3	0.132	$2.75^{+0.24}_{-0.29}$	ζ Oph	Crane *et al.* 1986, *Ap.J.*, 309, 822
227.3	0.132	2.83±0.09	HD21483	Meyer *et al.* 1989, *Ap.J. Lett.*, 343, L1
227.3	0.132	2.832±0.072	HD154368	Palazzi *et al.* 1990, *Ap.J.*, 357, 14

Table 4. CMB temperatures as measured through the molecules CN.

At the same time that the science portion of the team was developing, the management, engineering, technical and other mission support personnel were also developed. On November 18, 1989, after long preparation and delays, NASA launched its first satellite, the COBE, dedicated to cosmological observations. COBE had three scientific instruments:

(1) the Far-Infrared Absolute Spectrophotometer (FIRAS) to measure the CMB spectrum over the wavelength range 100 μm $< \lambda <$ 1 cm with a 7° resolution in order to investigate the blackbody nature of the CMB spectrum. It is designed to compare the spectrum of the CMB with that of a precise blackbody to measure tiny deviations from a blackbody spectrum.

(2) the Diffuse Infrared Background Experiment (DIRBE) to map the spectrum over the wavelength range 1.25μm $< \lambda <$ 240μm in 10 broad frequency bands with a 0.7° resolution to carry out a search for the cosmic

infrared background (CIB). CIB measurements are designed to measure visible light from very distant, unresolved galaxies. Light from all such distant galaxies is red shifted due to the cosmological expansion of the universe. The visible light from the galaxies is red shifted into the near infrared bands or absorbed by dust and reradiated in the far infrared and red shifted into the sub-millimeter band. The CIB measurements constrain models of the cosmological history of star formation and the buildup over time of dust and elements heavier than hydrogen.

(3) the Differential Microwave Radiometers (DMR) to map the CMB anisotropy in three frequency bands, 31.5, 53 and 90 GHz, with a 7° resolution and a sensitivity better than one part in 100,000 of the cosmic background temperature (see Figure 3). The primitive form of the DMR was utilized in the 1940's by Robert Dicke at Princeton University. The DMR does not measure the absolute temperature of a given direction of sky. Instead it measures the difference of temperatures of two different directions. A symmetrical DMR is one where two antennas pick signals from different directions and measure the difference between them. The two antennas quickly interchange positions and repeat the measurement. If the signals are of instrumental origin, they do not depend on direction, the difference would not change its sign, but if the signals are of cosmological origin, the difference should change its sign when the antennas views are swapped. This operational scheme greatly reduces the systematic problems and improves reliability. The anisotropy map, a map of temperature difference, provides a snapshot of the universe at the time of recombination about 380,000 years after Big Bang. The map shows the primordial structures which could not have been affected by any physical process no faster than the speed of light by the recombination era. Many cosmological parameters that describe the dynamics and geometry of universe and initial conditions of Big Bang cosmology can be estimated from the anisotropy map.

1.13 COBE Results

The COBE project was remarkably successful. FIRAS measurements corroborated the blackbody nature of the CMB spectrum (see Figure 8), giving the background temperature [45], $T_0 = 2.726 \pm 0.010$K, 95% CL. The DIRBE instrument provided absolutely calibrated maps of the sky at many wavelengths, gave an unparalleled picture of our own galaxy, and a good measurement of the cosmic infrared background radiation which is the radiation from the first generation of stars. Most of this radiation comes from early, bright, dusty galaxies.

Figure 6. Photograph of the COBE Science Working Group (minus Eli Dwek) taken during a meeting in 1988. From left to right: back row: Ed Cheng, Rick Schafer, Stephan Meyer, Mike Janssen, John Mather, Ned Wright, George Smoot, Tom Kelsall, middle row: Dave Wilkinson, Tom Murdock, Chuck Bennett, Bob Silverberg, Harvey Moseley, Michael Hauser, Rainer Weiss, front row: Nancy Boggess, Sam Gulkis, Phil Lubin.

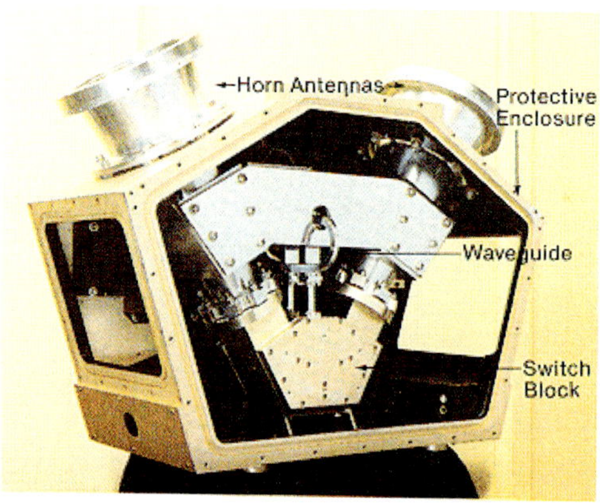

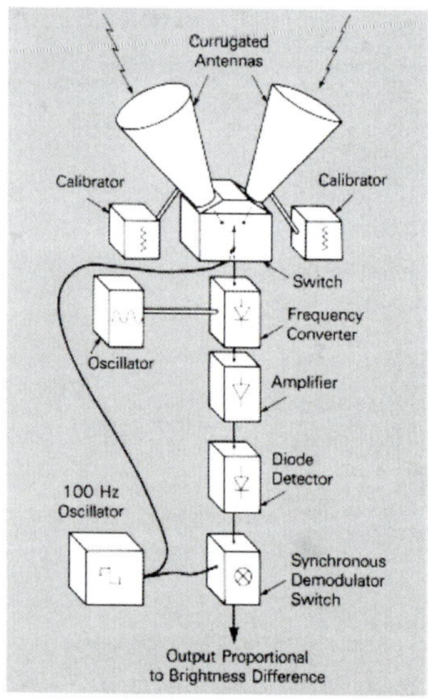

Figure 7. The DMR instrument design. Left: photograph of the DMR instrument, right: schematic of the DMR. The DMR has played the key role in the CMB anisotropy experiments, including the U-2 flight experiments and the COBE DMR. The WMAP also adopted differential radiometers as the basic apparatus for the CMB observations [66].

The FIRAS observations established very strongly that the CMB is the relic thermal radiation from the Big Bang and the DIRBE results revealed more about the later universe. The DMR discovery of CMB anisotropies got the most attention and has led to a whole area of activity with much theory, additional experiments and space missions. The discovery of CMB anisotropies was a many step process that involved the development of the instrumentation and techniques including calibration as well as the development of software and personnel to construct the instrument, carry out the observations, process and then finally analyze the data. To this point we have seen the historical development of some of these. For the COBE DMR the development of the experiment including instrumentation, calibration, software and personnel was highly integrated though often in time order due to the long period which COBE took. The outstanding instrumentation portion of the team, e.g. Roger Ratliff (microwave components), John Maruschak (component testing and verification), Robert Patschke (31.4 GHz), Maria Leche (53 GHz), Larry Hillard (90 GHz), Cathy Richards (receiver upgrade), Rick Mills (test chamber and organization), Peter Young (test results), Marco Toral (antennas), Gene Gochar (mechanical engineering), Frank Kirschman (thermal design), Dave Amason (command procedures), Chris Witebsky (90 GHz design), Dave Nace and Bernie Klein (instrument engineer), Dick Weber (systems engineer) and so on, tested and assembled components, subsystems and then the full set of DMRs were tested and calibrated. Goddard had been fairly restricted in hiring and chose to do COBE as an in-house project and thus was able to make additional hires so many young people were hired and trained on this project. It was very rewarding working with these people and to see their development and eagerness to take on challenging work and significant responsibilities.

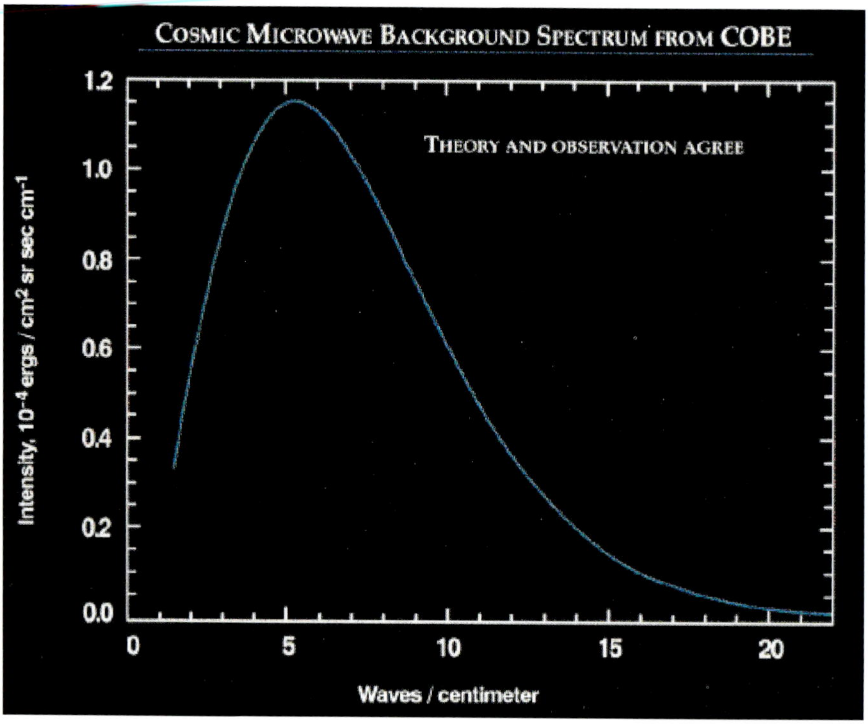

Figure 8. The solid (blue) curve shows the expected intensity from a single temperature blackbody spectrum, as predicted by the hot Big Bang theory. The FIRAS data fit with the expected blackbody curve with $T = 2.726$ K so precisely that the uncertainties are smaller than the width of the blackbody curve, the data points are covered by the curve and not visible [45].

Figure 9. False-color image of the near-infrared sky map (1.25, 2.2 and 3.5 μm composited) mapped by DIRBE. The dominant sources of infrared light at these wavelengths are stars in the Milky Way, as shown by the thin disk and the bulge at the center. Scattered bright sources can also be seen off the galactic disk [66].

During the end of this period, the data processing and analysis team began building a data processing pipeline with simulations and systematic checks and reviews. Sergio Torres, Jon Aymon, Charles Backus, Laurie Rokke, Phil Keegstra, Chuck Bennett, Luis Tenorio, Ed Kaita, R. Hollenhorst, Dave Hon, Qui Hui Huang, Al Kogut, Gary Hinshaw, Robert Kummerer, Jairo Santana, Kris Gorski, Tony Banday, Charley Lineweaver, Giovanni De Amici, Pete Jackson, Kevin Galuk,Vijay Kumar, and Karen Loewenstein were the many people who worked on the DMR data processing and analysis. This is, of course, a very small fraction of the people who worked on COBE over the many years.

We started with a small team of five post docs and one graduate student developing the software about one and half years before launch. We built up the team slowly as launch approached and passed. About half an hour after COBE was launched it flew over Antarctica and the extra ice-reflected sunlight produced excess solar electric power and mission control asked to turn on the DMRs earlier than their original timeline to absorb some of the excess power. I quickly approved and the DMR instruments were commanded on. As soon as the DMRs power came on, the internal noise calibrators fired as they were programmed, producing a nice strong but on scale signal, then as the COBE space craft rotated, the DMR beams swept over the moon (lucky alignment) and we could tell from our real-time software plots that the DMRs had survived the launch and were operating apparently nominally given the observed signals and apparent noise. Only time would tell if the instrument would remain sufficiently stable and reliable to make the required observations.

At the end of the first day, Sergio Torres and Dave Hon ran the full map-making pipeline and it worked well. We had DMR maps covering half the sky. They were noisy but still the best maps any one had to that point. We were all in good spirits and filled with optimism for the long period of observations and processing that was going to be necessary. After six months we had a substantial amount of data and things were working well enough that we made a full, carefully documented map pipeline run and produced maps covering essentially the full sky. These maps were quite good and had what we know in retrospect was just enough sensitivity to detect the primordial CMB anisotropies. We did note that there were small significant effects in the data but chose to publish only an upper limit on anisotropies to be conservative since we had had only limited time to understand fully the experiment and there were possible systematics in the instrument, our estimate of the possible astrophysical foregrounds, or in the data processing and analysis. We knew that we could make significant improvements in the processing of the data and correct for small residual effects. We wanted also to do sensitivity studies to ensure that we understood the experiment well.

During my training as a young scientist, my mentors, including Dave Frisch, Luis Alvarez, and their colleagues, repeatedly emphasized the importance of maintaining high scientific standards, both in choosing good projects and in executing them well. I heard of how important it was in this kind of experiment to set up systems and approaches so that one does not mislead oneself and assume that one has things right. Often even the most careful person lets down their vigilance when the results are coming out as expected and pays more attention when the result is different or unexpected. This is simply human nature and one must be careful not to fall in this trap. There was a famous paper showing the measurements of the speed of light with error estimates over time. It systematically changes with time but almost always within the estimated errors even though over many experiments, it changes by many standard deviations. This was unequivocal proof of this tendency and thus the need to be rigorous and systematic regardless of the result one was obtaining. One should work until confident in how well one has done before looking at the result and start believing in it. In many fields it is essential procedure to do the experiment in a double-blind method so as to guard against this very expectation effect.

So as soon as we had revamped and retested our data processing/map-making pipeline, we carefully validated it and then processed the full first-year data set. When the maps were made, there were clearly some residual signals above the noise. A sure-fire indicator was that when we made maps at each DMR frequency (31.4, 53, and 90 GHz) to get the best sensitivity we co-added the data from each of the two (A and B) channels at each frequency. However, we could also difference the A and B maps which should subtract any sky signal and leave only the independent instrumental noise. The variances and the smoothed A+B maps looked significantly different from the A–B maps. The A–B maps were consistent with what was expected from the simple Gaussian fluctuations of the receiver white noise. The question was, were these extra signals from the CMB, from astrophysical sources, from the instrument and experimental procedures, or from the data processing? At one of our COBE science team meetings I reported to the full SWG that we saw some effects but that it was too soon to be sure that it was not some artifact. I set the goal for the DMR analysis team of systematically going through all our steps and processes to show that we have each thing accounted for properly since in this kind of experiment one attributes to the CMB everything that is not attributed to anything else. We set up a plan and schedule to work through everything systematically and have each study confirmed and checked. The team fell into hard focused work on this program not jumping right to the conclusion that this was CMB anisotropy but that assuming that this was the correct approach. About six months (August 1991) into this, Ned Wright turned part of his attention to the DMR and noted that the one-year DMR maps had fluctuations in them consistent with Cold Dark Matter (CDM) cosmological model. This did not deflect the DMR team much as they were still trying to get done the projects set for them by their deadlines. In October Ned Wright did another bit of processing of the one-year maps with software he wrote and concluded that the data fit a Harrison-Zeldovich spectrum with a corresponding quadrupole amplitude of about 15 μK. We found about 30 μK in the DMR official software. Ned was now excited about the DMR potential results and wanted to bring it up at the next Science Team meeting. We agreed to have it at a more private special evening meeting after dinner at Al and Nancy Boggess's house as it would be a sensitive issue. The Science Working Group (as I argued) thought it appropriate to be conservative, educate all the members of the team on the status and reliability of the results and have the DMR team prepare and present to the full science team a review of all portions of the DMR experiment. This review was set for the end of the year 1991.

This extra attention from the SWG began to affect the DMR analysis team. They did not want to miss out on the discovery and especially the fun and excitement of the science analysis and the cosmological implications. I tried to have no one work on science analysis papers (myself included) until the checks were done. This was for the very point of ensuring that we had done a job that we were very confident was as good as possible. I worked hard to keep them on track and careful about getting the correct results and went so far as to offer two free tickets anywhere in the world to any one that could find an effect from anything (other than the CMB) what would explain a significant portion of what we were seeing. We divided into specialists in the areas of instrumental errors (led by the outstandingly able Alan Kogut who had to deal with very many potential issues), galactic foregrounds (Chuck Bennett and Gary Hinshaw), software (Sergio Torres for the DMR official code, George Smoot and my graduate students Luis Tenorio and Charley Lineweaver creating independent code and tests and Ned Wright with his own map processing software), and so forth. David Wilkinson was a particularly determined skeptic keeping everyone on their toes.

We held the review just after New Year's and presented all the tests, checks, calibrations and results in a day long review with extra material to the full DMR and SWG. There were a few things left to follow up but with them underway, we turned to analyzing the maps for their scientific results. There are a number of interesting things here including the low quadrupole and so forth but the most important issue was following a rigorous procedure. This was evident to me when we had the last SWG meeting to review the results and papers and discuss going public. At that meeting we also heard from the MIT-Princeton Team with a balloon-borne bolometer experiment whose data seemed to support the DMR results though were not as strong as the DMR. That seemed consistent but not conclusive. Then Phil Lubin got up and argued against going public stating that his group's data from their South Pole experiment was inconsistent and had a lower upper limit than the DMR results. This caused concern among some of the SWG but I found myself confidently arguing for going forward and publishing. I felt that the DMR team had done about a good a job as one could and were very likely to be correct.

In April 23, 1992, the COBE team announced the historical discovery of the anisotropies of cosmic microwave background radiation with characteristic anisotropy $\Delta T/T \approx 10^{-5}$ or $\Delta T \approx 30\mu K$ on angular scales larger than $\sim 7°$ at the annual meeting of American Physical Society in Washington D.C [47]. The CMB anisotropy provides very rich information on the early universe, allowing the calculation of cosmological parameters and discrimination of various detailed models of the Big Bang. The anisotropy map produced by the COBE DMR is composed of 6,144 pixels each 2.6° x 2.6°. This can be compared to the size of a patch of sky subtended by light that had been traveling since the beginning of the universe, about 1°. That is, the perturbations detected by DMR are directly from the primordial state set at the beginning. After 4 years of measurements by the COBE DMR, the typical signal-to-noise ratio in a 10° smoothed frequency-averaged map rose to ~2, so the anisotropy could be seen by eye. The DMR found the CMB thermodynamic temperature of $T_0 = 2.725 \pm 0.020$ K which was well consistent with the result of COBE FIRAS. If we subtract this temperature from the map and change the scale by a factor of 1,000, the CMB dipole looms out of the uniform background with amplitude 3.358 ± 0.024 mK toward the Galactic coordinates (l^G, b^G) = (264.31° ± 0.16°, 48.05° ± 0.10°) reconfirming the discovery of U-2 flight experiments. When the scale is increased to 100,000, higher multipoles ($l \geq 2$) can be seen. Figure 10 shows these features at these increasing scale factors. The quadrupole amplitude was estimated between 4μK and 28μK. The analysis of multipoles with $l > 2$ showed that the fluctuations are consistent with scale invariant (an $n = 1$ power-law) fluctuation spectrum as predicted by models of the inflationary Big Bang (Figure 11). Another important test for the inflationary Big Bang theory is the Gaussian distribution of the primordial temperature fluctuations. Our COBE DMR data found no evidence for deviations from a Gaussian distribution [49] as have all the experiments that have followed thus far.

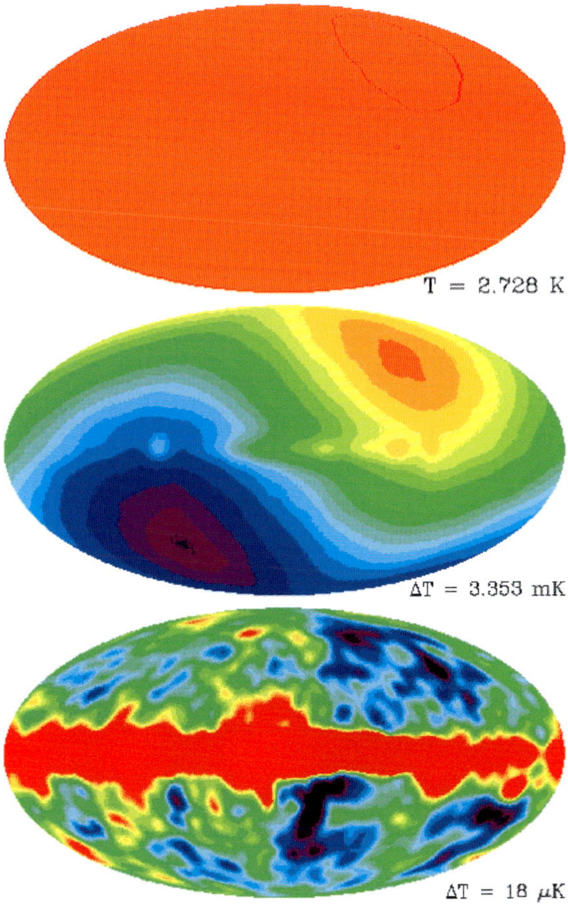

Figure 10. The breakdown of the CMB sky map produced by COBE DMR [66], monopole (top), dipole (middle) and multipole (bottom). At $T \sim 3$K level (top), we have uniform radiation from every direction. When the sensitivity is changed to ~ mK level, the dipole pattern shows up, which is due to the peculiar motion of our solar system relative to the CMB rest frame. As the scale is refined to $\Delta T \sim 10\mu$K, after removal of the dipole, the multipole features of anisotropy become evident. These very tiny fluctuations give us the information about the early universe.

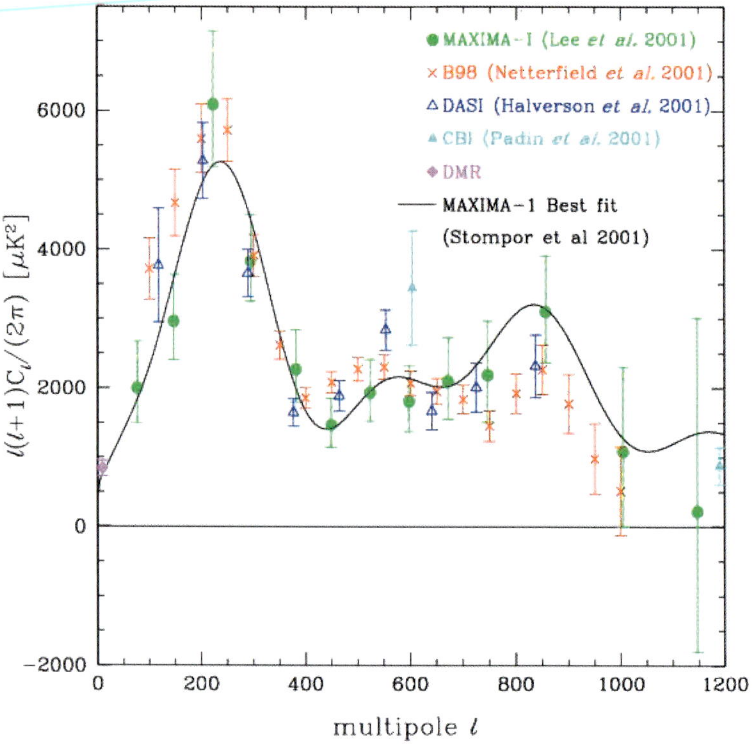

Figure 11. By 2001 the first acoustic peak was observed and there was evidence mounting for a second peak in the CMB angular power spectrum based upon the observations obtained from COBE DMR, MAXIMA, BOOMERanG, DASI, and CBI. In a couple more years much more data were available and the WMAP first year data then traced out the plateau and first peaks. The solid blue line is the prediction curve of a model with $n = 1$, $H = 50$km/s/Mpc and cold dark matter [67].

2 FORGING THE STANDARD MODEL OF COSMOLOGY: ΛCDM

2.1 The Suborbital CMB experiments

The CMB anisotropy is the most important cosmological observable to date, so many more ground based and balloon-borne CMB experiments followed the COBE mission in 1990's [52]. While some experiments focused on large angular scales at frequency bands not used in the COBE DMR, most of the projects worked on the smaller angular scales which were not explored by the COBE DMR. These small angular scale experiments put very tight constraints on cosmological models. Results from some representative experiments on the angular power spectrum are presented in Figure 12. Supernovae observations startled the cosmology community with the discovery of accelerated expansion of space [56, 57]. This brought back the cosmological constant to the Einstein equations, introduced originally with different motivation and then discarded soon after. Observations other than the CMB experiments including galaxy surveys have provided valuable information and allowed cross checks with the results from the CMB experiments. Some recent results of suborbital experiments/observations are presented in Table 5, Figure 12 and Figure 17.

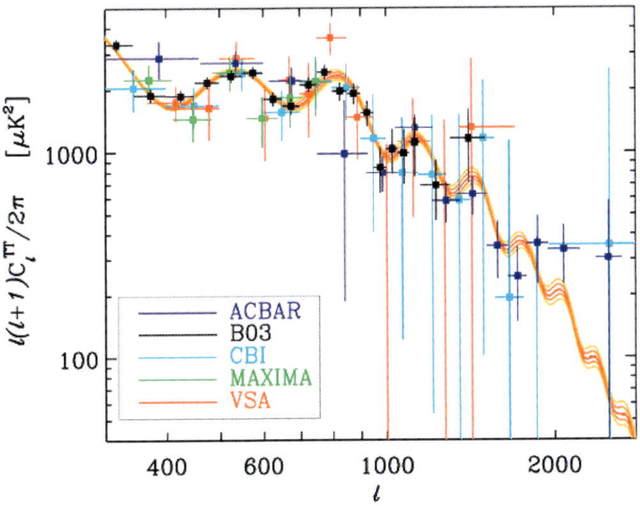

Figure 12. The best fit curve (orange) of small-scale angular power spectrum for ΛCDM model estimated with WMAP data and measurements carried out by ground-based and balloon-borne CMB experiments [51]. Notice that this shows the greater details of angular power spectrum at the range $l = 300$.

2.2 *Physics from CMB Anisotropy Power Spectrum*

The observations of the CMB anisotropy power spectrum have turned us from speculations about the Universe with meager measurements to a working cosmological model in which the Universe is spatially flat, consists mainly of dark matter and small fraction of ordinary matter just sufficient to produce the light element primordial abundances. All the complex (cosmic web) structure of matter formed from primordial adiabatic fluctuations believed to be the result of quantum mechanical fluctuations from when the Universe was a tiny fraction of a second old – the Inflationary epoch. Observations to date have achieved of order a few per cent accuracy on the key cosmological parameters and the coming decade is likely to see this accuracy improve to less than a per cent. In large part, this is due to the expected improvement in CMB temperature and polarization anisotropies. That change will make a very substantial difference in our understanding and in our ability not only to determine those parameters but also to test the key elements of the assumed physics of our cosmological model and even probe what were the natural and minimum initial conditions.

These new results, their analysis and interpretation, will have profound implications for the physics and astronomy. The current successful models call for four major new pieces of physics: dark energy, dark matter, Inflation (or alternative), and the matter-antimatter asymmetry and assume four other major items: no other significant relics of the early universe, there are no significant extra dimensions, the fundamental constants do not vary with time, there are no other significant exotic forces in play. All of these will be tested and probed in a precise way.

An important reason for this is the physical simplicity of the early universe and processes that leave their imprint on the CMB. The CMB is a very rich source of information because all the oscillations are still linear and the physics of the fluid is well understood and there are many possible features to probe and observe.

The early universe is dominated by the cosmic background radiation (CBR) photons which interact strongly with the electrons (electrically coupled to the protons and helium nuclei) to make what is called the photon-baryon plasma. This plasma undergoes simple acoustic oscillations until the universe cools enough for neutral atoms to form and the baryons and photons lose their tight coupling. The photons are then free to propagate across the universe from this last epoch plasma which marks the cosmic photosphere very much like the apparent surface of the Sun (solar photosphere). The photons that were set free in the transition from a cosmic plasma to stable atoms make up the cosmic microwave background radiation which is present everywhere in the cosmos. As we observe this radiation in the sky, we are practically looking at a snapshot of the early universe.

It is easy to calculate that this de-ionization happened when the universe was 1,089 times smaller than the present at a time about 379,000 years from the beginning of the Big Bang. The process is set by the adiabatic expansion of the universe which gives the ratio of CBR temperatures being the inverse ratio of universe scale sizes. Thus when the universe was 1,089 times smaller than now, the CBR temperature was 1,089 times higher or about 3,000 K which because of the large ratio of photons to baryons is sufficient to ensure that the universe is fully ionized. (Go to the equilibrium equations of Saha and put in a hydrogen ionization energy of 13.6 eV and the temperature of 3,000 K or equivalent mean energy about 1/4 eV per photon provides enough photons above 13.6 eV to ionize all the hydrogen and helium.) This happens fairly rapidly but not instantaneously and thus extends over about $\Delta z \approx 70$ compared to 1,089 which is similar to the skin on an apple. The finite thickness of the region of last scattering of photons with the primordial plasma is an important reason that there is a damping of the visibility of small features or the damping of the high spatial frequencies (high l). Now we can understand the roll off (diminishing of signal) at small angular scales.

2.2.1 *The Geometry of Space Time*

We can probe the geometry of space time by observing sound waves in the primordial plasma. The CBR dominates the early universe and thus the photons (and other relativistic particles) moving at the speed of light in random directions are the particles that make up most of any sound wave in the early universe. Because we live in three dimensions, the speed of sound will be the speed of light divided by root 3 ($v_s = c/\sqrt{3}$).*

According to Jeans' theory, first derived to explain the formation of the solar system and oscillations of stars under the influence of gravity, if the free fall time is shorter than the sound crossing time, gravitational collapse occurs. If the sound crossing time is shorter than the free fall time, then the system can adjust and undergoes acoustic oscillations. The Sun oscillates as a result of any perturbation. So does the early universe, since the sound speed is so high and the gravitational perturbations are so small (10^{-5}) that gravity is weak and the free fall time is very long. Primordial perturbations all start near time equals zero and oscillate acoustically - i.e. as sound waves. The CBR is freed at $1 + z = 1{,}089$ or about 379,000 years so that the CMB fluctuations image them at that epoch. Primordial perturbations have had essentially an elapsed time of 379,000 years to oscillate at the speed of sound.

At the close of those 379,000 years, the largest possible coherent acoustic oscillations had a spatial extent of roughly 220,000 lightyears (or 70,000 parsec). There was simply no time for more: With the speed of sound at 0.6 light speed and a time of 379,000 years, the largest regions in which coherent oscillations could develop had a spatial extent of 0.6 x 379,000 = 220,000 light years. This upper limit is called the "sound horizon". One reason these oscillations are of great interest to cosmology is that CMB anisotropy observations can determine the apparent angular size of the sound horizon in the sky.

When the first stable atoms formed, the sound waves in the cosmic plasma caused tiny fluctuations in the temperature of the cosmic microwave background. The cosmic microwave background fluctuations were there before the atoms formed. Satellites WMAP and Planck and other suborbital instruments can map those temperature differences with high precision.

However, what is observed is not the absolute size of the sound horizon, but its angular size in the sky. We do know the absolute size already, namely the 220,000 lightyears mentioned above. By comparing the angular and the absolute sizes, we can determine the curvature of our cosmos – whether space is flat, or has a sphere-like or saddle-like shape. In ordinary Euclidean space ("flat space"), we are well acquainted with the relationship: The angular size (the "apparent size") of a given object decreases linearly with the distance – at least for far-away objects. The following image shows the relationship between the spatial extent L of a measuring rod, its distance and its angular size

* At the last scattering surface the baryon density has increased to the point that it slightly affects the speed of sound. The baryon-photon momentum density ratio $R = (\rho_b + p_b)/(\rho_\gamma + p_\gamma) \approx 0.6(\Omega_b h^2 / 0.02091000/1 + z)$ so that $v_s = c/\sqrt{3(1+R)} \simeq c/\sqrt{3(1+3\rho_B/4\rho_\gamma)}$ and this causes a small correction to this estimate.

α. Shown in red are light rays connecting us, the observer, with both ends of the measuring rod:

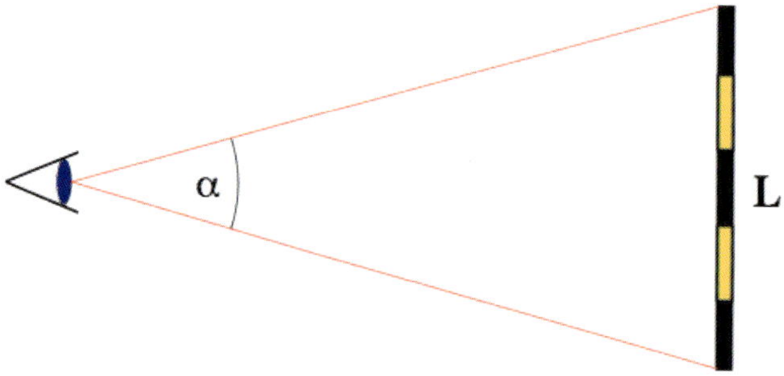

Figure 13. Flat or Euclidean geometry.

Matters are somewhat different in a space of positive curvature, the three-dimensional analogue of a spherical surface. In such a space, light does not travel along straight lines. Instead, light rays converge, as can be seen in the next image. The angular size (the "apparent size") of a given object decreases more slowly than linearly with the distance – at least for far-away objects. In such a space, the same measuring rod at the same distance from us will have a larger apparent size, corresponding to a larger angle α.

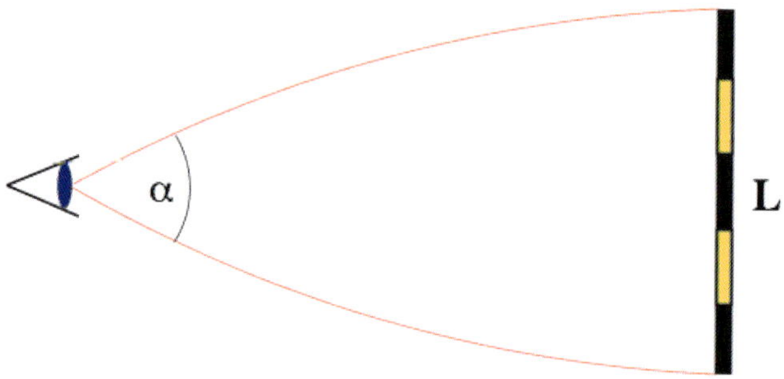

Figure 14. Closed or positively curved geometry.

In a space with negative curvature, it is the other way around: Light rays diverge, and as a result the same measuring rod, still at the same distance, will have a smaller apparent size, corresponding to smaller α.

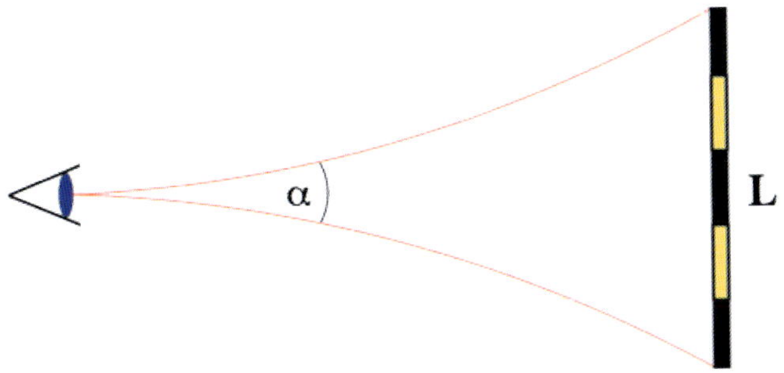

Figure 15. Open or negatively curved geometry.

Consequently, it is simple to read off the curvature of space from the cosmic microwave background. We know the absolute size of the largest sound waves in the early universe, and we can measure their apparent size in the sky. The distance of the microwave background can be calculated. We know the temperature at which it was formed, and we can measure its temperature today. The temperature difference between then and now is directly related to the amount by which the universe has expanded from then until now, and this, in turn, is directly related to the distance. Comparing the apparent and true sizes at the calculated distance, we obtain the curvature of space. In this way, we have determined that, to high precision, space in our cosmos is flat.

2.2.2 *Acoustic Oscillations*

So why do the acoustic peaks show up? What do they tell us?

The geometry of spacetime is most precisely determined by measuring the angular scale of the first acoustic oscillation peak in the angular power spectrum. With observations of several peaks we are able to tie down several parameters and determine the geometry of spacetime even more precisely. How can we understand the plateau and the acoustic peaks?

That again is conceptually simple. If a physical system is hit by a spike of short duration – the extreme version is a Dirac delta function, i.e. an impulse of zero time duration – then all possible oscillation modes are excited equally on average but stochastically. This would be equivalent for the early universe of equal power or identical rms perturbation amplitudes into all plane wave for flat spacetime or appropriate modes. So, if we choose the correct plotting of the angular power spectrum, the initial perturbations or excitation would be basically independent of scale and particularly angular frequency. The finite thickness of last scattering will damp out the high angular frequencies. Another effect which is leakage of photons out of high frequency perturbations because of minor edge effects adds to this effect. Because of the geometry of spacetime and the contents of the universe, this otherwise nearly scale invariant power spectrum has a set of acoustic peak at a fundamental angular frequency and its harmonics. These are really three-dimensional perturbations that are then manifested upon the apparently spherical sky.

On the largest angular scales, e.g. those first observed by the COBE DMR, we observe what are the basically initial conditions. These scales are so large that even moving at the full speed of sound ($c/\sqrt{3}$) the change is small compared to the size of the structure and elements that large have only undergone a very small fraction of an oscillation from the beginning until the last scattering surface. Perturbations that are much larger than the sound horizon remain little changed and will show up in the power spectrum as simply the low angular frequency (large angular scale) plateau that reflect the primordial perturbations. Without the acoustic oscillations, the angular power spectrum would remain flat until the high angular frequency (small physical scales) damping cuts it off.

Why is there a well-defined first peak, if all scales are simply oscillating and there is little damping on the larger scales?

However, when one reaches the scale of about 220,000 light years at the surface and time of last scattering, the oscillation has had time to just fully compress. Compression makes that region hotter and that is what shows up in the snap shot provided by the CMB.

At half that size an acoustic oscillation has time to compress and then expand and reach maximum rarefaction. It would appear to be cooler than the average. In an angular power spectrum this would appear as the second peak, since taking the power rectifies (squares) the variation which was to cooler.

At one third that size the acoustic oscillation has time to compress, rarify, and reach maximum compression again. This is the source of the third acoustic. And so on for higher spatial frequency peaks. It will appear warmer than the average. At $3/4^{\text{ths}}$ that size the oscillation has compressed and then expanded back to the starting place for no contrast to the average. Thus measuring the mean square fluctuations on each scale or the power spectrum, one would anticipate peaks at the sound horizon, half the sound horizon, one third the sound horizon, etc. and then nulls mid way in between. However, there is an additional effect from the motion of the oscillations – at the half way points the sound wave is moving with maximum speed and the Doppler effect produces a small secondary peak exactly out of phase with the compression and rarefaction. These somewhat fill in between the acoustic peaks so that there is not a precise null.

The key points are that: (1) Compressed fluid is hotter. (2) Oscillation frequency scales inversely with size. (3) Oscillations stop at last scattering. (4) Nulls of oscillations are when Doppler effect is maximal from the motion of oscillation. (5) Fluctuations are imprinted on CMB at the last scattering surface.

All these effects are readily calculable as they depend only on the geometry and the speed of sound. There are small corrections for the speed of sound near last scattering due to the baryon loading (baryons are relatively more important as the CMB is cooled by the expansion of the universe) and also the effect of the dark matter and even a small correction for dark energy. One finds that the height of the first peak is proportional to the total matter content of the universe while the second (rarefaction) peak is set more by the baryon content and so on. As a result, by carefully fitting to the amplitudes and l values, one can precisely constrain the physical density of baryons and dark matter as well as the geometry of space-time.

The first acoustic peak occurs at $l \sim 220$ or about 0.9° angular scale for a flat universe. The early observations quickly confirmed this value indicative of a universe near flat. Now observations are getting sufficiently precise that one must take into account the small corrections that come from the possible variation of the contents of the universe. One must take into account that the photon-baryon fluid is not purely photon-dominated at the last scattering surface. Baryons add to the mass of the photon-baryon plasma without adding pressure. Also the universe is not exactly matter dominated and there is some dark energy content. The dark energy content comes to dominate in later epochs and changes the distance to the last scattering surface somewhat and thus changes the apparent angles on the sky a small amount. All of these effects are small but are important at the percent to few percent level. By far the most dominant factor is the geometry of the universe and via General Relativity the total energy content of the universe is very close to the critical value. In current models this means $\Omega_{total} = \Omega_m + \Omega_\Lambda = 1.02 \pm 0.02$.

2.2.3 The Dark Matter and the Baryon Content of the Universe

By the epoch of last scattering the energy density in dark matter gets to be larger than the energy density in photons. As a result, the inertia and gravitational potential of the matter is the source of the restoring force (gravity) for the acoustic oscillations and thus affects the amplitude of the oscillations directly. Thus the height of the acoustic peaks is proportional to the total mass density which is primarily dark matter. By measuring the height of the peaks, particularly the first peak, the physical density of matter $\Omega_m h^2$ is well determined.

Baryons (or ordinary matter) load down the photon-baryon plasma and add inertial (and gravitational) mass to the oscillating system. The effect on the acoustic peaks is easy to understand. If one adds mass to a spring and lets it fall in the gravitational field, with more mass loading the spring, it falls further before pulled back by the spring. Then it rebounds to the same position it started from. That is the maximum compression increases but the rarefaction is unchanged.

Since the odd numbered (first, third, fifth...) acoustic peaks are associated with how much the plasma compresses (falls into the gravitational potential wells), they are enhanced by an increase in the amount of baryons in the universe. The even numbered peaks (second, fourth, sixth) are associated with how far the plasma rarefies (rebounds in the gravitational field). The addition of baryons enhances the odd peaks over the even peaks. Added baryons make the first acoustic peak much larger than the second. The more baryons the more the second peak is relatively suppressed

If the baryons contribute a negligible amount of mass to the plasma, the CMB temperature at the bottom of the potential well oscillates symmetrically around zero. With more baryons in the system, the plasma is loaded down. The plasma compresses further inside the potential well before pressure can reverse the motion. The oscillation is now asymmetric in that the extrema that represent compressions inside potential wells are increased over those that represent rarefactions. The power spectrum doesn't care about the sign and so take the absolute value of the temperature fluctuation. Now we see that the first and third peaks are enhanced over the second peak. When we do the full calculation of the power spectrum, the basic physics of a mass on the spring appears as advertised. The odd numbered acoustic peaks in the power spectrum are enhanced in amplitude over the even numbered ones as we increase the baryon density of the universe.

There are second order effects. Since adding mass to a spring slows the oscillation down, adding baryons to the plasma decreases the frequency of the oscillations pushing the position of the peaks to slightly higher multipoles l. (More bayons means slightly slower sound speed so a given number of oscillation cycles must take place over a small size. $v_s\downarrow \Rightarrow l\uparrow$) Baryons also affect how sound waves damp, which affects how spectrum falls off at high l. The many ways that baryons show up in the angular power spectrum provides many independent checks on the baryon density of the universe. The baryon density is a quantity that the CMB can measure to exquisite precision. The CMB observations agree with Big Bang Nucleosynthesis (BBN) at about 4.4 ± 0.4% of critical density. The BBN estimate comes from nuclear physics and conditions in the first few minutes of the Big Bang. The CMB acoustic peaks ratio comes from atomic physics at 380,000 after the Big Bang. As a consequence there has been no significant change in the baryon–photo ratio from an energy scale of an MeV to an eV and probably to now with an energy scale of 1/4000 eV which is a substantial portion of the expansion history of the universe.

2.2.4 Other Cosmological Parameters including Dark Energy, Equation of State

The CMB anisotropies are sensitive in varying degrees to a number of other cosmological quantities including the dark energy density, its equation of state, the age of the universe, the optical depth to the re-ionization of the universe, and the slight tilt of the primordial perturbation spectrum. For some of these observations of the CMB angular power spectrum turn out to be extremely sensitive for some others there are degeneracies or near degeneracies that require the CMB observations to be combined with other cosmological observations to get a truly accurate result. Needless to say, the CMB anisotropies do provide a substantial amount of information. In principle, measuring the power spectrum multipoles provides up to 3,000 independent numbers; while we believe that our standard model of cosmology can be well-described by less than 20 parameters. Thus we have substantial redundancy going from millions of pixels in our maps, by assuming rotational invariance down to about 3,000 numbers in the angular power spectrum and then by cosmological model fitting down to a couple of dozen parameters.

2.3 Wilkinson Microwave Anisotropy Probe

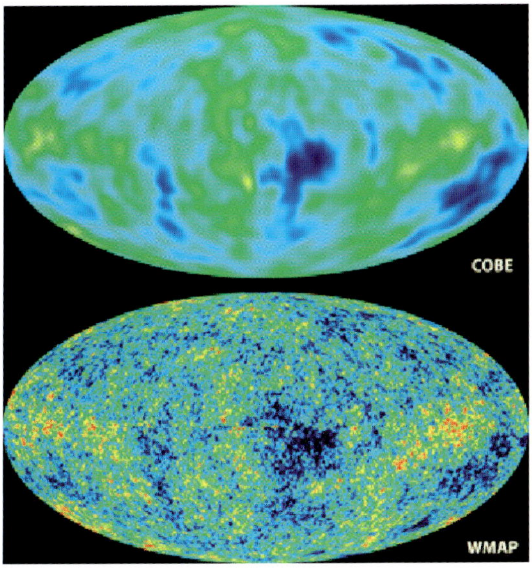

Figure 16. A visual comparison of COBE DMR and WMAP. The COBE was able to show only the super-horizon scale (~ 10°) structures while the WMAP has sufficient angular resolution to show more detailed structures down to sub-horizon scale (~ 0.3°).

The Wilkinson Microwave Anisotropy Probe (WMAP) mission was the second, more extensive satellite-borne CMB project also by NASA, which followed the COBE. The skymap data taken by WMAP have 45 times the sensitivity and 33 times the angular resolution of the COBE DMR mission. WMAP used five separate frequency bands from 23 to 94 GHz. The goal of WMAP was to map the relative CMB temperature over the full sky with an angular resolution ~ 0.3°, a sensitivity of 20 μK per 0.3° x 0.3° square pixel, with systematic artifacts limited to 5 μK per pixel. In February 2003, the WMAP first-year data and results were released and three-year data were released in March 2006 [68]. The results strongly support the inflationary Big Bang models. WMAP also set tight constraints on the cosmological models, among which the ΛCDM model (nearly flat universe with dark energy ~ 70%, cold dark matter ~ 25%, baryonic matter ~ 5%) fits best with WMAP along with various independent experiments (see Table 5 and Figure 17).

Cosmological Parameters for ΛCDM Model

parameter	WMAP 3-year	All combined[1]
h_0 (Hubble parameter)	73.5 ± 3.2km/s/Mpc	$70.8^{+1.5}_{-1.6}$ km/s/Mpc
h (reduced Hubble parameter)	0.735 ± 0.032	$0.708^{+0.015}_{-0.016}$
n_s (scalar spectral index)[2]	0.951 ± 0.016	$0.938^{+0.015}_{-0.014}$
$\Omega_b h^2$ (baryon density)	$0.0223^{+0.00075}_{-0.00073}$	$0.02193^{+0.00067}_{-0.00068}$
Ω_Λ (dark energy density)	0.763 ± 0.034	0.738 ± 0.016
Ω_m (matter density)	0.237 ± 0.034	0.262 ± 0.016
σ_8 (matt. fluc. On $8h^{-1}$ Mpc)	0.742 ± 0.051	$0.751^{+0.032}_{-0.031}$
t_0 (age of the universe)	$13.73^{+0.16}_{-0.15}$ Gyr	13.84 ± 0.14Gyr
τ (reionization optical depth)	$0.088^{+0.029}_{-0.030}$	$0.070^{+0.027}_{-0.028}$

Table 5. ΛCDM parameters computed using WMAP 3-year data only (middle column) and various data sets combined (right column). Ω refers to the density relative to the critical density.

[1] Results with combined data from the experiments WMAP, 2df, SDSS, BOOMERanG, ACBAR, CB I, VSA, SN Astier, SN gold, WL and BAO.
[2] At k = 0.002/Mpc.

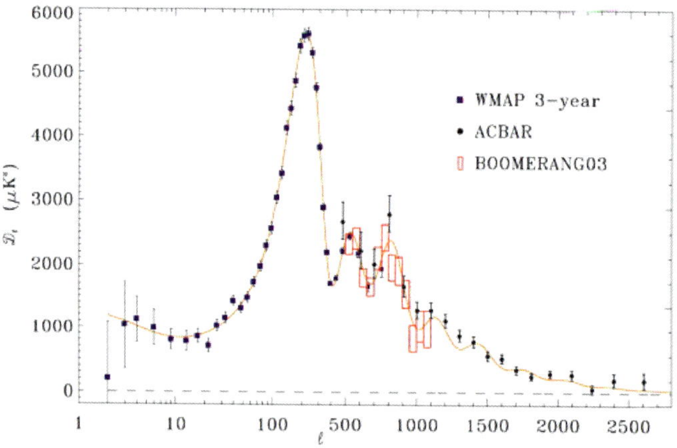

Figure 17. Power spectrum predicted by ΛCDM model and plots by WMAP 3-year data experiments, ACBAR and BOOMERANG03. Solid (yellow) curve is the predicted power spectrum of ΛCDM model.

2.4 PLANCK: The Third Generation CMB Project

Figure 18. An artist's rendering of the Planck satellite in space [69].

The Planck mission is the third generation space mission for the CMB experiments. Conducted by the European Space Agency (ESA), it is scheduled to be launched in 2008. In 1992, two potential space-based CMB experiments, COBRAS and SAMBA, were proposed to ESA and adopted as a combined mission called COBRAS/SAMBA. This project was later renamed as *PLANCK* in honor of the German physicist Max Planck. The Planck mission will sweep the full sky with frequency windows from 30 to 857 GHz, mapping precise and extensive CMB anisotropies with angular resolution down to 5' and sensitivity $\Delta T/T \sim 2 \times 10^{-6}$ [58]. Planck will improve on WMAP with advanced features, 10 times the sensitivity, 2 or 3 times better angular resolution and 6 times the frequency coverage of WMAP. The resolutions of COBE, WMAP and Planck are visually compared in Figure 19. Planck measurements are expected to set constraints on physics at energies greater than 10^{15} GeV and its precise measurements of the CMB anisotropy with angular resolution of 10' will revolutionize cosmology.

Figure 19. The progression of satellite CMB map resolutions going from the COBE DMR discovery maps to the WMAP and anticipated Planck map [69].

3 CONCLUDING REMARKS

The demonstration that we have understood the CMB as a relic from the early Universe in a simple configuration has led us toward fulfilling the promise that its existence and small deviations are a unique probe of cosmology propelling us into an era which we now call 'Precision Cosmology'. It is that precision that makes cosmology a true physical science and now promises to provide a means to know the universe better and to test our very assumptions. The soon to arrive era of the Planck mission, with its related ground-based CMB observations along with other cosmological observations, promises to move us to the 1% or better level on all key cosmological parameters. As one reaches this level of precision with cross-constraining observations one not only determines the parameters of the universe but also strongly tests the assumptions one has made in determining those parameters.

The cosmic microwave background has many more features yet to be explored fully, and one of the most important topics is the CMB polarization. Another key topic is the Sunyaev-Zel'dovich effect via clusters of galaxies. Linear polarization of the CMB arises via CMB photon scattering with free electrons as long as there is a net quadrupole anisotropy seen by the free electrons. So as Rees predicted in 1968, linear polarization should arise from the anisotropies produced by the primordial perturbations that eventually lead to large scale structure formation. Since Thomson scattering of an anisotropic radiation field also generates linear polarization, the CMB is predicted to be polarized at the roughly 5% level [83]. This polarization has been observed by the DASI [84] group and a few following instruments.

In 2003 the WMAP experiment demonstrated that it was able to measure the TE cross-correlation power spectrum [64]. The cross-correlation power spectrum provides supporting evidence of the adiabatic nature of the perturbations, as well as directly constraining the thickness of the last scattering surface. Since the polarization anisotropies are generated in this scattering surface, the existence of correlations at angles far larger than about a degree demonstrate that there were super-Hubble horizon length fluctuations at the recombination epoch. We are just now in the early phases of observing and exploiting the polarization of the CMB.

However, polarization of the CMB can arise from various sources not yet observed such as re-scattering of the CMB photons during the re-ionization and gravitational waves from inflation and small scale anisotropies. Polarization measurements are anticipated to significantly improve the accuracy of cosmological parameters and measurements will, in addition to temperature anisotropy measurements, provide an independent test for cosmological models. The B-mode (due to gravitational waves) polarization in particular will give substantial information on the energy scale of inflation. The CMB is also a favored subject in which one can probe theorized remnants of Big Bang inflation such as topological defects [67] which are directly related to physics in very early universe ($T \sim 10^{16}$ GeV).

ACKNOWLEDGEMENTS

I would like to thank and acknowledge the many people who contributed to this work, particularly my many colleagues and coworkers over the years and particularly the COBE team. The development of cosmology has been a significant intellectual achievement powered by the work of very many scientists, engineers, and staff over many decades. A number of slides and figures that I used in my Nobel Lecture were provided by others and many names were cut off; unfortunately, in the rush of connecting my laptop, the wrong resolution was used and the edges of the slides not projected. Among the many slides were those from Don Groom, Al Kogut, Wayne Hu, Ned Wright, WMAP+SDSS, Max Tegmark, David Wilkinson, Julian Borrill and others which were there but not shown. I thank a number of colleagues, especially my graduate students and post docs for their discussions and comments both at present and over the many years as well as for their great excitement and intellectual efforts on the various projects. Eunhwa Jeong and Bruce Grossan were directly helpful with this write up. This work was supported by the Lawrence Berkeley National Laboratory and the Department of Physics at the University of California, Berkeley.

REFERENCES

[1] A. A. Penzias, R. W. Wilson, Bell Telephone Labs, Inc., Holmdel, N.J., "A Measurement of Excess Antenna Temperature at 4080 mc/s," *Ap. J.*, vol. 142, pp. 419-421, 1965.

[2] R. H. Dicke, P. J. E. Peebles, P. S. Roll, and D. T. Wilkinson, "Cosmic Black-Body Radiation," *Ap. J.*, vol. 142, p. 414, 1965.

[3] R. A. Alpher, H. Bethe, and G. Gamow, "The Origin of Chemical Elements," *Physical Review*, vol. 73, issue 7, pp. 803-804, 1948.
R. A. Alpher, R. Herman, and G. A. Gamow, "Thermonuclear Reactions in the Expanding Universe," *Physical Review*, vol. 74, issue 9, pp. 1198-1199, 1948.
R. A. Alpher and R. C. Herman, "The Origin and Abundance Distribution of the Elements," *Annual Review of Nuclear and Particle Sciences*, vol. 2, pp.1-40, 1953.
R. A. Alpher, J. W. Follin, and R. C. Herman, "Physical Conditions in the Initial Stages of the Expanding Universe," *Physical Review*, vol. 92, issue 6, pp. 1347-1361, 1953.
R. A. Alpher, G. Gamow, and R. Herman, "Thermal Cosmic Radiation and the Formation of Protogalaxies," *Proceedings of the National Academy of Sciences of the United States of America*, vol. 58, issue 6, pp. 2179-2186, 1967.
R. A. Alpher and R. C. Herman, "Reflections on early work on 'big bang' cosmology", *Physics Today*, vol. 41, no. 8, p. 24, 1988.
R. A. Alpher and R. Herman, "Early work on 'big-bang' cosmology and the cosmic blackbody radiation," *Modern Cosmology in Retrospect*, Edited by B. Bertotti, R. Balbinot, and S. Bergia. Cambridge University Press, p. 129, 1990.

[4] A. G. Doroshkevich and I. G. Novikov, *Sov. Phys-Doklady*, vol. 9, p. 11, 1964.

[5] R. H. Dicke and P. J. Peebles, *Space Sci. Rev.*, vol. 4, p. 419, 1965.

[6] R. B. Partridge "3K: The Cosmic Microwave Background Radiation," *Cambridge Astrophysics Series Cambridge, University Press*, 1995.

[7] R. K. Sachs and A. M. Wolfe, "Perturbations of a Cosmological Model and Angular Variations of the Microwave Background," *Ap. J.*, vol. 147, p. 73, 1967.

[8] J. Silk, "1967 Fluctuations in the Primordial Fireball," *Nature*, vol. 215, pp. 1155-1156, 09 September 1967.

[9] J. Silk, "Cosmic Black-Body Radiation and Galaxy Formation," *Ap. J.*, vol. 151, pp. 459-471, 1968.

[10] S. W. Hawking and G. F. R. Ellis, "The Cosmic Black-Body Radiation and the Existence of Singularities in our Universe," *Ap. J.*, vol. 152, p. 25, 1968.

[11] D. W. Sciama, "Peculiar Velocity of the Sun and the Cosmic Microwave Background," *Phys. Rev. Lett.*, vol. 18, p. 1065, 1967.

[12] R. B. Partridge and D. T. Wilkinson, *Phys. Rev. Lett.*, vol. 18, p. 557, 1967.

[13] M. Rees, "Polarization and Spectrum of the Primeval Radiation in an Anisotropic Universe," *Ap. J.*, vol. 153, p. L1, 1968.

[14] R. A. Sunyaev and Y. Zel'dovich, *Comm. Astrophys. Sp. Phys.*, vol. 4, p. 173, 1972.
R. A. Sunyaev and Y. B. Zel'dovich, *Astrophysics & Space Sci.*, vol. 7, p. 3, 1970.

[15] M. Birkinshaw, *Phys. Rep.*, vol. 310, p. 98, 1999.

[16] S. W. Hawking, "On the Rotation of the Universe", *Mon. Not. R. Astr. Soc.*, vol. 142, pp. 129-141, 1969.

[17] O. Fackler, D. Frisch, J. Martin, G. Smoot, and L. Sompayrac, "Test of the $\Delta S = \Delta Q$ Rule in Ke3 Decay," *Phys. Rev. Lett.*, vol. 31, pp. 847–850, 1973.
G. F. Smoot, O. D. Fackler, D. H. Frisch, J. F. Martin, and L. M. Sompayrac, "K^0 production by K^+ on platinum at 3 GeV/c," *Nuclear Physics B*, vol. 96, issue 3, pp. 379-400.

[18] P. J. E. Peebles, *Physical Cosmology*, Princeton, NJ: Princeton U. Press, 1971.
P. J. E. Peebles, *Principles of Physical Cosmology*, Princeton, NJ: Princeton U. Press, p. 168, 1993.

[19] J. M. Stewart and D. W. Sciama, "Peculiar Velocity of the Sun and its relation to the Cosmic Microwave Background," *Nature*, vol. 216, p. 748, 1967.

[20] G. F. Smoot *et al.*, "Detection of Anisotropy in the Cosmic Blackbody Radiation," *Phys. Rev. Lett.*, vol. 39, p. 898 1977.

[21] M. V. Gorenstein and G. F. Smoot, "Large-angular-scale anisotropy in the cosmic background radiation," *Ap. J.*, part 1, vol. 244, pp. 361–381, Mar. 1, 1981. Results of an extended series of airborne measurements of large-angular-scale anisotropy in the 3-K cosmic background radiation are reported. A dual-antenna microwave radiometer operating at 33 GHz flown aboard a U-2 aircraft to 20-km altitude on 11 flights between December 1976 and May 1978 measured differential intensity between pairs of directions distributed over most of the Northern Hemisphere. Measurements show clear evidence of anisotropy that is readily interpreted as due to the solar motion relative to the sources of the radiation. The anisotropy is well fitted by a first order spherical harmonic of amplitude 3.6 + or -0.5 mK, corresponding to a velocity of 360 + or - 50 km/s toward the direction 11.2 + or -0.5 hours of right ascension and 19 deg + or - 8 deg declination.

[22] M. V. Gorenstein, R. A. Muller, G. F. Smoot, A. Tyson, "Radiometer system to map the cosmic background radiation," *Rev. Sci. Instrum.*, vol. 49(4), Apr. 1978.

[23] G. F. Smoot and P. M. Lubin, "Southern Hemisphere measurements of the anisotropy in the cosmic microwave background radiation," *Ap. J.*, Part 2 - Letters to the Editor, vol. 234, p. L83-L86, Dec. 1, 1979.

[24] G. F. Smoot, "Aether Drift and the isotropy of the universe: A measurement of anisotropes in the primordial black-body radiation," *Final Report*, 1 Nov. 1978–31 Oct. 1980. University of California, Berkeley. Lawrence Berkeley Lab. Large-angular-scale anisotropies in the 3 K primordial black-body radiation were detected and mapped with a sensitivity of 2 x 10^{-4} K and an angular resolution of about 10 deg. The motion of the Earth with respect to the distant matter of the Universe ("Aether Drift") was measured and the homogeneity and isotropy of the Universe (the "Cosmological Principle") was probed. The experiment uses two Dicke radiometers, one at 33 GHz to detect the cosmic anisotropy, and one at 54 GHz to detect anisotropies in the residual oxygen above the detectors. The system was installed in the NASA-Ames Earth survey aircraft (U-2), and operated successfully in a series of flights in both the Northern and Southern Hemispheres. Data taking and analysis to measure the anisotropy were successful.

[25] M. A. Janssen, S. M. Bednarczyk, S. Gulkis, H. W. Marlin, and G. F. Smoot, "Pattern measurements of a low-sidelobe horn antenna," *IEEE Transactions on Antennas and Propagation*, vol. AP-27, pp. 551-555, July 1979.

[26] M. A. Toral, R. B. Ratliff, M. C. Lecha, J. G. Maruschak, C. L. Bennett, "Measurements of very low-sidelobe conical horn antennas," *IEEE Transactions on Antennas and Propagation*, vol. 37, pp. 171-177, Feb. 1989.

[27] G. P. Nanos Jr., "Polarization of the blackbody radiation at 3.2 centimeters," *Ap. J.*, Part 1, vol. 232, pp. 341–347, Sept. 1, 1979.

[28] P. M. Lubin, G. F. Smoot, "Search for linear polarization of the cosmic background radiation," *Physical Review Letters*, vol. 42, issue 2, pp. 129-132, Jan. 8, 1979.

[29] P. Lubin, P. Melese, and G. F. Smoot, "Linear and circular polarization of the cosmic background radiation," *Astrophys. J. Ltrs.*, vol. 273, pp. L51-L54, Oct. 15, 1983.

[30] P. M. Lubin, G. L. Epstein, and G. F. Smoot, "3-mm Anisotropy Measurement and the Quadrupole Component in the Cosmic Background Radiation," *Physical Review Letters*, vol. 50, p. 616, 1983.
P. Lubin, T. Villela, G. Epstein, and G. Smoot, "A map of the cosmic background radiation at 3 millimeters," *Astrophysical Journal Letters*, vol. 298, pp. L1-L5, 1985.

[31] P. M. Lubin, G. L. Epstein, and G. F Smoot, "Four-Way Stretch Universe Denied," *Science News*, vol. 123, p. 126, 1983.
A. A. Friedmann, P. Boynton, F. Melchiori, and G. F. Smoot, "A Possible 4-WAY Stretch in the Universe" *Science News*, vol. 117, p. 54, 1980.

[32] P. Lubin, T. Villela, G. Epstein, and G. F. Smoot, "A map of the cosmic background radiation at 3 millimeters," *Astrophys. J. Ltrs.*, vol. 298, pp. L1-L5, Nov. 1, 1985.

Data from a series of balloon flights covering both the Northern and Southern Hemispheres, measuring the large angular scale anisotropy in the cosmic background radiation at 3.3 mm wavelength are presented. The data cover 85 percent of the sky to a limiting sensitivity of 0.7 mK per 7-deg field of view. The data show a 50-sigma (statistical error only) dipole anisotropy with an amplitude of 3.44 +/- 0.17 mK and a direction of alpha= 11.2 h +/- 0.1 h, and delta = -6.0 deg +/- 1.5 deg. A 90 percent confidence level upper limit of 0.00007 is obtained for the rms quadrupole amplitude. Flights separated by six months show the motion of Earth around the Sun. Galactic contamination is very small, with less than 0.1 mK contribution to the dipole quadrupole terms. A map of the sky has been generated from the data.

[33] P. J. E. Peebles, *Principles of Physical Cosmology*, Princeton, NJ: Princeton U. Press, p. 168 (1993).

[34] P. G. Roll and D. T. Wilkinson, "Cosmic Background Radiation at 3.2 cm –Support for Cosmic Black-body Radiation," *Phys. Rev. Letters*, vol. 16, p. 405, 1966.
P. G. Roll and D. T. Wilkinson, "Measurement of Cosmic Background Radiation at 3.2 cm Wavelength," *Annals of Physics*, vol. 44, p. 289, 1967.

[35] R. A. Stokes, R. B. Partridge, and D. T. Wilkinson, "New Measurements of the Cosmic Microwave Background at $\lambda = 3.2$ cm and $\lambda = 1.58$ cm-Evidence in Support of a Blackbody Spectrum," *Phys. Rev. Lett.*, vol. 19, pp. 1199–1202, 1967.
D. T. Wilkinson, "Measurement of the Cosmic Microwave Background at 8.56 mm Wavelength," *Phys. Rev. Lett.*, vol. 19, pp. 1195-1198, 1967.
P. E. Boynton, R. A. Stokes, and D. T. Wilkinson, "Primeval Fireball Intensity at $\lambda = 3.3$ mm," *Phys. Rev. Lett.*, vol. 21, p. 462, 1968.

[36] L. Danese and G. de Zotti, "The relic radiation spectrum and the thermal history of the Universe," *Nuovo Cimento, Rivista, Serie 2*, vol. 7, pp. 277-362, July-Sept. 1977.

[37] L. Danese and G. F. De Zotti, "Double Compton process and the spectrum of the microwave background," *Astronomy and Astrophysics*, vol. 107, p. 39, 1982.

[38] G. F. Smoot, G. de Amici, S. D. Friedman, C. Witebsky, N. Mandolesi, R. B. Partridge, G. Sironi, L. Danese, and G. de Zotti, "Low-frequency measurement of the spectrum of the cosmic background radiation," *Physical Review Letters*, vol. 51, p. 1099-1102, Sept. 19, 1983.
G. F. Smoot, G. de Amici, S. D. Friedman, C. Witebsky, N. Mandolesi, R. B. Partridge, G. Sironi, L. Danese, and G. de Zotti, "New Multifrequency Measurements of the Spectrum of the Cosmic Background Radiation General Relativity and Gravitation," vol. 1, Classical Relativity. Proceedings of the 10th International Conference on General Relativity and Gravitation, held July 4-9, 1983, in Padova, Italy. Edited by B. Bertotti, F. de Felice and A. Pascolini. Published by Consiglio Nazionale delle Ricerche, p. 865, Rome, 1983.

[39] G. F. Smoot, G. de Amici, S. D. Friedman, C. Witebski, N. Mandolesi, R. B. Partridge, G. Sironi, L. Danese, and G. de Zotti, "New multifrequency measurements of the spectrum of the cosmic background radiation," COSPAR, IAU, ESA, *et al.*, Symposium on High-Energy Astrophysics and Cosmology, Pamporovo, Bulgaria, July 18-23, 1983. *Advances in Space Research*, vol. 3, no. 10-12, pp. 465-467, 1984.

[40] G. F. Smoot, M. Bensadoun, M. Bersanelli, G. de Amici, A. Kogut, S. Levin, and C. Witebsky, "Long-wavelength measurements of the cosmic microwave background radiation spectrum," *Astrophys. J. Ltrs.*, vol. 317, pp. L45-L49, June 15, 1987.

[41] G. F. Smoot, S. M. Levin, C. Witebsky, G. de Amici, and Y. Rephaeli, "An analysis of recent measurements of the temperature of the cosmic microwave background radiation." *Ap. J.*, Part 1, vol. 331, pp. 653-659, Aug. 15, 1988.

[42] D. G. Johnson and D. T. Wilkinson, "A 1% Measurement of the Temperature of the Cosmic Microwave Radiation at $\lambda = 1.2$ Centimeters," *Astrophys. J. Ltrs.*, vol. 313, pp. L1-L4, Feb. 1, 1987.

[43] S. Staggs, *et al.*, "A Measurement of the Cosmic Background Spectrum at 1.4 GHz," *Astro. Ltrs. and Communications*, vol. 32, pp. 3-6, 1995.

[44] D. J. Fixsen, A. Kogut, S. Levin, M. Limon, P. Lubin, P. Mirel, M. Seiffert, and E. Wollack, "The Temperature of the CMB at 10 GHz," *Ap. J.*, vol. 612, pp. 86-95, 2004. The measured radiometric temperature of the CMB is 2.721 +/- 0.010 K at 10 GHz and 2.694 +/- 0.032 K at 30 GHz.

[45] J. C. Mather, *et al.*, "Measurement of the Cosmic Microwave Background Spectrum by the COBE FIRAS Instrument," *Ap. J.*, vol. 420, pp. 439-444, 1994.

[46] G. Smoot, C. Bennett, R. Weber, J. Maruschak, R. Ratliff, M. Janssen, *et al.*, "COBE Differential Microwave Radiometers - Instrument design and implementation," *Ap. J.*, Part 1, vol. 360, pp. 685-695, Sept. 10, 1990.

[47] G. F. Smoot, *et al.*, "Structure in the COBE Differential Microwave Radiometer First-Year Maps," *Ap. J.*, vol. 396, pp. L1-L5, 1992.

[48] R. L. Davis, H. M. Hodges, G. F. Smoot, P. J. Steinhardt, and M. S. Turner, "Cosmic microwave background probes models of inflation," *Physical Review Letters*, vol. 69, no. 13, p. 1856-1859, 1993.
G. F. Smoot and P. J. Steinhardt, "Gravity's rainbow," *Class Quant. Grav.*, vol. 10, pp. S19-S32, 1993.
G. F. Smoot and P. J. Steinhardt, "Gravity's rainbow (microwave anisotropy and gravitational waves)," *Classical and Quantum Gravity*, vol. 10, pp. S19-S32, 1993.

[49] G. F. Smoot, *et al.*, "Statistics and Topology of the COBE DMR First Year Maps," *Ap. J.*, vol. 437, pp. 1-11, 1994.

[50] S. Torres, J. Aymon, C. Backus, C. Bennett, S. Gulkis, R. Hollenhorst, D. Hon, Q. H. Huang, M. Janssen, S. Keihm, L. Olson, G. F. Smoot, and E. Wright, "Cosmic Background Explorer/COBE/Satellite Anisotropy Experiment Data Analysis Techniques," Proceedings of the Third International Workshop on Data Analysis in Astronomy, held June 20-27, 1988, in Erice, Sicily, Italy. Editors, V. Di Gesu, L. Scarsi, P. Crane, J.H. Friedman, S. Levialdi, M.C. Maccarone; Publisher, Plenum Press, New York, NY, 1989.

[51] D. N. Spergel, *et al.*, "Wilkinson Microwave Anisotropy Probe (WMAP) three year results: Implications for cosmology," astro-ph/0603449, 2006.

[52] G. F. Smoot, "The CMB Anisotropy Experiments," astro-ph/9705135.

[53] U. Seljak, M. Zaldarriaga, *Ap. J.*, vol. 469, p. 437, 1996.

[54] P. B. Keegstra, G. F. Smoot, K. M. Gorski, G. Hinshaw, and L. Tenorio, "Generalized Spherical Harmonics for All-Sky Polarization Studies astronomical Data Analysis Software and Systems VI," A.S.P. Conference Series, vol. 125, Gareth Hunt and H. E. Payne, eds., p. 198, 1997.

[55] A. Kogut, A. J. Banday, C. L. Bennett, K. M. Gorski, G. Hinshaw, P. D. Jackson, P. Keegstra, C. Lineweaver, G. F. Smoot, L. Tenorio, and E. L. Wright, "Calibration and Systematic Error Analysis for the COBE DMR 4 Year Sky Maps," *Ap. J.*, vol. 470, p. 653, 1996.

[56] S. Perlmutter, *et al.*, "Measurements of Omega and Lambda from 42 high redshift supernovae," *Ap. J.*, vol. 517, pp. 565-586, 1999.

[57] A. G. Riess *et al.*, "Observational evidence from supernovae for an accelerating universe and a cosmological constant," *Astron. J.*, vol. 116, pp. 1009-1038, 1998. astro-ph/9805201.

[58] Planck Science Team, PLANCK BLUEBOOK, available on line, http://www.rssd.esa.int/Planck/

[59] G. F. Smoot and K. Davidson, *Wrinkles In Time*, New York: Avon Books, 1993.

[60] G. F. Smoot, "COBE Observations of The Early Universe," *J. Roy. Astron. Soc. Can.*, vol. 88, no. 2, 1994.

[61] E. L. Wright *et al.*, "Interpretation of the Cosmic Microwave Background Radiation Anisotropy detected by the COBE Differential Microwave Radiometer," *Ap. J.*, vol. 396, pp. L13-L18, 1992.

[62] N. W. Boggess *et al.*, "The COBE Mission: It's design and performance two years after launch," *Ap. J.*, vol. 397, pp. 420-429, 1992.

[63] W. Hu and S. Dodelson, *Ann. Rev. Astron. & Astrophys*, vol. 40, p. 171, 2002.
J. E. Carlstrom, G. P. Holder, and E. D. Reese, *Ann. Rev. Astron. & Astrophys*, vol. 40, p. 643, 2002.

[64] A. Kogut *et al.*, *Ap. J.*, vol. S148, p. 161, 2003.

[65] W. Hu and S. Dodelson, *Ann. Rev. Astron. & Astrophys*, vol. 40, p. 171, 2002.

[66] http://lambda.gsfc.nasa.gov/product/cobe/

[67] http://cosmos.lbl.gov/

[68] http://map.gsfc.nasa.gov/

[69] http://www.rssd.esa.int/Planck/

[70] E. R. Harrison, "Fluctuations at the threshold of classical cosmology," Phys. Rev. D1, p. 2726, 1970.
P. J. E. Peebles and J. T. Yu, "Primeval adiabatic perturbation in an expanding universe," *Astrophysical Journal*, vol. 162, p. 815, 1970.
Ya. B. Zel'dovich, "A hypothesis, unifying the structure and entropy of the universe," *Monthly Notices of the Royal Astronomical Society*, 160, 1972.

[71] E. S. Cheng, P. R. Saulson, D. T. Wilkinson, and B. E. Corey, "Large-Scale Anisotropy in the 2.7 K Radiation," *Astrophys. J. Ltrs.*, vol. 232, pp. L139-L143, September 15, 1979.

[72] S. P. Boughn, E. S. Cheng, and D. T. Wilkinson, "Dipole and Quadrupole Anisotropy of the 2.7 K Radiation," *Astrophys. J. Ltrs.*, vol. 243, pp. L113-L117, 1981.

[73] D. J. Fixsen, E. S. Cheng, and D. T. Wilkinson, "Large-Scale Anisotropy in the 2.7 K Radiation with a Balloon-Borne Maser Radiometer at 24.5 GHz," *Phys. Rev. Ltrs.*, vol. 50, no. 8, pp. 620-622, Feb. 21, 1983.

[74] J. M. Uson and D. T. Wilkinson, "Small-Scale Isotropy of the Cosmic Microwave Background," *Ap. J.*, vol. 283, pp. 471-478, 1984.
J. M. Uson and D. T. Wilkinson, "Improved Limits on Small-Scale Anisotropy in Cosmic Microwave Background," *Nature*, vol. 312, pp. 427-429, 1985.

[75] P. T. Timbie and D. T. Wilkinson, "A Search for Anisotropy in the Cosmic Microwave Radiation at Medium Angular Scales," *Ap. J.*, vol. 353, pp. 140-144, April 10, 1990.

[76] E. K. Conklin and R. N. Bracewell, *Phys. Rev. Lett.*, vol. 18, p. 614, 1967.

[77] D. J. Fixsen *et al.*, *Ap. J.*, vol. 473, p. 576, 1996.

[78] P. J. E. Peebles and J. T. Yu, *Ap. J.*, vol. 162, p. 815, 1970.

[79] D. W. Sciama, "Eppur Si Muove," *Comments on Astrophysics and Space Physics*, vol. 4, p. 35, March 1972.

[80] J. Carlstrom, M. Joy, L. Grego, G. Holder, W. L. Holzapfel, S. LaRoque, J. J. Mohr & E. D. Reese "The Sunyaev-Zel'dovich Effect: Results and Future Prospects", IAP Conference Proceedings, 2000.

[81] South Pole Telescope, http://spt.uchicago.edu/
http://cosmology.berkeley.edu/group/swlh/sp/_telescope/

[82] N. Jarosik, *et al.*, "Three-Year Wilkinson Microwave Anisotropy Probe (WMAP) Observations: Beam Profiles, Data Processing, Radiometer Characterization and Systematic Error Limits," astro-ph/0603452.

[83] W. Hu and M. White, *Ap. J.*, vol. 435, p. 568, 1997.

[84] J. M. Kovac, E. M. Leitch, C. Pryke, J. E. Carlstrom, N. W. Halverson, and W. L. Holzapfel, "Detection of polarization in the cosmic microwave background using DASI," *Nature*, vol. 420, pp. 772-787, Dec. 19, 2002.
E. M. Leitch, J. M. Kovac, N. W. Halverson, J. E. Carlstrom, C. Pryke, and M. W. E. Smith, "Degree Angular Scale Interferometer 3 Year Cosmic Microwave Background Polarization Results," *Ap. J.*, vol. 624, part 1 pp. 10-20, 2005.

Portrait photo of Professor Smoot III by photographer Jerry Bauer.

Physics 2007

Albert Fert and Peter A. Grünberg

"for the discovery of Giant Magnetoresistance"

THE NOBEL PRIZE IN PHYSICS

Speech by Professor Börje Johansson of the Royal Swedish Academy of Sciences.
Translation of the Swedish text.

Your Majesties, Your Royal Highnesses, Ladies and Gentlemen,

This year's Nobel Prize in Physics concerns magnetism and electricity, more specifically how an external magnetic field can influence the electrical resistance of a metal.

According to Alfred Nobel's will, the Nobel Prize in Physics is to be awarded for the discovery or invention that has conferred the greatest benefit on mankind. In their discovery of Giant Magnetoresistance Albert Fert and Peter Grünberg have provided us with an invention whose importance cannot be doubted. One can in all justice say that this effect is one prerequisite for the modern IT revolution.

Uses of electricity are among the applications of physics that have most clearly served mankind throughout modern history. When, in the early 19th century, the British physicist Faraday made his first hesitant experiments involving electricity during his public demonstrations, one of those watching was Gladstone, at that time Chancellor of the Exchequer. When he asked if the experiment could lead to any practical benefits Faraday's immediate answer was "One day you will be able to tax it". Few findings can have pleased as many ministers of finance through the ages as those related to the movements of electrons.

In traditional electronic applications, it is the charge of the electrons that is exploited. But electrons have another feature, namely their own rotation or spin. This spin brings with it an element of magnetism. The electron is like a tiny magnet that either points upwards or downwards. In what we call ferromagnetic metals, such as iron, there are considerably more electrons with magnetism upwards than downwards. The overall effect is what gives iron its magnetic qualities.

When an electric current is transmitted through a metal or another conductor – irrespective of whether it is magnetic or not – the movement of the electrons is impeded by atomic irregularities in the material. It is these obstructions at atomic level that result in what we call the electrical resistance. In a ferro-magnetic metal like iron, however, we encounter something specific. The resistance for spin-up electrons differs from the resistance for spin-down electrons. This means that the resistance of the material can be influenced magnetically.

The discovery, or invention, of Giant Magnetoresistance is an unusually clear example of the interaction of technological developments and fundamental scientific insights. One prerequisite for the discovery of this new phenomenon was the technology that developed in the 1980s in the wake of the emerging semi-conductor industry: nascent nanotechnology.

Peter Grünberg and Albert Fert both used the art of creating nanometre thin layers of only a few atoms' thickness in order to learn more about how magnetic materials behave. They discovered simultaneously, but independently of each other, that an external magnetic field could have a very great impact on electrical resistance if the magnetic layers alternated with non-magnetic layers. The concept of Giant Magnetoresistance was coined to refer to this new phenomenon.

Retrieving data from hard disks is one of the most important applications of Giant Magnetoresistance, since the information is stored on the hard disks in the form of small magnetic areas. Giant Magnetoresistance can be used by the read-out head that scans the hard disk and converts the magnetic information into electrical current.

But this is merely one of a horde of potential application areas. Giant Magnetoresistance also means the beginning of an entirely new form of electronics, spintronics, in which both the spin and the charge of electrons can be exploited.

Cher Professeur Fert: votre découverte de la magnétoresistance géante a déjà transformé le monde de l'informatique et les promesses pour le futur sont importantes et nombreuses.

Lieber Herr Professor Grünberg: Ihre Entdeckung des Riesenmagnetowiderstands hat bereits die Welt der Informationstechnologie verwandelt und die Verheißungen für die Zukunft sind groß.

Professor Fert and Professor Grünberg,

You have been awarded the 2007 Nobel Prize in Physics for your discovery of Giant Magnetoresistance. It is an honour for me to convey to you the warmest congratulations of the Royal Swedish Academy of Sciences. I now ask you to step forward to receive your Nobel Prizes from the hands of His Majesty the King.

ALBERT FERT

I was born in March 1938 in the small town of Carcassonne in the south of France. Later, and until I was two years old, I lived in Toulouse, another city in Southern France, where my parents were high school teachers, my father in physics and my mother in economics. But war loomed on the horizon. My father was mobilized into the army in June 1939, two months before the birth of my brother André. In 1940, he was captured and made a prisoner of war by the Germans; he returned home only in 1945. For the duration of the war, my brother and I were sent to live at our grandparents' farm in Montclar, a very small village in the foothills of the Pyrénées. My mother, who continued to teach in Toulouse, came to see us every weekend and so, until the age of seven, I lived the life of a country boy, milking the goats, harvesting grapes in my grand father's vineyards, setting snares for rabbits or hares. I lived very close to the world of plants and animals and very far removed from the world of physics.

In June 1945, my father returned from prisoner of war camp and our family was reunited in Toulouse. I became a city boy with a longing for life in the countryside. My father, while continuing to teach high school, prepared his doctoral thesis. Eventually, he was promoted to the rank of Professsor at the University of Toulouse where he made important contributions to the development of electron microscopy. My brother and I applied ourselves studiously to our primary and secondary school work. Our father followed our progress in the sciences. Unquestionably, his penchant for rigorous thinking had a major influence on my approach to mathematics and physics. I earned good grades in the sciences. During my adolescent years, I also developed a great interest in literature, the arts and sports. At the age of fourteen, I began playing rugby and I was very proud of being selected to play in our high school team. At the age of seventeen, having completed my studies for the baccalauréat, I found myself drawn to Paris and above all to the Ecole Normale Supérieure (ENS), not only because it was a prestigious institution but also because it had the added attraction of being in the center of Paris, close to the Latin Quarter and St. Germain des Près, and at the heart of the city's intellectual life. I worked hard to prepare for the competitive entrance examinations at the ENS, in mathematics and physics. I was accepted. In September 1957, I lightheartedly left Toulouse for Paris and the Ecole Normale Supérieure.

My six years at the Ecole Normale Supérieure between the ages of 19 and 24 were a very intense period of my youth. The richness of my life on the small campus of the rue d'Ulm came from daily contacts with students working in a broad spectrum of disciplines, the sciences, philosophy, literature, history etc. In addition Paris was everywhere around us, with its museums, exhibitions, cinemas, concert halls and jazz clubs. I became a passionate fan of jazz, photography, film. I actually wrote a script and made a film.

It is possible that I might have been diverted from the pursuit of a career in science had I not been studying with excellent physics instructors. In particular, Jacques Friedel, who had established a physics program at the masters degree level, exposed me to the most up to date developments in condensed matter physics and emphasized as well an in depth teaching of quantum mechanics and statistical mechanics. It was this program which drew me into the study of condensed matter physics. When I returned from my military service in 1965, I began work on my doctoral thesis under the direction of Ian Campbell, in the Laboratory for Solid States Physics at Orsay which Jacques Friedel directed. I had the good luck of having the thesis topic: "Testing the suggestion of Neville Mott (future Noble Prize laureate) that the mobility of electrons in a ferromagnetic metal depends on the orientation of their spin in relation to their magnetic orientation". In the course of the research for my thesis, I became hooked on physics. I discovered that the work of a young inexperienced researcher can open many doors – provided that his experiments are well executed and rigorously interpreted. My findings in my thesis, as well as my in- depth knowledge of the physics of electron transport, made it possible for me to return to this subject fifteen years later and to discover the giant magnetoresistance. I am still surprised when I see that my findings during that earlier period comprise a large part of the basis for spintronics today.

However, when I defended my thesis in 1970, the available technologies did not allow for further advancement in the direction of giant magnetoresistance and spintronics. Although a concept similar to that of GMR already had emerged from my thesis experiments on ternary alloys, it proved impossible to extend it to the case of multilayers because it was impossible at that time to produce sufficiently thin layers. Thus, the exploitation of the influence of the spin on the mobility of electrons in ferromagnetic metals had to be postponed until the mid 1980s. Following a post-doctoral appointment at the University of Leeds, I returned to Orsay to take up a position as Assistant Professor at the University of Paris-Sud and to supervise a small group of doctoral students and post-docs at the Laboratory of Solid State Physics. I was promoted to the rank of professor in 1976. Marie-Josée and I lived in Paris; Ariane and Bruno, born in 1968 and 1971, were growing up and I divided my time between family life and a very sustained effort in research and teaching. During our summer vacations, we returned to the Mediterranean and the Pyrénées, to the home of my parents in Banyuls sur Mer, close to the Spanish frontier. My research activities during the 1970s and at the beginning of the 1980s touched on a variety of subjets and yielded fruitful results. The thesis of my first doctoral student, Alain Friederich, shed new light on problems relating to the Hall effect and to the anisotropy of magnetoresistance linked to spin-orbit coupling. Today the Spin Hall effect has become a hot topic in physics. I think that some of our results from that period should have an impact on spintronics today. In 1975, in order to interpret our results on the Hall effect and the anisotropy of interactions between electrons and magnetic impurities, I began a collaboration with Peter M. Levy, a theorist at New York University. Over the course of several years, I spent many summer months at New York University and learned to love the city, Greenwich Village and its jazz clubs, SoHo and its art galleries. My collaboration with Peter Levy on theoretical problems encompassed numerous subjects and, among other results, led to the discovery of the existence of Dzyaloshinsky-Moriya interactions in spin glasses. One consequence of these interactions is the triad anisotropy in spin glasses, which was confirmed experimentally in the thesis of my Ph.D student Dimitri Arvanitis.

In the mid 1980s, it seemed likely that techniques developed in the field of microelectronics would make it possible to grow magnetic multilayers built up from layers of nanometer thickness. It was precisely at this time that my former student, Alain Friederich, was developing Molecular Beam Epitaxy (MBE) in his research group of the Thomson-CSF company. While attending a meeting in San Diego in 1985, Alain Friederich and I were discussing our work at a bar next to the swimming pool, under the palm trees and the stars and we decided to collaborate on the growth and study of magnetic multilayers. With help from the expert on MBE at Thomson-CSF, Patrick Etienne, as well from my doctoral students, Frédéric Nguyen Van Dau, Frédéric Petroff and Agnès Barthélémy, and two post-docs, Mario Baibich and Jean-Marc Broto, our collaboration resulted in the discovery of giant magnetoresistance at the beginning of 1988.

On the day we discovered giant magnetoresistance, we were measuring several multilayers of Fe/Cr, one after the other. We were not altogether certain about the quality of the multilayers that contained the thinnest layers of chromium. Exercising caution, we began by measuring multilayers in which the layers of chromium were the thickest. We determined that there was indeed a magnetoresistive effect. We then moved on to multilayers in which the chromium layers were progressively less thick. And wonder of wonders! The greater the decrease in the thickness of the chromium, the larger the magnetoresistance. The final measurement, which we took on 0.9 nm of chromium, was stupefying. The resistance increase between the parallel and anti-parallel configurations of the layers of iron went up to 80%! We had just participated in the birth of the phenomenon which we called giant magnetoresistance (GMR).

The International Conference on Magnetism which took place in Paris at the beginning of July 1988 offered us our first occasion to present the results of our work on giant magnetoresistance in Fe/CR multilayers. However, my Ph.D student Frédéric Nguyen Van Dau had only a few minutes in which to make his presentation and therefore his paper did not provoke much of a response. Peter Grünberg did not attend the conference. It was only the following week, at the International Conference on Magnetic Films and Surfaces (ICMFS) at Le Creusot in Burgundy, that Peter Grünberg and I were able to present the details of our respective results and to take the measure of the considerable interest which our findings provoked among members of the magnetism community. At this conference, I also was able to explain my earlier results which showed the influence of spin on the mobility of electrons in ferromagnetic metals and to present the interpretation of "GMR" on this basis. Interest within the magnetism community in this new type of phenomenon which exploited the spin of electrons in magnetic nanostructures grew exponentially following the publication of our article by Physical Review Letters in December 1988. I have some heartwarming memories of this period. As I neared the end of my first paper on this subject presented at a conference in the United States, I was anxious to know whether I would be able to complete my presentation within the designated time frame. I asked Bret Heinrich, the Chair, what time was left. He told me that, given the level of interest in the work I was presenting, I could continue for as long as I wanted. Every researcher dreams of such a response from the chair. Thanks Bret!

Beginning in 1989, the physics of multilayers and GMR became very hot topics and subjects of research in an increasing number of laboratories. In addition, my own research activity was expanding on many fronts. Apart from experimental work on the Fe/Cr systems in the theses of Frédéric Nguyen Van Dau, Agnès Barthélémy and Frédéric Petroff, I developed, with Agnès Barthélémy, a semi-classical theory of GMR which complemented that of Camley and Barnas. I also collaborated with Peter Levy and Shufeng Zhang of New York University on the development of the first quantum model of GMR. In 1990, we began to study multilayers prepared by sputtering in a collaboration with the team of Peter Shroeder, Jack Bass and Bill Pratt at Michigan State University (MSU). This collaboration led in 1991 to the first observation of GMR and of oscillations in the interlayer coupling for the Co/Cu system by my doctoral student Dante Mosca. Stuart Parkin obtained similar results at approximately the same time in Almaden. The system Co/Cu later became the archetypical GMR system. The first observations of inverse GMR in the thesis of Jean-Marie George constitute another important result of my team in the early 1990s. Beginning in 1991, I became interested as well in results of the team at MSU on GMR in the geometry where the current is perpendicular to the plane of the layers (CPP-GMR) . In collaboration with a young researcher at Thomson-CSF, Thierry Valet, I developed a theory of the CPP-GMR based on the concept of spin accumulation. Today, this theory has become essential for many current developments in spintronics. The 1993 Valet-Fert paper in Physical Review is my second most frequently cited paper, coming just after the paper in which we presented the discovery of GMR. Beginning in 1994, I collaborated with Luc Piraux at the University of Louvain, on studies of CPP-GMR using multilayered nanowires. These studies made it possible to extend the results of MSU to much greater thicknesses and to confirm the length scale linked to the effects of spin accumulation in the Valet-Fert model.

Thus, the beginning of the 1990s was for me and my research team a period of intense and fruitful activity during which we made important contributions to the development of the physics of GMR and to the establishment of essential concepts in spintronics. Peter Grünberg and I, in collaboration with firms like Thomson-CSF, Philips and Siemens, have participated as well in several projects financed by the European Community in order to further the development of new applications. The year 1994 brought us international recognition, with the 'International Union of Pure and Applied Physics' presentation of its Magnetism Award to Peter Grünberg and myself, and the American Physical Society's award of the International Prize for New Materials to Peter Grünberg, Stuart Parkin and me. It also strucks me at this time that GMR was only a first step in the exploitation of spin in magnetic nanostructures and that this first step had opened the door to a much larger field of research – the field we now call spintronics. Alain Friederich, the director of the physics group at the laboratories of Thomson-CSF, proposed to me that we create a new laboratory which would bring together the CNRS and Thomson-CSF for a joint exploration of the perspectives opened up by GMR. We proposed a research program which moved us into a large number of new areas: magnetic tunnel junctions, half-metallic ferromagnets, spin injection phenomena, spin transport in semi-conductors... The project was accepted. The new laboratory, called the Unité Mixte de Physique (UMP) CNRS-Thomson-CSF, (which later became the UMP CNRS-Thales) was established in the spring of 1995, at Corbeville, a few kilometers from the University of Paris-Sud. An agreement joined the laboratory to the University, in which I continued my teaching. From the beginning, the laboratory brought together my team at Orsay, the team of Thomson-CSF with which we were collaborating and several researchers and engineers at CNRS: Jean-Marie George, Annie Vaurès, Jean-Luc Maurice (shortly thereafter, a team for the study of superconductivity joined the laboratory). Since then, it is within this Unité Mixte CNRS–Thales that my research activity has progressed in many directions in the field of spintronics. The team has been progressively enlarged with the arrival of new researchers from CNRS and assistant professors from the University of Paris-Sud – Abdelmadjid Anane, Manuel Bibes, Vincent Cros, Cyrile Deranlot, Julie Grollier, Henri Jaffrès, Richard Mattana and Pierre Seneor.

It is difficult for me to describe my research activities at the CNRS /Thales laboratory between 1995 and 2007 precisely because this activity has moved in so many directions and has explored very diverse dimensions of spintronics. I can only refer the reader to my Nobel Lecture for details about the results of this activity. It is also the case that the relatively young members of my team are achieved their independence and have progressively developed their own research programs. Numerous and diversified studies of magnetic tunnel junctions have been undertaken by Frederic Petroff, Agnès Barthélémy and Vincent Cros. Agnès Barthélémy, assisted by Manuel Bibes, is now responsible for all research on the applications of magnetic oxides and multiferroics in spintronics. Our very important activity on the phenomena of spin transfer is being developed, for the most part, by Vincent Cros and Julie Grollier, with the help of Abdelmadjid Anane and Henri Jaffres. Spintronics involving semiconductors is being studied by Jean-Marie George, Henri Jaffres, Abdelmadjid Anane and Richard Mattana. Single electron spintronics has become the domain of Pierre Seneor and Frédéric Petroff, while Frédéric Nguyen Van Dau directs the development of various applications and Jean-Luc Maurice is the expert in structural studies. An activity in molecular spintronics is in the process of being launched by Pierre Seneor, Frédéric Petroff, Jean-Marie George and Richard Mattana. All of them are currently participating in the development of nanotechnologies, with the valuable assistance from Cyrile Deranlot and from Karim Bouzehouane, Stéphane Fusil and Eric Jacquet of the superconductivity team. As for me, I am making every effort to participate in most of these current activities. I also have been able to devote myself a little more to theory: the modeling of single-electron spintronics with Jozef Barnas at the University of Poznan; the theory of the injection of spin into semiconductors with Henri Jaffrès; the theory of the phenomenons of spin transfer with Peter Levy, Henri Jaffrès, Jozef Barnas.

The year 2007 marked a high point in the activity of the UMP CNRS/Thales. It was rich in new results and publications and, for me, abundant in scientific prizes: the Japan Prize in April, in Tokyo; the Wolf Prize in Physics in May, in Jerusalem and the Nobel Prize in December, in Stockholm. The Nobel Prize, of course, has changed my life. I have received innumerable requests; and new responsibilities are on the horizon. In addition, I am eager to return to my research projects and to concretize my recent ideas. A difficult challenge! I also hope that the Nobel Prize will facilitate the entire team's energetic communication of its ideas and messages.

In conclusion, I want to thank all those who have helped me to become who I am today: first of all my parents and Marie-Josée; also Jacques Friedel and Ian Campbell, my first guides in my trajectory as a physicist; Alain Friederich, with whom I was able to launch a collaboration which was decisive for the discovery of GMR; the team at the Unité Mixte CNRS/Thales, and, finally, all those who have worked with me during my forty year-long career as a physicist.

THE ORIGIN, DEVELOPMENT AND FUTURE OF SPINTRONICS

Nobel Lecture, December 8, 2007

by

ALBERT FERT

Unité Mixte de Physique CNRS/Thales, 91767, Palaiseau, and Université Paris-Sud, 91405, Orsay, France.

OVERVIEW

Electrons have a charge and a spin, but until recently, charges and spins have been considered separately. In conventional electronics, the charges are manipulated by electric fields but the spins are ignored. Other classical technologies, magnetic recording for example, are using the spin but only through its macroscopic manifestation, the magnetization of a ferromagnet. This picture started to change in 1988 when the discovery[1-2] of the Giant Magneto-Resistance (GMR) of the magnetic multilayers opened the way to an efficient control of the motion of the electrons by acting on their spin through the orientation of a magnetization. This rapidly triggered the developments of a new field of research and technology, today called spintronics and, like the GMR, exploiting the influence of the spin on the mobility of the electrons in ferromagnetic materials. Actually, the influence of the spin on the mobility of the electrons in ferromagnetic metals, first suggested by Mott[3], had been experimentally demonstrated and theoretically described in my Ph.D. thesis more than ten years before the discovery of 1988. The GMR was the first step on the road of the exploitation of this influence to control an electrical current. Its application to the read head of hard discs greatly contributed to the fast rise in the density of stored information and led to the extension of the hard disk technology to consumer's electronics. Then, the development of spintronics revealed many other phenomena related to the control and manipulation of spin currents. Today this field of research is extending considerably, with very promising new axes like the phenomena of spin transfer, spintronics with semiconductors, molecular spintronics or single-electron spintronics.

FROM SPIN DEPENDENT CONDUCTION IN FERROMAGNETS TO GIANT MAGNETORESISTANCE

GMR and spintronics take their roots in previous researches on the influence of the spin on the electrical conduction in ferromagnetic metals[3-5]. The spin

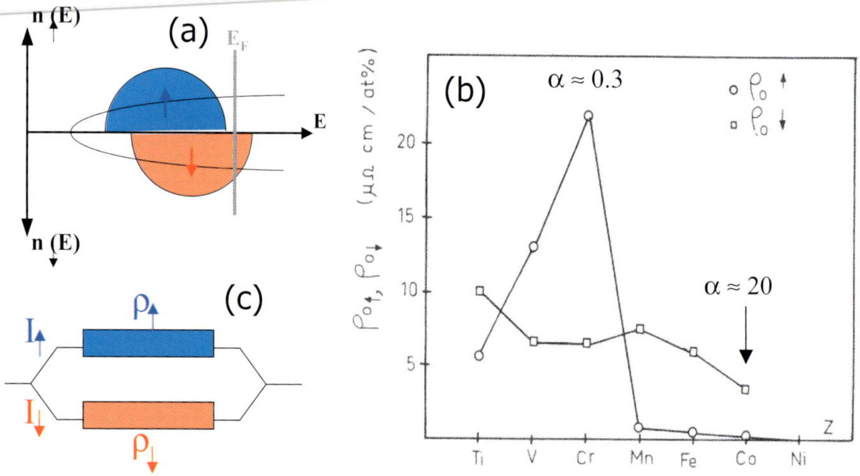

Figure 1. Basics of spintronics. (a) Schematic band structure of a ferromagnetic metal showing the energy band spin-splitting . (b) Resistivities of the spin up and spin down conduction channels for nickel doped with 1% of several types of impurity (measurements at 4.2 K)[4]. The ratio α between the resistivities $\rho_{0\downarrow}$ and $\rho_{0\uparrow}$ of the spin ↓ and spin↑ channels can be as large as 20 (Co impurities) or, as well, smaller than one (Cr or V impurities). (c) Schematic for spin dependent conduction through independent spin ↓ and spin↑ channels in the limit of negligible spin mixing ($\rho_{\uparrow\downarrow} = 0$ in the formalism of Ref.[4]).

dependence of the conduction can be understood from the typical band structure of a ferromagnetic metal shown in Fig.1a. The splitting between the energies of the "majority spin" and "minority spin" directions (spin up and spin down in the usual notation) makes that the electrons at the Fermi level, which carry the electrical current, are in different states for opposite spin directions and exhibit different conduction properties. This spin dependent conduction was proposed by Mott[3] in 1936 to explain some features of the resistivity of ferromagnetic metals at the Curie temperature. However, in 1966, when I started my Ph.D. thesis, the subject was still almost completely unexplored. My supervisor, Ian Campbell, proposed that I investigate it with experiments on Ni- and Fe-based alloys and I had the privilege to be at the beginning of the study of this topic. I could confirm that the mobility of the electrons was spin dependent and, in particular, I showed that the resistivities of the two channels can be very different in metals doped with impurities presenting a strongly spin dependent scattering cross-section[4]. In Fig.1b, I show the example of the spin up (majority spin) and spin down (minority spin) resistivities of nickel doped with 1% of different types of impurities. It can be seen that the ratio α of the spin down resistivity to the spin up one can be as large as 20 for Co impurities or, as well, smaller than one for Cr or V impurities, consistently with the theoretical models developed by Jacques Friedel for the electronic structures of these impurities. The two current conduction was rapidly confirmed in other groups and, for example, extended to Co-based alloys by Loegel and Gautier[5] in Strasbourg.

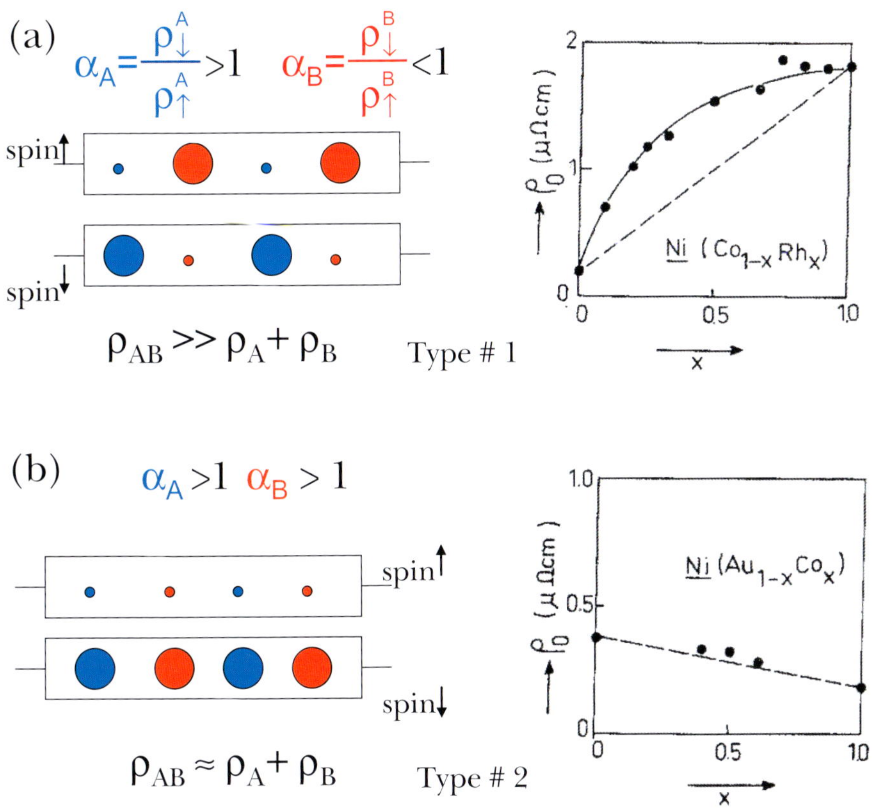

Figure 2. Experiments on ternary alloys based on the same concept as that of GMR[4]. (a) Schematic for the spin dependent conduction in alloys doped with impurities of opposite scattering spin asymmetries ($\alpha_A = \rho_{A\downarrow}/\rho_{A\uparrow} > 1$, $\alpha_B = \rho_{B\downarrow}/\rho_{B\uparrow} < 1$, which leads to $\rho_{AB} >> \rho_A + \rho_B$) and experimental results for $Ni(Co_{1-x}Rh_x)$ alloys. (b) Same for alloys doped with impurities of similar scattering spin asymmetries ($\alpha_A = \rho_{A\downarrow}/\rho_{A\uparrow} > 1$, $\alpha_B = \rho_{B\downarrow}/\rho_{B\uparrow} > 1$, whitch leads to $\rho_{AB} \approx \rho_A + \rho_B$) and experimental results for $Ni(Au_{1-x}Co_x)$ alloys. In GMR the impurities A and B are replaced by multilayers, the situation of a (b) corresponding to the antiparallel (parallel) magnetic configurations of adjacent magnetic layers.

In my thesis, I also worked out the so-called two current model[4] for the conduction in ferromagnetic metals. This model is based on a picture of spin up and spin down currents coupled by spin mixing, i.e. by momentum exchange. Spin mixing comes from spin-flip scattering, mainly from electron-magnon scattering which increases with temperature and equalizes partly the spin up and spin down currents above room temperature in most ferromagnetic metals. The two-current model is the basis of spintronics today, but, surprisingly, the interpretation of the spintronics phenomena is generally based on a simplified version of the model neglecting spin mixing and assuming that the conduction is by two independent channels in parallel, as illustrated by Fig. 1c. It should be certainly useful to revisit the interpretation of many recent experiments by taking into account the spin mixing contributions (note that the mechanism of spin mixing should not be confused with the relaxation of spin accumulation by other types of spin-flips[6]).

As a matter of fact, some experiments of my thesis with metals doped with two types of impurities[4] were already anticipating the GMR. This is illustrated by Fig. 2. Suppose, for example, that nickel is doped with impurities of Co which scatter strongly the electrons of the spin down channel and with impurities of rhodium which scatter strongly the spin up electrons. In the ternary alloy Ni(Co + Rh), that I call of type #1, the electrons of both channels are strongly scattered either by Co or by Rh, so that the resistivity is strongly enhanced. In contrast, there is no such enhancement in alloys of type #2 doped with impurities (Co and Au for example) scattering strongly the electrons in the same channel and leaving the second channel open. The idea of GMR is the replacement of the impurities A and B of the ternary alloy by layers A and B in a multilayer, the antiparallel magnetic configuration of the layers A and B corresponding to the situation of an alloy of type #1, while the configuration with a parallel configuration corresponds to type #2. This brings the possibility of switching between high and low resistivity states by simply changing the relative orientation of the magnetizations of layers A and B from antiparallel to parallel. However, the transport equations tell us that the relative orientation of layers A and B can be felt by the electrons only if their distance is smaller than the electron mean free path, that is, practically, if they are spaced by only a few nm. Unfortunately, in the seventies, it was not technically possible to make multilayers with layers as thin as a few nm. I put some of my ideas in the fridge and, in my team at the Laboratoire de Physique des Solides of the Université Paris-Sud, from the beginning of the seventies to 1985, I worked on other topics like the extraordinary Hall effect, the spin Hall effect, the magnetism of spin glasses and amorphous materials.

In the mid-eighties, with the development of techniques like the Molecular Beam Epitaxy (MBE), it became possible to fabricate multilayers composed of very thin individuals layers and I could consider trying to extend my experiments on ternary alloys to multilayers. In addition, in 1986, I saw the beautiful Brillouin scattering experiments of Peter Grünberg and coworkers[7] revealing the existence of antiferromagnetic interlayer exchange couplings in Fe/Cr multilayers. Fe/Cr appeared as a magnetic multilayered system in which it was possible to switch the relative orientation of the magnetization in adjacent magnetic layers from antiparallel to parallel by applying a magnetic field. In collaboration with the group of Alain Friederich at the Thomson-CSF company, I started the fabrication and investigation of Fe/Cr multilayers. The MBE expert at Thomson-CSF was Patrick Etienne, and my three Ph.D. students, Frédéric Nguyen Van Dau first and then Agnès Barthélémy and Frédéric Petroff, were also involved in the project. This led us in 1988 to the discovery[1] of very large magnetoresistance effects that we called GMR (Fig. 3a). Effects of the same type in Fe/Cr/Fe trilayers were obtained practically at the same time by Peter Günberg at Jülich[2] (Fig. 3b). The interpretation of the GMR is similar to that described above for the ternary alloys and is illustrated by Fig. 3c. The first classical model of the GMR was published in 1989 by Camley and Barnas[8] and I collaborated with Levy and Zhang for the first quantum model[9] in 1991.

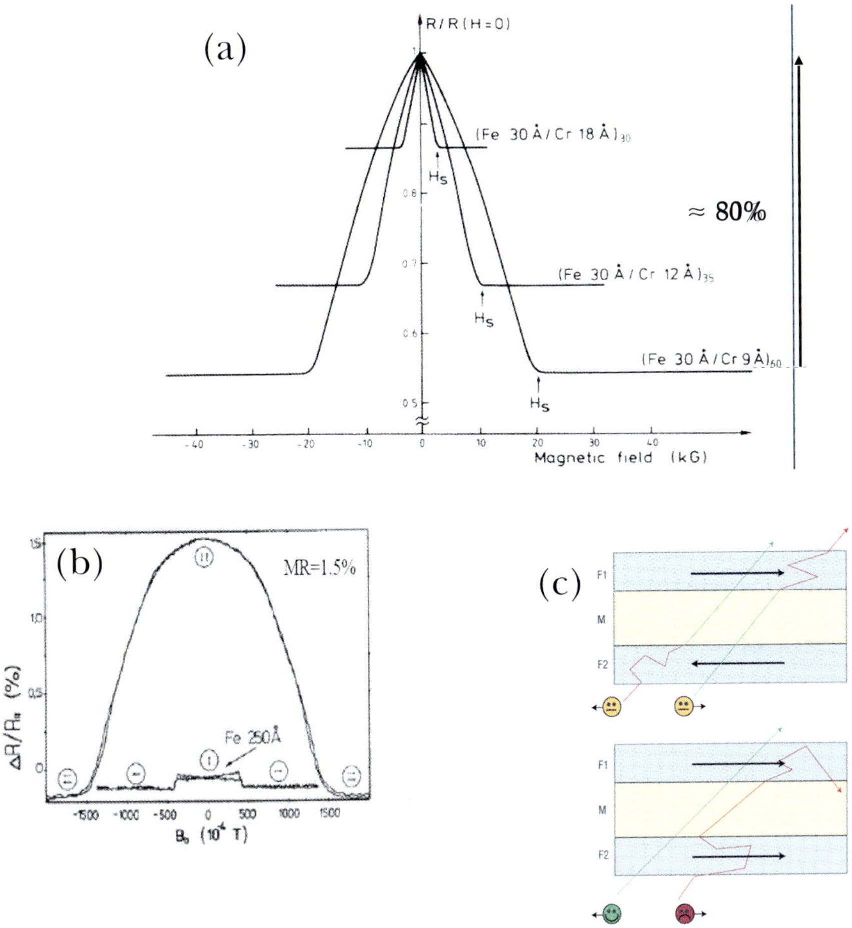

Figure 3. First observations of giant magnetoresistance. (a) Fe/Cr(001) multilayers[1] (with the current definition of the magnetoresistance ratio, MR = 100(R_{AP}-R_P)/R_p, MR = 80% for the (Fe 3nm/Cr 0.9nm) multilayer). (b) Fe/Cr/Fe trilayers[2]. (c) Schematic of the mechanism of the GMR. In the parallel magnetic configuration (bottom), the electrons of one of the spin directions can go easily through all the magnetic layers and the short-circuit through this channel lead to a small resistance. In the antiparallel configuration (top), the electrons of each channel are slowed down every second magnetic layer and the resistance is high (figure from Ref.[18]).

I am often asked if I was expecting such large MR effects. My answer is yes and no: on one hand, a very large magnetoresistance could be expected from an extrapolation of my preceding results on ternary alloys, on the other hand one could fear that the unavoidable structural defects of the multilayers, interface roughness for example, might introduce spin-independent scatterings cancelling the spin-dependent scattering inside the magnetic layers. The good luck was finally that the scattering by the roughness of the interfaces is also spin dependent and adds its contribution to the "bulk" one (the "bulk" and interface contributions can be separately derived from CPP-GMR experiments).

THE GOLDEN AGE OF GMR

Rapidly, our papers reporting the discovery of GMR attracted attention for their fundamental interest as well as for the many possibilities of applications, and the research on magnetic multilayers and GMR became a very hot topic. In my team, reinforced by the recruitment of Agnés Barthélémy and Frédéric Petroff, as well as in the small but rapidly increasing community working in the field, we had the exalting impression of exploring a wide virgin country with so many amazing surprises in store. On the experimental side, two important results were published in 1990. Parkin *et al.*[10] demonstrated the existence of GMR in multilayers made by the simpler and faster technique of sputtering (Fe/Cr, Co/Ru and Co/Cr), and found the oscillatory behaviour of the GMR due to the oscillations of the interlayer exchange as a function of the thickness of the nonmagnetic layers. Also in 1990 Shinjo and Yamamoto[11], as well as Dupas *et al.*[12], demonstrated that GMR effects can be found in multilayers without antiferromagnetic interlayer coupling but composed of magnetic layers of different coercivities. Another important re-

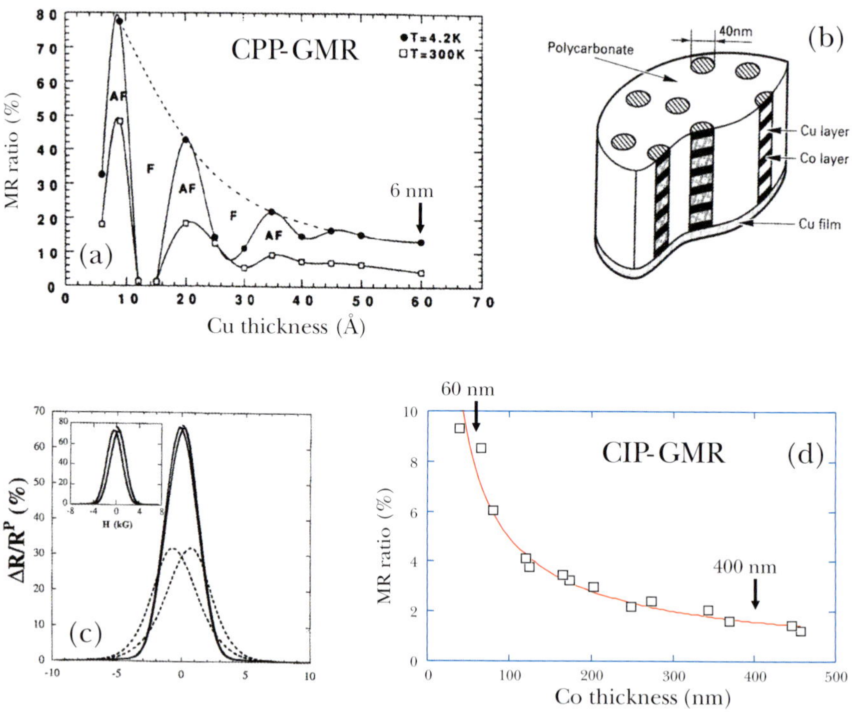

Figure 4. (a) Variation of the GMR ratio of Co/Cu multilayers in the conventional Current In Plane (CIP) geometry as a function of the thickness of the Cu layers[13]. The scaling length of the variation is the mean free path (short). (b) Structure of multilayered nanowires used for CPP-GMR measurements [21]. (c) CPP-GMR curves for (Permalloy 12 nm/ Copper 4 nm) multilayered nanowires (solid lines) and (Cobalt 10 nm/Copper 5nm) multilayered nanowires (dotted lines)[21]. (d)) Variation of the CPP-GMR ratio of Co/Cu multilayered nanowires as a function of the thickness of the Co layers[21]. The scaling length of the variation is the spin diffusion length (long).

sult, in 1991, was the observation of large and oscillatory GMR effects in Co/Cu, which became the archetypical GMR system (Fig. 4a). The first observations[13] were obtained in my group by my Ph. D. student Dante Mosca with multilayers prepared by sputtering at Michigan State University and at about the same time in the group of Stuart Parkin at IBM[14]. Also in 1991, Dieny *et al.*[15] reported the first observation of GMR in spin-valves, i.e. trilayered structures based on a concept of my co-laureate Peter Grünberg[16] in which the magnetization of one of the two magnetic layers is pinned by coupling with an antiferromagnetic layer while the magnetization of the second one is free. The magnetization of the free layer can be reversed by very small magnetic fields, so that the concept is now used in most applications.

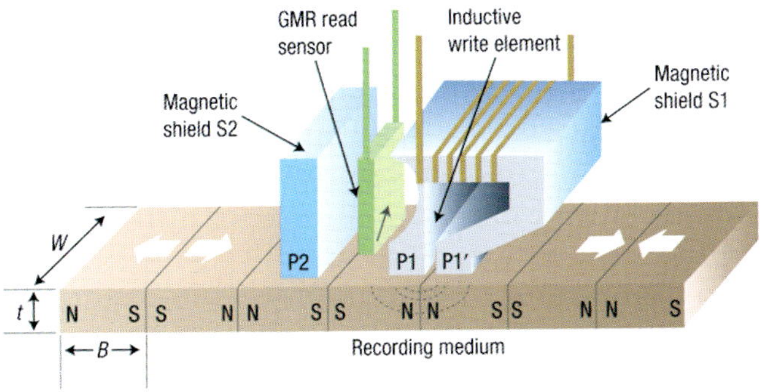

Figure 5. GMR head for hard-disc. Figure from Chappert *et al.*[18].

Other developments of the research on magnetic multilayers and GMR at the beginning of the seventies are described in the Nobel lecture of my co-laureate Peter Grünberg, with, in particular, a presentation of the various devices bases on the GMR of spin valve structures[17-18]. In the read heads (Fig.5) of the Hard Disc Drives (HDDs), the GMR sensors based on spin-valves have replaced the AMR (Anisotropic Magnetoresistance) sensors in 1997. The GMR, by providing a sensitive and scalable read technique, has led to an increase of the areal recording density by more than two orders of magnitude (from ≈ 1 to ≈ 600 Gbit/in^2 in 2007). This increase opened the way both to unprecedented drive capacities (up to 1 terabyte) for video recording or backup and to smaller HDD sizes (down to .85-inch disk diameter) for mobile appliances like ultra-light laptops or portable multimedia players. GMR sensors are also used in many other types of application, mainly in automotive industry and bio-medical technology[19].

CPP-GMR AND SPIN ACCUMULATION PHYSICS

During the first years of the research on GMR, the experiments were performed only with currents flowing along the layer planes, in the geometry we call CIP (Current In Plane). It is only in 1993 that experiments of CPP-

GMR begun to be performed, that is experiments of GMR with the Current Perpendicular to the layer Planes. This was done first by sandwiching a magnetic multilayer between superconducting electrodes by Bass, Pratt and Shroeder at Michigan State University[20], and, a couple of years after, in a collaboration of my group with Luc Piraux at the University of Louvain, by electrodepositing the multilayer into the pores of a polycarbonate membrane[21] (Fig. 4b-d). In the CPP-geometry, the GMR is not only definitely higher than in CIP (the CPP-GMR will be probably used in a future generation of read heads for hard discs), but also subsists in multilayers with relatively thick layers, up to the micron range[21], as it can be seen in Fig. 4c-d. In a theoretical paper with Thierry Valet[22], I showed that, owing to spin accumulation effects occurring in the CPP-geometry, the length scale of the spin transport becomes the long spin diffusion length in place of the short mean free path for the CIP-geometry. Actually, the CPP-GMR has revealed the spin accumulation effects which govern the propagation of a spin-polarized current through a succession of magnetic and nonmagnetic materials and play an important role in all the current developments of spintronics. The diffusion current induced by the accumulation of spins at the magnetic/nonmagnetic interface is the mechanism driving a spin-polarized current at a long distance from the interface, well beyond the ballistic range (i.e. well beyond the mean free path) up to the distance of the spin diffusion length (SDL). In carbon molecules for example, the spin diffusion length exceeds the micron range and, as we will see in the Section on molecular spintronics, strongly spin-polarized currents can be transported throughout long carbon nanotubes.

The physics of the spin-accumulation occurring when an electron flux crosses the interface between a ferromagnetic and a nonmagnetic material is explained in Fig. 6. Far from the interface on the magnetic side, the current is larger in one of the spin channels (spin up on the figure), while, far from the interface on the other side, it is equally distributed in the two channels. With the current direction and the spin polarization of the figure, there is accumulation of spin up electrons (and depletion of spin down for charge neutrality) around the interface, or, in other word, a splitting between the Fermi energies (chemical potentials) of the spin up and spin down electrons. This accumulation diffuses from the interface in both directions to the distance of the SDL. Spin-flips are also generated by this out of equilibrium distribution and a steady splitting is reached when the number of spin-flips is just what is needed to adjust the incoming and outgoing fluxes of spin up and spin down electrons. To sum up, there is a broad zone of spin accumulation which extends on both sides to the distance of the SDL and in which the current is progressively depolarized by the spin-flips generated by the spin accumulation.

Figure 6 is drawn for the case of spin injection, i.e. for electrons going from the magnetic to the nonmagnetic conductor. For electrons going in the opposite direction (spin extraction), the situation is similar except that a spin accumulation in the opposite direction progressively polarizes the current in

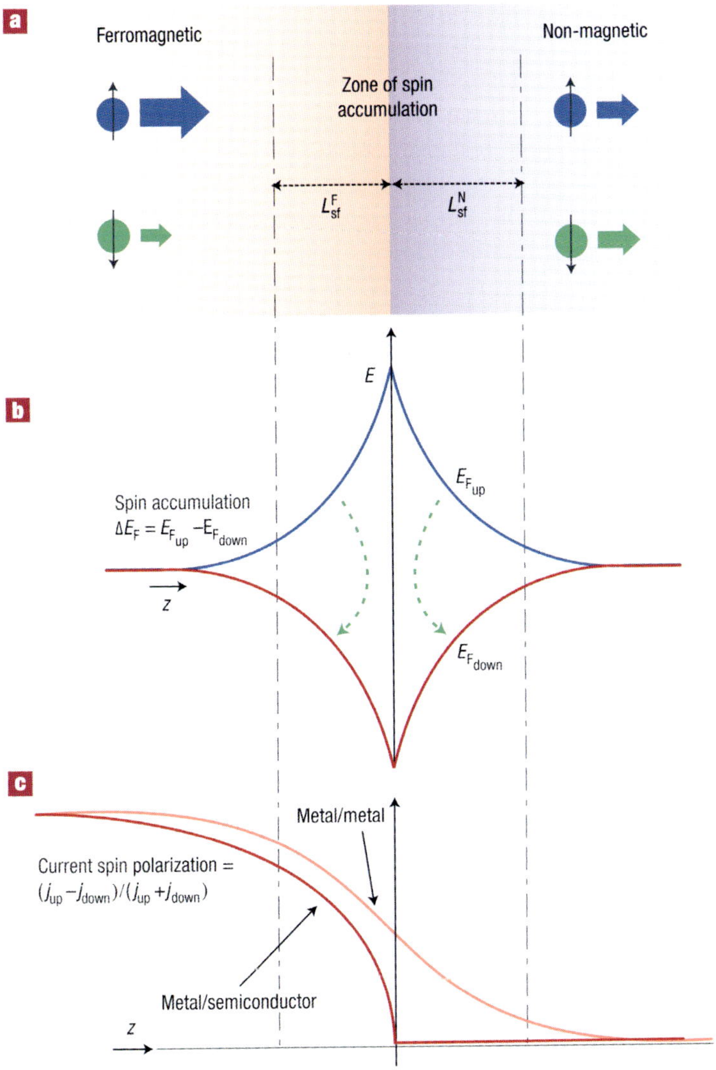

Figure 6. Schematic representation of the spin accumulation at an interface between a ferromagnetic metal and a non magnetic layer. (a) Spin-up and spin-down currents far from an interface between ferromagnetic and nonmagnetic conductors (outside the spin-accumulation zone). (b) Splitting of the chemical potentials $E_{F\uparrow}$ and $E_{F\downarrow}$ at the interface. The arrows symbolize the spin flips induced by the spin-split out of equilibrium distribution. These spin-flips control the progressive depolarization of the electron current between the left and the right. With an opposite direction of the current, there is an inversion of the spin accumulation and opposite spin flips, which polarizes the current when it goes through the spin-accumulation zone. (c) Variation of the current spin polarization when there is an approximate balance between the spin flips on both sides (metal/metal) and when the spin flips on the left side are predominant (metal/semiconductor without spin-dependent interface resistance, for example). Figures from Ref.[18].

the nonmagnetic conductor. In both the injection and extraction cases, the spin-polarization subsists or starts in the nonmagnetic conductor at a long distance from the interface. This physics can be described by new types of transport equation[22] in which the electrical potential is replaced by a spin and position dependent electro-chemical potential. These equations can be applied not only to the simple case of a single interface but to multi-interface systems with overlap of the spin accumulations at successive interfaces. They can also be extended to take into account band bending and high current density effects[23-24].

The physics of spin accumulation plays an important role in many fields of spintronics, for example in one of the most active field of research today, spintronics with semiconductors. In the case of spin injection from a magnetic metal into a nonmagnetic semiconductor (or spin extraction for the opposite current direction), the much larger DOS in the metal makes that similar spin accumulation splittings on the two sides of the interface, as in Fig. 6, lead to a much larger spin accumulation density and to a much larger number of spin flips on the metallic side. The depolarization is therefore faster on the metallic side and the current is almost completely depolarized when it enters the semiconductor, as shown in Fig. 6c. This problem has been first raised by Schmidt and coworkers[25]. I came back to the theory with my coworker Henri Jaffrès to show that the problem can be solved by introducing a spin dependent interface resistance, typically a tunnel junction, to introduce a discontinuity of the spin accumulation at the interface, increase the proportion of spin on the semiconductor side and shift the depolarization from the metallic to the semiconductor side (the same conclusions appear also in a paper of Rashba)[26-27]. Spin injection through a tunnel barrier has now been achieved successfully in several experiments but the tunnel resistances are generally too large for an efficient transformation of the spin information into an electrical signal[24].

MAGNETIC TUNNEL JUNCTIONS AND TUNNELLING MAGNETORESISTANCE (TMR)

An important stage in the development of spintronics has been the research on the Tunnelling Magnetoresistance (TMR) of the Magnetic Tunnel Junctions(MTJ). The MTJ are tunnel junctions with ferromagnetic electrodes and their resistance is different for the parallel and antiparallel magnetic configurations of their electrodes. Some early observations of TMR effects, small and at low temperature, had been already reported by Jullière[28] in 1975, but they were not easily reproducible and actually could not be really reproduced during 20 years. It is only in 1995 that large ($\approx 20\%$) and reproducible effects were obtained by Moodera's and Miyasaki's groups on MTJ with a tunnel barrier of amorphous alumina[29-30]. From a technological point of view, the interest of the MTJ with respect to the metallic spin valves comes from the vertical direction of the current and from the resulting possibility of a reduction of the lateral size to a submicronic scale by lithographic

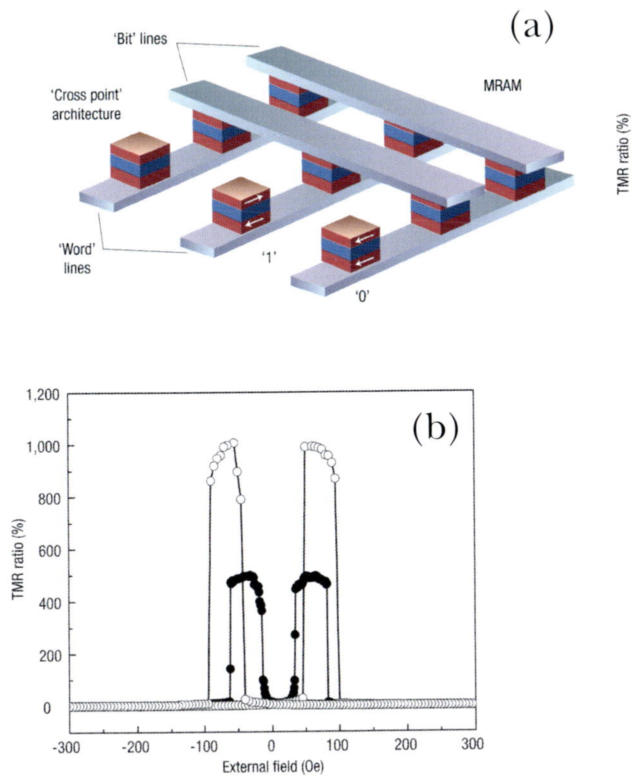

Figure 7. (a) Principle of the magnetic random access memory MRAM in the basic "cross point" architecture. The binary information "0" and "1" is recorded on the two opposite orientations of the magnetization of the free layer of magnetic tunnel junctions (MTJ), which are connected to the crossing points of two perpendicular arrays of parallel conducting lines. For writing, current pulses are sent through one line of each array, and only at the crossing point of these lines the resulting magnetic field is high enough to orient the magnetization of the free layer. For reading, one measures the resistance between the two lines connecting the addressed cell. Schematic from Ref.[18]. (b) High magnetoresistance, $TMR=(R_{max}-R_{min})/R_{min}$, measured by Lee *et al.*[34] for the magnetic stack: $(Co_{25}Fe_{75})_{80}B_{20}$(4nm)/MgO(2.1nm)/$(Co_{25}Fe_{75})_{80}B_{20}$(4.3nm) annealed at 475°C after growth, measured at room temperature (black circles) and low temperature (open circles).

techniques. The MTJ are at the basis of a new concept of magnetic memory called MRAM (Magnetic Random Access Memory) and shematically represented in Fig. 7a. The MRAMs are expected to combine the short access time of the semiconductor-based RAMs and the non-volatile character of the magnetic memories. In the first MRAMs, put onto the market in 2006, the memory cells are MTJs with an alumina barrier. The magnetic fields generated by "word" and "bit" lines are used to switch their magnetic configuration, see Fig. 7a. The next generation of MRAM, based on MgO tunnel junctions and switching by spin transfer, is expected to have a much stronger impact on the technology of computers.

The research on the TMR has been very active since 1995 and the most important step was the recent transition from MTJ with amorphous tunnel barrier (alumina) to single crystal MTJ and especially MTJ with MgO barrier. In the CNRS/Thales laboratory we founded in 1995, the research on TMR was one of our main projects and, in collaboration with a Spanish group, we obtained one of the very first TMA results[31] on MTJ with epitaxial MgO. However our TMR was only slightly larger than that found with alumina barriers and similar electrodes. The important breakthrough came in 2004 at Tsukuba[32] and IBM[33] where it was found that very large TMR ratios, up to 200% at room temperature, could be obtained from MgO MTJ of very high structural quality. TMR ratios of about 600% have been now reached[34] (Fig. 7b). In such MTJ, the single crystal barrier filters the symmetry of the wave functions of the tunnelling electrons[35-37], so that the TMR depends on the spin polarization of the electrodes for the selected symmetry.

The high spin polarization obtained by selecting the symmetry of the tunnelling waves with a single crystal barrier is a very good illustration of what is under the word "spin polarization" in a spintronic experiment. In the example of Fig. 8, taken from an article by Zhang and Butler[37], one sees the density of states of evanescent waves functions of different symmetries, Δ_1, Δ_5, etc, in a MgO(001) barrier between Co electrodes. The key point is that, at least for interfaces of high quality, an evanescent wave function of a given symmetry is connected to the Bloch functions of the same symmetry at the Fermi level of the electrodes. For Co electrodes, the Δ_1 symmetry is well represented at the Fermi level in the majority spin direction sub-band and not in the minority one. Consequently, a good connection of the slowly decaying channel Δ_1 with both electrodes can be obtained only in their parallel magnetic configuration, which explains the very high TMR. Other types of barrier can select other symmetries than the symmetry Δ_1 selected by MgO(001). For example, a $SrTiO_3$ barrier predominantly selects evanescent wave functions of Δ_5 symmetry, which are connected to minority spin states of cobalt[38]. This explains the negative effective spin-polarization of cobalt we had observed in $SrTiO_3$-based MTJ[39]. This finally shows that there no intrinsic spin polarization of a magnetic conductor. The effective polarization of a given magnetic conductor in a MTJ depends on the symmetry selected by the barrier and, depending on the barrier, can be positive or negative, large or small. In the same way the spin polarization of metallic conduction depends strongly on the spin dependence of the scattering by impurities, as illustrated by Fig. 1b.

There are other promising directions to obtain large TMR and experiments in several of them are now led by Agnès Barthélémy (much more than by myself) in the CNRS/Thales laboratory. First, we tested ferromagnetic materials which were predicted to be half-metallic, i.e. metallic for one of the spin direction and insulating for the other one, in other words 100% spin-polarized. Very high spin polarization (95%) and record TMR (1800%) have been obtained by our Ph.D. student Martin Bowen with $La_{2/3}Sr_{1/3}MnO_3$ electrodes[40] but the Curie temperature of this manganite (around 350K) is

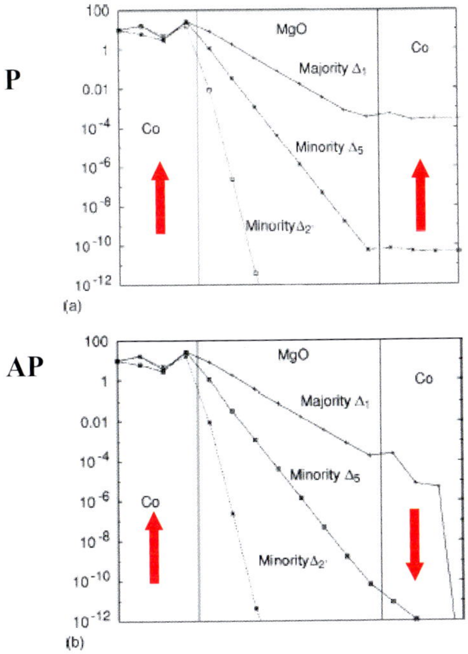

Figure 8. Physics of TMR illustrated by the decay of evanescent electronic waves of different symmetries in a MgO(001) layer between cobalt electrodes calculated by Zhang and Butler[37]. The Δ_1 symmetry of the slowly decaying tunnelling channel is well represented at the Fermi level of the spin conduction band of cobalt for the majority spin direction and not for the minority spin one, so that a good connection by tunnelling between the electrodes exists only for the parallel magnetic configuration when a Δ_1 channel can be connected to both electrodes (above). In the antiparallel configuration (below), both the spin up and spin down Δ_1 channels are poorly connected on one of the sides. This explains the very high TMR of this type of junction.

too low for applications. It now turns out from recent results in Japan[41] that ferromagnets of the family of the Heusler alloys also present very large TMR ratios with still 90% at room temperature[41]. Another interesting concept that we are exploring is spin filtering by tunnelling through a ferromagnetic insulator layer[42-43]. This can be described as the tunnelling of electrons through a barrier of spin dependent height if the bottom of the conduction band is spin-split, which gives rise to a spin-dependence of the transmission probability (spin filtering). Very high spin filtering coefficients have been found at low temperature with Eus barriers[42] at the MIT and at Eindhoven. Promising results with insulating ferromagnets of much higher Curie temperature have been recently obtained, see, for example Ref. [43]. Some of the magnetic barriers we have recently tested in MTJ are also ferroelectric, so that the MTJ present the interesting property of four states of resistance corresponding to the P and AP magnetic configurations and to the two orientations of the ferroelectric polarization[44], as shown in Fig. 9.

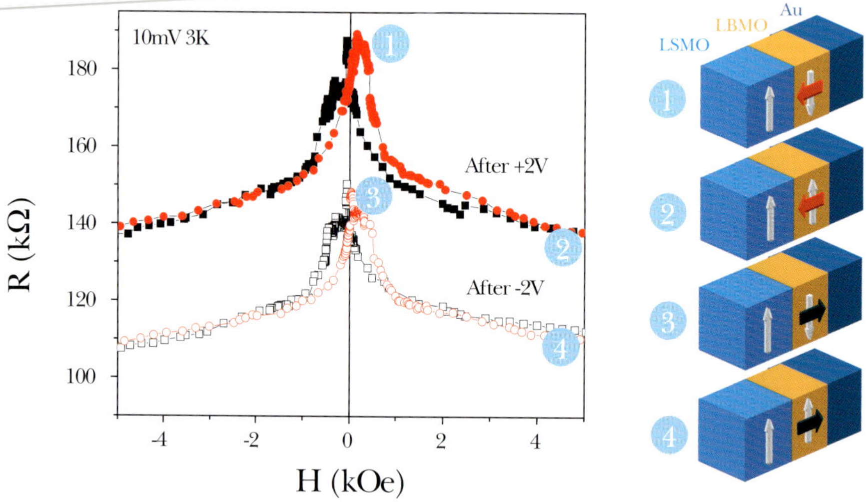

Figure 9. Four state resistance of a tunnel junction composed of a biferroic tunnel barrier ($La_{0.1}Bi_{0.9}MnO_3$) between a ferromagnetic electrode of $La_{2/3}Sr_{1/3}MnO_3$ and a nomagnetic electrode of gold. The sates 1-4 correspond to the magnetic (white arrows) and electric (black arrows) polarizations represented on the right of the figure. From Gajek *et al.*[44].

MAGNETIC SWITCHING AND MICROWAVE GENERATION BY SPIN TRANSFER

The study of the spin transfer phenomena is one of the most promising new directions in spintronics today and also an important research topic in our CNRS/Thales laboratory. In spin transfer experiments, one manipulates the magnetic moment of a ferromagnetic body without applying any magnetic field but only by transfer of spin angular momentum from a spin-polarized current. The concept, which has been introduced by John Slonczewski[45] and appears also in papers of Berger[46], is illustrated in Fig. 10. As described in the caption of the figure, the transfer of a transverse spin current to the "free" magnetic layer F_2 can be described by a torque acting on its magnetic moment. This torque can induce an irreversible switching of this magnetic moment or, in a second regime, generally in the presence of an applied field, it generates precessions of the moment in the microwave frequency range.

The first evidence that spin transfer can work was indicated by experiments of spin injection through point contacts by Tsoi *et al.*[47] but a clear understanding came later from measurements[48-49] performed on pillar-shaped metallic trilayers (Fig. 11a). In Fig.11b-c, I present examples of our experimental results in the low field regime of irreversible switching, for a metallic pillar and for a tunnel junctions with electrodes of the dilute ferromagnetic semiconductor $Ga_{1-x}Mn_xAs$. For metallic pillars or tunnel junctions with electrodes made of a ferromagnetic transition metal like Co or Fe, the current density needed for switching is around 10^6-10^7 Amp/cm^2, which is still slightly too high for applications, and an important challenge is the reduction of this current density. The switching time has been measured in other groups and

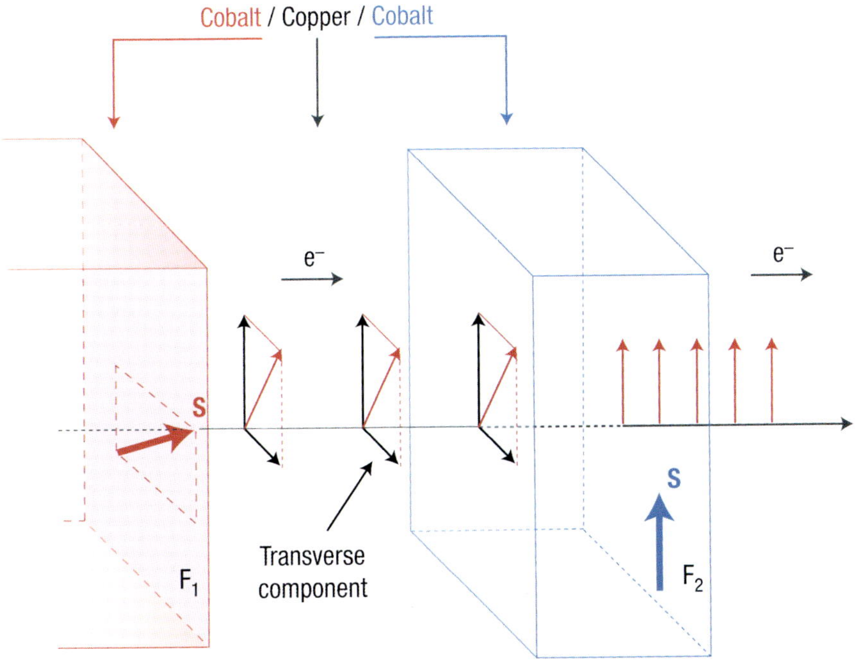

Figure 10. Illustration of the spin transfer concept introduced by John Slonczewski[45] in 1996. A spin-polarized current is prepared by a first magnetic layer F with an obliquely oriented spin-polarization with respect to the magnetization axis of a second layer F_2. When this current goes through F_2, the exchange interaction aligns its spin-polarization along the magnetization axis. As the exchange interaction is spin conserving, the transverse spin-polarization lost by the current has been transferred to the total spin of F_2, which can also be described by a spin-transfer torque acting on F_2.This can lead to a magnetic switching of the F_2 layer or, depending on the experimental conditions, to magnetic oscillations in the microwave frequency range. Figure from Ref.[18].

can be as short as 100 ps, which is very attractive for the switching of MRAM. For the tunnel junction of Fig. 11c, the switching current is only about 10^5 Amp./cm^2 and smaller than that of the metallic pillar by two orders of magnitude. This is because a smaller number of individual spins is required to switch the smaller total spin momentum of a dilute magnetic material.

In the presence of a large enough magnetic field, the regime of irreversible switching of the magnetization of the "free" magnetic layer in a trilayer is replaced by a regime of steady precessions of this free layer magnetization sustained by the spin transfer torque[52]. As the angle between the magnetizations of the two magnetic layers varies periodically during the precession, the resistance of the trilayer oscillates as a function of time, which generates voltage oscillations in the microwave frequency range. In other conditions, the spin transfer torque can also be used to generate an oscillatory motion of a magnetic vortex.

The spin transfer phenomena raise a series of various theoretical problems. The determination of the spin transfer torque is related to the solution of

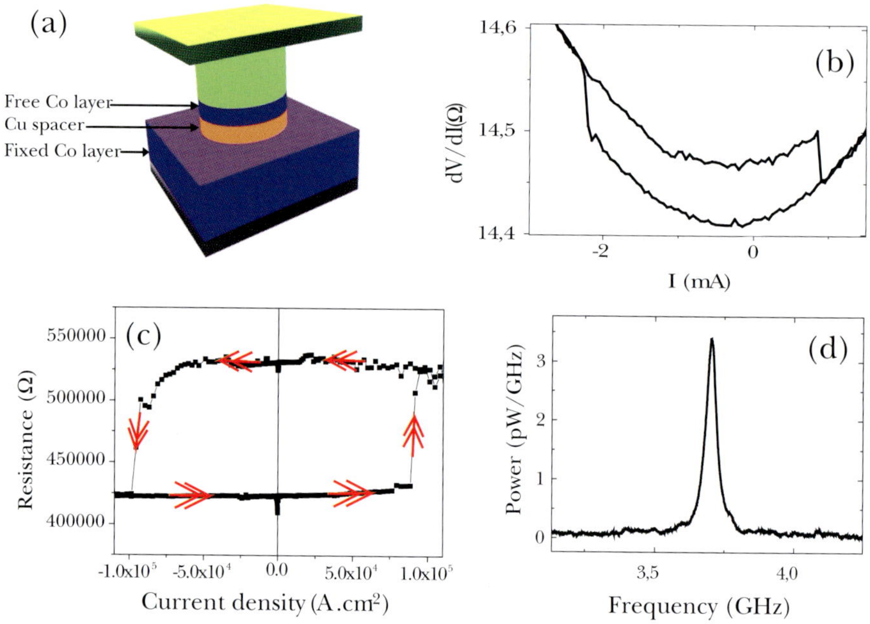

Figure 11. Experiments of magnetic switching and microwave generation induced by spin transfer from an electrical DC current in trilayered magnetic pillars. (a) Schematic of a trilayered magnetic pillar. (b) Switching by spin transfer between the parallel and antiparallel magnetic configurations of a Co/Cu/Co metallic pillar[49]. The switching between parallel and antiparallel orientations of the magnetizations of the two magnetic layers of the trilayer is detected by irreversible jumps of the resistance at a critical value of the current. The critical current density is of the order of 10^7 A/cm^2. (c) Switching by spin transfer of a pillar-shaped tunnel junction composed of electrodes of the dilute ferromagnetic semiconductor GaMnAs separated by a tunnel barrier of InGaAs[50]. The critical current is about hundred times smaller than in the Py/Cu/Py pillar. Similar results have been obtained by Hayakawa *et al.*[51]. (d) Typical microwave power spectrum of a Co/Cu/Py pillar (Py =permalloy)[57].

spin transport equations[53-56], while the description of the switching or precession of the magnetization raises problems of non-linear dynamics[53]. All these problems are interacting and, for example, some of our recent results show that it is possible to obtain very different dynamics (with, for applications, the interest of oscillations without applied field) by introducing strongly different spin relaxation times in the two magnetic layers of a trilayer to distort the angular dependence of the torque[57].

The spin transfer phenomena will have certainly important applications. Switching by spin transfer will be used in the next generation of MRAM and will bring great advantages in terms of precise addressing and low energy consumption. The generation of oscillations in the microwave frequency range will lead to the design of Spin Transfer Oscillators (STOs). One of the main interests of the STOs is their agility, that is the possibility of changing rapidly their frequency by tuning a DC current. They can also have a high quality factor. Their disadvantage is the very small microwave power of an individual STO, metallic pillar or tunnel junction. The solution is certainly the synchronization of a large number of STOs. The possibility of synchronization has been already demonstrated for two nano-contacts inducing spin transfer excitations in the same magnetic layer[58-59]. In our laboratory we are exploring theoretically and experimentally a concept of self-synchronization of a collection of electrically connected STOs by the RF current components they induce[60].

SPINTRONICS WITH SEMICONDUCTORS AND MOLECULAR SPINTRONICS

Spintronics with semiconductors[61-62] is very attractive as it can combine the potential of semiconductors (control of current by gate, coupling with optics, etc) with the potential of the magnetic materials (control of current by spin manipulation, non-volatility, etc.). It should be possible, for example, to gather storage, detection, logic and communication capabilities on a single chip that could replace several components. New concepts of components have also been proposed, for example the concept of Spin Field Effect Transistors (Spin FETs) based on spin transport in semiconductor lateral channels between spin-polarized sources and drains with control of the spin transmission by a field effect gate[63]. Some nonmagnetic semiconductors have a definite advantage on metals in terms of spin-coherence time and propagation of spin polarization on long distances[61-62]. However, as it will be discussed below, the long standing problem of the Spin FET it still far from being solved.

Spintronics with semiconductors is currently developed along several roads.

i) The first road is by working on hybrid structures associating ferromagnetic metals with nonmagnetic semiconductors. As this has been mentioned in the Section on spin accumulation, Schmidt *et al.*[25] have raised the problem of "conductivity mismatch" to inject a spin-polarized current from a magnetic metal into a semiconductor. Solutions have been proposed by the theory[26-27] and one knows today that the injection/extraction of a spin-polarized current into/from a semiconductor can be achieved with a spin-dependent interface resistance, typically a tunnel junction. Spin injection/extraction through a tunnel contact has been now demonstrated in spin LEDs and magneto-optical experiments[61-62,64].

ii) Another road for spintronics with semiconductors is based on the fabrication of ferromagnetic semiconductors. The ferromagnetic semiconductor $Ga_{1-x}Mn_xAs$ ($x \approx$ a few %) has been discovered[65] by the group of Ohno in Sendai in 1996, and, since this time, has revealed very interesting properties, namely the possibility of controlling the ferromagnetic properties with a gate voltage, and also large TMR and TAMR (Tunnelling Anisotropic Magnetoresistance) effects. However its Curie temperature has reached only 170 K, well below room temperature, which rules out most practical applications. Several room temperature ferromagnetic semiconductors have been announced but the situation is not clear on this front yet.

iii) The research is now very active on a third road exploiting spin-polarized currents induced by spin-orbit effects, namely the Spin Hall[66-68], Rashba or Dresselhaus effects. In the Spin Hall Effect (SHE), for example, spin-orbit interactions deflect the currents of the spin up and spin down channels in opposite transverse directions, thus inducing a transverse spin current, even in a nonmagnetic conductor. This could be used to create spin currents in structures composed of only nonmagnetic conductors. Actually the SHE can be also found in nonmagnetic metals[69-70] and the research is also very active in this field. May I mention that, already in the seventies, I had found very large SHE induced by resonant scattering on spin-orbit-split levels of nonmagnetic impurities in copper[71].

Several groups have tried to probe the potential of spintronics with semiconductors by validating experimentally the concept of Spin FET[63] described above. Both ferromagnetic metals and ferromagnetic semiconductors have been used for the source and the drain, but the results have been relatively poor. In a recent review article, Jonker and Flatté[61] note that a contrast larger than about 1% (i. e. $[R_{AP}\text{-}R_P]/R_P > 1\%$) has never been observed between the resistances of the parallel and antiparallel magnetic orientations of the source and the drain, at least for lateral structures. We have recently proposed[24] this can be understood in the models[27] I had developed with Henri Jaffrès to describe the spin transport between spin-polarized sources and drains. In both the diffusive and ballistic regimes, a strong contrast between the conductances of the two configurations can be obtained only if the resistances of the interfaces between the semiconductor and the source or drain are not only spin dependent but also chosen in a relatively narrow window. The resistances must be larger than a first threshold value for spin injection/extraction from/into a metallic source/drain, and smaller than a second threshold value to keep the carrier dwell time shorter than the spin lifetime. For vertical structures with a short distance between source and drain, the above conditions can be satisfied more easily and relatively large magnetoresistance can be observed, as illustrated by the results I present in Fig. 12. However the results displayed in Fig. 12c show that the magnetoresistance drops rapidly when the interface resistance exceeds some threshold value. This can be explained by the increase of the dwell time above the spin lifetime. Alternatively, the magnetoresistance also drops to zero when an increase of temperature shortens the spin lifetime and increases the ratio of the dwell time to the spin lifetime. For most experiments on lateral structures, it turns out that a part of the difficulties comes from too large interface resistances giving rise to too short dwell times. Min *et al.*[73] have arrived at similar conclusions for the particular case of silicon-based structures and propose interesting solutions to lower the interface resistances by tuning the work function of the source and the drain.

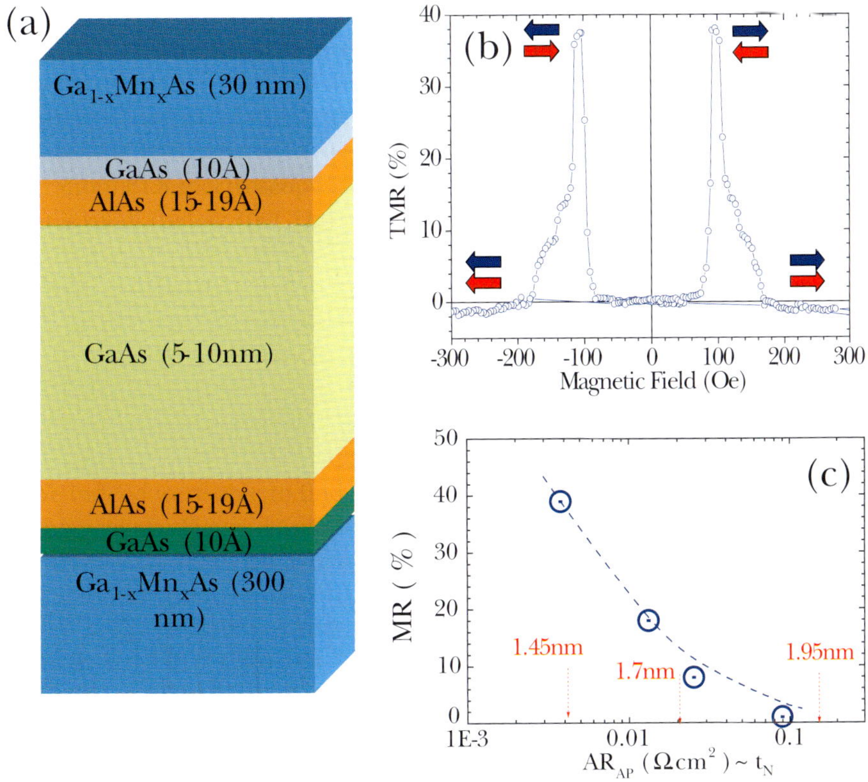

Figure 12. Spintronics with semiconductors illustrated by experimental results[24,72] on the structure represented on the right and composed of a GaAs layer separated from the GaMnAs source and drain by tunnel barriers of AlAs. (a) MR curve at 4.2 K showing a resistance difference of 40 % between the parallel and antiparallel magnetic configurations of the source and the drain. (b) MR ratio as a function of the resistance of the tunnel barriers.

A recently emerging direction is spintronics with molecules. Very large GMR- or TMR-like effects are predicted by the theory, especially for carbon-based molecules in which a very long spin lifetime is expected from the small spin-orbit coupling. Promising experimental results have been published during the last years on spin transport in carbon nanotubes[74-75]. In particular, my recent work[75] in collaboration with a group in Cambridge on carbon nanotubes between ferromagnetic source and drain made of the metallic manganite $L_{1/3}Sr_{1/3}MnO_3$ has shown that the relative difference between the resistances of the parallel and antiparallel configurations can exceed 60-70%, well above what can be obtained with semiconductor channels. This can be explained not only by the long spin lifetimes of the electrons in carbon nanotubes but also by their short dwell time related to their high Fermi velocity (a definite advantage on semiconductors). The research is currently very active in this field and, in particular, graphene-based devices are promising.

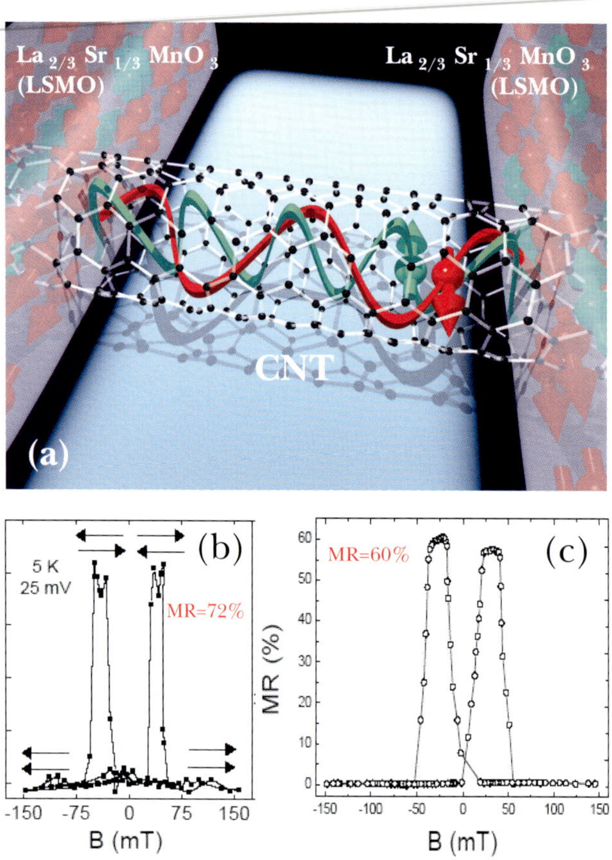

Figure 13. Spintronics with molecules illustrated by, (a): Artistic view of spin transport through a carbon nanotube between magnetic electrodes (courtesy of T. Kontos). (b) and (c): Magnetoresistance experimental results[75] at 4.2 K on carbon nanotubes between electrodes made of the ferromagnetic metallic oxide $La_{2/3}Sr_{1/3}MnO_3$. A contrast of 72% and 60% is obtained between the resistances for the parallel (high field) and antiparallel (peaks) magnetic configurations of the source and drain.

CONCLUSION

In less than twenty years, we have seen spintronics increasing considerably the capacity of our hard discs, extending the hard disc technology to mobile appliances like cameras or portable multimedia players, entering the automotive industry and the bio-medical technology and, with TMR and spin transfer, getting ready to enter the RAM of our computers or the microwave emitters of our cell phones. The researches of today on the spin transfer phenomena, on multiferroic materials, on spintronics with semiconductors and molecular spintronics, open fascinating new fields and are also very promising of multiple applications. Another perspective, out of the scope of this lecture, should be the exploitation of the truly quantum mechanical nature of spin and the long spin coherence time in confined geometry for quantum computing in an even more revolutionary application. Spintronics should take an important place in the science and technology of our century.

REFERENCES

[1] M.N. Baibich, J.M. Broto, A. Fert, F. Nguyen Van Dau, F. Petroff, P. Etienne, G. Creuzet, A. Friederich, and J. Chazelas, Phys. Rev. Lett. **61**, 2472 (1988).

[2] G. Binash, P. Grünberg, F. Saurenbach, and W. Zinn, Phys. Rev. B **39**, 4828 (1989).

[3] F. Mott, Proc. Roy. Soc. A**153**, 699 (1936).

[4] A. Fert and I. A. Campbell, Phys. Rev. Lett. **21,** 1190 (1968); A. Fert and I. A. Campbell, J. Physique **32,** C1-46 (1971); A. Fert and I. A. Campbell, J. Phys. F **6**, 849 (1976).

[5] B. Loegel and F. Gautier, J. Phys. Chem. Sol. **32**, 2723 (1971).

[6] The contribution of spin-flips to spin mixing (i.e. to momentum exchange between the two channels, mainly through magnon scattering[4]) should not be confused with the contribution of spin-flips to the relaxation of spin accumulation effects (mainly through spin-lattice relaxation by spin-orbit interactions).

[7] P. Grünberg, R. Schreiber, Y. Young, M. B. Brodsky, H. Sowers, Phys. Rev. Lett. **57**, 2442 (1986).

[8] R. E. Camley, J. Barnas, Phys. Rev. Lett. **63**, 664 (1989).

[9] P. M. Levy, S. Zhang, A. Fert, Phys. Rev. Lett. **65**, 1643 (1990).

[10] S. S. P. Parkin, N. More, K.P. Roche, Phys. Rev.Lett. **64,** 2304 (1990).

[11] T. Shinjo and H. Yamamoto, J. Phys. Soc. Jpn. **59**, 3061 (1990).

[12] C. Dupas, P. Beauvillain, C. Chappert, C. Chappert, J.P. Renard, F. Trigui, P. Veillet, E. Velu, D. Renard, J. Appl. Phys. **67,** 5680 (1990).

[13] D. H. Mosca, F. Petroff, A. Fert, P. A. Schroeder, W. P. Pratt, R. Loloee, J. Magn. Magn. Mater. **94**, L1 (1991).

[14] S. S. P. Parkin, R. Bhadra, K. P. Roche, Phys. Rev. Lett. **66**, 2152 (1991).

[15] B. Dieny, V. S. Speriosu, S. S. P. Parkin, B. A. Gurney, D. R. Wilhoit, D. Mauri, Phys. Rev. B **43**, 1297 (1991).

[16] P. Grünberg, Magnetic field sensor with ferromagnetic thin layers having magnetically antiparallel polarized components, US patent 4,949,039 (1990).

[17] S. S. P. Parkin, in Spin Dependent Transport in Magnetic nanostructures (Edited by S. Maekawa and T. Shinjo, Taylor and Francis 2002), p. 237.

[18] C. Chappert, A. Fert, F. Nguyen Van Dau, Nature Materials, vol. 6, 813 (2007).

[19] P. P. Freitas, H. Ferreira, D. Graham, L. Clarke, M. Amaral, V. Martins, L. Fonseca, J. S. Cabral, Europhysiscs News 34/6, 225 (2003).

[20] W. P. Pratt *et al.*, Phys. Rev. Lett. **66**, 3060 (1991); J. Bass and W. P. Pratt, J. Magn. Magn. Mater. **200,** 274 (1999).

[21] L. Piraux, J-M. George, C. Leroy, R. Legras, A. Fert, Appl. Phys. Lett. **65**, 2484 (1994); A. Fert and L. Piraux, J. Magn. Magn. Mater. **200**, 338 (1999).

[22] T. Valet and A. Fert, Phys. Rev. B **48**, 7099 (1993).

[23] Z. G. Yu and M. E. Flatté, Phys. Rev. B **66**, 201202 (2002).

[24] A. Fert, J. M. George, H. Jaffrès and R. Mattana, IEEE Transactions on Electron Devices **54**, 921 (2007).

[25] G. Schmidt *et al.,* Phys. Rev. B **62,** 4790 (2000).

[26] E. I. Rashba, Phys. Rev. B **62,** 16267 (2000).

[27] A. Fert and H. Jaffrès, Phys. Rev. B **64,** 184420 (2001).

[28] Jullière, Phys. Lett. **54A**, 225 (1975).

[29] J. S. Moodera, L. R. Kinder, T. M. Wong, R. Meservey, Phys. Rev. Lett. **74**, 3273 (1995).

[30] T. Miyazaki and N. Tezuka, J. Magn. Magn. Mater. **139** (1995) 231.

[31] M. Bowen, V. Cros, F. Petroff, A. Fert, A. Cebollada, F. Briones, Appl. Phys. Lett. **79**, 1655 (2001).

[32] Yuasa *et al.,* Nature Mater. **3**, 868 (2004).

[33] S. S. P. Parkin *et al.*, Nature Materials **3**, 862 (2004).

[34] Y. M. Lee, J. Hayakawa, S. Ikeda, F. Matsukura, H. Ohno, Appl. Phys. Lett. **90**, 212507 (2007).

[35] J. Mathon and A. Umerski ; Phys. Rev. B **60**, 1117 (1999).

[36] Ph. Mavropoulos, N. Papanikolaou and Ph. Dederichs, *Phys. Rev. Lett.* **85** 1088 (2000).

[37] X. G. Zhang and W. H. Butler, Phys. Rev. B **70**, 173407 (2004).

[38] S. Velev *et al.*, Phys. Rev. Lett. **95** (2005); M. Bowen *et al.*, Phys. Rev. B **73,** 140408 (2006).
[39] J. M. De Teresa, A. Barthélémy, A. Fert, J. P. Contour, F. Montaigne, A. Vaures; Science **286,** 507 (1999).
[40] M. Bowen, M. Bibes, A. Barthélémy, J. P. Contour, A. Anane, Y. Lemaitre, A. Fert ; Appl. Phys. Letters **82**, 233, (2003).
[41] T. Ishikawa *et al.*, Appl. Phys. Lett. **89,** 192505 (2006).
[42] P. Leclair, J. K. Ha, J. M.Swagten, J. T. Kohlhepp, C. H. Van de Vin and W. J. M. de Jonge; Appl. Phys. Lett. **80**, 625 (2002).
[43] A. V. Ramos *et al.*, Appl. Phys. Lett. **91,** 122107 (2007).
[44] M. Gajek, M. Bibes, S. Fusil, K. Bouzehouane, J. Fontcuberta, A. Barthélémy, A. Fert, Nature Materials **6**, 296 (2007).
[45] J. C. Slonczewski, J. Magn. Mat. **159**, L1 (1996).
[46] L. Berger, Phys. Rev. B **54**, 9353 (1996).
[47] M. Tsoi, A. G. M. Jansen, J. Bass, W. C. Chiang, V. Tsoi, M. Seck, P. Wyder, Phys. Rev. Lett. **80**, 4281 (1998).
[48] Katine *et al.*, Phys. Rev. Lett. **84**, 3149 (2000).
[49] J. Grollier, V. Cros, A. Hanzic, J. M. George, H. Jaffres, A. Fert, G. Faini, J. Ben Youssef, H. Le Gall., Appl. Phys. Lett. **78**, 3663 (2001). J. Grollier, Ph. D. thesis (Université Paris-Sud, 2003).
[50] M. Elsen, O. Boulle, J. M.George, H. Jaffrès, V. Cros, A. Fert, A. Lemaître, R. Giraud, G. Faini, Phys.Rev. B **73**, 035303 (2006).
[51] J. Hayakawa *et al.*, Jpn. J. Appl. Phys., Part 2 **44**, L1267 (2005).
[52] W. H. Rippart *et al.*, Phys. Rev. Lett. **92**, 027201 (2004).
[53] M. D. Stiles and J. Miltat in *Spin Dynamics in Confined Magnetic Structures, III,* edited by B. Hillebrands and A. Thiaville (Springer, Berlin, 2006).
[54] J. C. Slonczewski, J. Magn. Magn. Mat.,**247**, 324 (2002).
[55] A. A. Kovalev, A. Brattas, G. E. W. Bauer, Phys. Rev. B **66**, 224424 (2002).
[56] J. Barnas, A. Fert, M. Gmitra, I. Weymann, V. K. Dugaev, Phys. Rev. B **72**, 024426 (2005).
[57] O. Boulle, V. Cros, J. Grollier, L. G. Pereira, C. Deranlot, F. Petroff, G. Faini, J. Barnas, A. Fert, Nature Physics, **3,** 492 (2007), O. Boulle, Ph. D. Thesis (2006, Université Paris-Sud).
[58] S. Kaka *et al.*, Nature **437,** 389 (2005).
[59] F. B. Mancoff *et al.*, Nature **437,** 393 (2005).
[60] J. Grollier, V; Cros, A. Fert, Phys. Rev. B **73,** 060409 [R], 2006.
[61] B. T. Jonker and M. E. Flatté, in Nanomagnetism (edited by D.L. Mills and J.A.C. Bland, Elsevier, 2006), p. 227.
[62] D. D. Awschalom and M. E. Flatté, Nature Physics **3**, 153 (2007).
[63] S. Datta and B. Das, Appl. Phys. Lett. **56**, 665 (1990).
[64] Stephens *et al.*, PRL **93**, 097602 (2004).
[65] H. Ohno *et al.*, Appl. Phys. Lett. **69**, 363 (1996).
[66] Y. Kato, R. C. Myers, A. C. Gossard, D. D. Awschalom, Science **306**, 1910 (2004).
[67] S. Zhang, Phys. Rev. Lett. **85,** 393 (2000).
[68] M. Koenig, S. Wiedmann, C. Bruene, A. Roth, H. Buhmann, L. W. Molenkamp, X. L. Qi and S-C. Zhang, Science, **318,** 766 (2007).
[69] L. Vila, T. Kjimura, Y. Otani, Phys. Rev. Lett. **99,** 226604 (2007).
[70] T. Seki *et al.*, Nature Materials **7**, 125 (2008).
[71] A. Fert, A. Friederich, A. Hamzic, J. of Magn. Magn. Mat. **24,** 231 (1981).
[72] R. Mattana, J. M. George, H. Jaffrès, F. Nguyen Van Dau, A. Fert, B. Lépine, A. Guivarch, G. Jézéquel, Phys. Rev. Letters **90**, 166601, (2003).
[73] B. C. Min, K. Motohashi, C. Lodder, R. Jansen, Nature Mat. **5**, 817 (2006).
[74] A. Cottet, T. Kontos, S. Sahoo, H. T. Man, W. Belzig, C. Bruder, C. Schönenberger, Semicond. Sci.Technol. **21**, 578 (2006).
[75] L. E. Hueso, J. M. Pruneda, V. Ferrari, G. Burnell, J. P. Valdes-Herrera, B. D. Simmons, P. B. Littlewood, E. Artacho, A. Fert, N. D. Mathur, Nature **445,** 410 (2007).

Portrait photo of Albert Fert by photographer Ulla Montan.

Peter A. Grünberg

PETER A. GRÜNBERG

A typical question I am often asked as a Nobel Laureate in physics is "what brought you to physics?" I am not sure but I do know that in school in geography when looking at the presentations of the planets orbiting around the sun I asked myself: "What is the reason for this strange behavior?" It was a real revelation when my physics teacher Mr. Röderer explained to me that this is caused by the balance of the attraction between masses and centrifugal forces. This roused my enthusiasm and whetted my appetite. Still, during my last few years in high school, I spent more time with sports, boy scouting, alpinism, music etc. so my performance in school was reasonable but not more.

Perhaps in this context, it might be of interest to know that my father held a diploma in mechanical engineering from the Technical University of Bruenn (Cechia), and worked for the Skoda factory in Pilsen designing locomotives. He died during the last days of the war, so after our expatriation to the western part of Germany, my mother was left alone to take care of her two children, namely my two-year older sister and myself.

At the age of 19, I started to study physics at the University of Frankfurt and changed to the Technical University of Darmstadt three years later. I finished my diploma thesis there at the age of 26 and my PhD thesis at the age of 29. In both theses, I applied optical spectroscopy to determine crystal field split energy levels of rare earth ions in garnets. The director of the institute was Professor K. H. Hellwege. As I did a lot of computational work for the experiments, he used to say to me when he met me in the corridors: "Above all mathematics, don't neglect the physical background" which he called "Physikalischer Hosenboden" (lit. "physical bottom"). This must have made a lasting impression on me because I have a strong desire to explain new phenomena which I come across in simple physical pictures and do not feel comfortable with mathematical formalism alone. Today, I would like to pass this attitude on. In computer simulations in particular we would get lost in an abundance of results if we did not maintain a critical view for the source of the effects. This is especially true for phenomena caused by a constructive or destructive interaction of various mechanisms. My direct supervisor for my thesis work was Prof. Stefan Hüfner.

The topics of garnets and crystal fields also brought me to the laboratory of Prof. A. Koningstein in Ottawa, Canada, where I worked as a postdoc for a little less than three years. As was the case until then my goal remained the determination of crystal fields, but here I used the electronic Raman effect to determine the energy levels experimentally. Since Raman scattering is mainly caused by the optical phonons, these were also included in my investigations. A photo of me at the age of 33 is shown in Figure 1.

Figure 1. Peter A. Grünberg (PAG) at the age of 33 when he started his work at the research centre in Jülich.

In 1972, I was offered a position as a research scientist at the newly founded Institute for Magnetism at the research centre in Jülich. The director was Prof. Werner Zinn. His picture (marked with WZ) is included in the group photo of Figure 2. He was mainly interested in investigating the model magnetic semiconductors EuO and EuS. I also started to work on these compounds, applying optical absorption spectroscopy as well as Raman scattering (RS) from phonons. The RS experiments were performed at the Max Planck Institute for Solid State Physics in Stuttgart in the group headed by M. Cardona. My partner in Stuttgart was Gernot Güntherodt (marked GG in Figure 2) with whom I explored the magnon phonon interaction in EuO and EuS.

Figure 2. Group photo from the conference on thin magnetic films and surfaces in Le Creusot (France in 1988) when GMR was publically announced for the first time. This photo appears here because it shows many colleagues who are mentioned in the text that follows.

RS often causes a problem if we want to observe scattering with small frequency shifts below about 30 GHz (corresponding to 1 cm^{-1}). This is the domain of Brillouin lightscattering spectroscopy (BLS), where instead of a grating monochromator as in RS, a Fabry Perot interferometer is used as the dispersive element. In the early 1970s in BLS instrumentation, an interesting development occured due to the pioneering work of John Sandercock in

Zürich. He invented the multipass operation, and later on, showed how two multipass interferometers could be used in tandem. Following this example we assembled our own setup, namely a triple-pass spectrometer without tandem operation. A schematic is shown in Figure 1 of [1]. John not only pioneered the experimental technique, he also demonstrated the improvement with the first BLS measurements of various phenomena in solid state physics. This work had a great impact on ours. At the same time, he established his own company. Researcher, inventor, entrepreneur, salesman – all at the same time: following in the footsteps of Alfred Nobel! I have the greatest respect for such people. He has since been awarded various prices, of which the David Richardson Medal of the Optical Society of America (2005) is the latest.

One of his experiments grabbed our attention in particular. It was the first measurement of acoustic spinwaves in Yttrium Iron Garnet (YIG) and therefore the first time that acoustic spinwaves were measured in ferro- or ferrimagnets by means of LS. This result encouraged us to try the same experiment with EuO. However, as this compound only orders ferromagnetically below 60 K, cooling was necessary. Fortunately, a cryostat with a superconducting magnet was available from the optical experiments, so it was brought to the BLS setup. Before long, we were able to detect BLS from the bulk spinwaves. Additional scattering with a very strong Stokes/anti-Stokes (S/aS) anomaly could finally be identified. This was due to a spinwave propagating along the surface in one direction only and not the opposite. The successful observation of spinwaves in the bulk and at the surface of EuO is described in [1]. It has the character of an anecdote that the clue for the interpretation of a certain peak, caused by a surface spinwave, was uncovered when the experimental setup broke down and had to be repaired.

Further important results in this context were obtained by John Sandercock together with Wolfram Wettling. They measured the bulk and surface spinwaves from ferromagnetic metals in the shape of platelets with thicknesses of the order of mm. Standing spinwaves in thin films with thicknesses of the order of 100 nm or less were seen by Marcus Grimsditch and Alex Malozemoff (AM in Figure 2) for the first time. References to all of this work can be found in reference 2 of [1]. From this period, I also would like to mention fruitful contacts with Burkhard Hillebrands (BHi in Figure 2), who later extended BLS to thin film structures with lateral confinement.

Often in experiments with light, strong metallic optical absorption is considered a disadvantage because it significantly reduces the interaction volume. In LS experiments on standing spinwaves, however, it turns out to be of advantage because the wavevector of the incident light is smeared out due to strong optical absorption. As a result, a whole band of wavevectors is available to fulfill total momentum conservation, which is one of the required experimental conditions in BLS. A BLS spectrum from spinwaves showing the surface- and a standing mode is displayed in Figure 2 of [1]. To emphasise the last point: the standing modes are only observed because the wavevector of the incident light is sufficiently smeared out due to strong metallic optical absorption. In contrast, unpinned standing modes with antinodes at the surfaces are not seen by microwave absorption because the total precessing moment cancels out. This aspect was later discussed by John Cochran and Bret Heinrich. Both are seen in Figure 2 on the upper right-hand side, marked JC and BH, respectively. Next, we extended the LS investigations of spinwaves to multilayered structures. We concentrated on "magnetic double layers" i.e. two ferromagnetic films separated by a nonferromagnetic interlayer. One main interest was the coupling of DE modes to the collective dipolar modes of the structure. The most important results of these investigations are summarized in Figure 3 of [1].

At this point, I wish to mention that my meanwhile permanent staff position in a government-funded research laboratory allowed me to make long-term planning and to build up equipment for sophisticated sample preparation. At that time, such components were only partly available on the market. In this endeavour, I was assisted by my technician Reinert Schreiber who designed, assembled and operated the machine to prepare the samples. A particular feature which he installed was the "wedge technique". It provides the possibility of making thin-film samples with increasing thickness from one end to the other. Later, this turned out to be very helpful for the study of thickness dependencies. A picture of Reinert performing some leak testing is shown in Figure 3.

I did the first calculations on the dipolar-coupled DE modes, as displayed in Figure 3 of [1] myself. The extension to multilayers was done together with my colleague K. Mika from the mathematical department of our institute. Knowledge of the effect of dipolar coupling on the spinwave frequencies and the experimental result on standing modes was sufficient to be able to qualitatively predict their behaviour under the effect of ferro- or antiferromagnetic exchange coupling. This is explained in [1] in the context of Figure 3 and Figure 4.

Figure 3. My technician for many years, Reinert Schreiber, while occupied with leak testing.

Let us return to the "trilayer" or "magnetic double layer" structure as displayed in Figure 3 of [1]. It explains how F and AF coupling can be determined qualitatively from the frequency shift of the optic mode. What we observed at that time were only frequency upshifts. According to Figure 4 of [1], the coupling then is ferromagnetic.

In the early 1980s, it was believed that there were mainly two mechanisms that explained coupling. These were pinhole coupling via magnetic bridges across holes in the interlayers and "orange peel" or Néel-type coupling due to meandering interlayers. Remagnetization curves which can reveal antiferromagnetic-type coupling yield no information when the coupling is ferromagnetic. For these reasons, the situation concerning coupling was still rather uncertain in the early '80s. Under these conditions, we started a systematic search using the spinwave method. We also concentrated on the more unique case of antiferromagnetic (AF) type coupling, trying to find an example.

While on a year's sabbatical leave at Argonne National Laboratory in the U.S. as a guest of Mervin Brodsky, this search was finally succesfull. We detected AF coupling in Fe/Cr/Fe structures on cleaved substrates of rocksalt. Details are found in [1]. Two of the colleagues I met at Argonne, with whom I worked on other topics namely Ivan Schuller (IS) and Sam Bader (SB), can also be seen in Figure 2. When I came back from the U.S., I continued working on the reproducibility of the effect, now using also epipolished GaAs substrates. At that time, Jozef Barnas, a theoretician from Poznan in Poland, joined my group and we started to work on a quantitative theory of the effect of coupling on spinwaves. A photo of Jozef is shown in Figure 4, and some results of our joint efforts have been published in ref. 2 of [1].

Figure 4. Jozef Barnas from Poznan in Poland worked in Jülich on the theory of spinwaves as modified by coupling, and on the theory of GMR.

Furthermore, at that time, anisotropic magnetoresistance (AMR) was widely discussed by the community for applications in sensors for hard disk drives. So "something was already in the air" regarding MR effects and we decided to complement the available experimental techniques with magnetoresistance. Jean Pierre Renard (JR in Figure 2) had found an effect which later on turned out to be related to GMR. In our laboratory, Gaby Binasch installed the equipment during her diploma work and did the first measurements on permalloy films. Since Reinert Schreiber was now also able to make AF coupled Fe/Cr/Fe structures in a form suitable for measuring of electrical resistivity, GMR was seen before long as discussed in [1] and published in ref. 8 thereof. Gaby finished her thesis and received her diploma in physics (equivalent to masters). After this, she left Jülich for an interesting job in industry. She can be seen in Figure 5 at the celebration for her diploma.

Figure 5. Gaby Binasch celebrating her diploma and receiving a handmade medal. Here she is just listening to Peter A. Grünberg who is reading the dedication to her.

Before I come back to GMR, let me add a few more details on coupling. As mentioned in [1], at the same time that we were working on Fe/Cr/Fe, other groups found similar phenomena for rare earth layers (Gd, Y) separated by Y layers. A RKKY-type mechanism was proposed by Yako Yafet (marked YY in Figure 2) as an explanation. Even oscillatory behaviour as a function of the interlayer (Y) thickness had already been reported. However, it seems that only after the discovery of GMR, coupling per se also received general attention. And sure there were many results and surprises. The RKKY-type mechanism was applied to structures with magnetic 3d- transition metals by George Mathon (GM in Figure 2), David Edwards and Patrick Bruno (PB in Figure 2). With the permission of the colleagues involved, I may at this point tell the following anecdote. In the course of his search for AF couplings Stuart Parkin from IBM San Jose (SP in Figure 2) had found it also for sputtered Co films with [111] texture interspaced by Cu. The Dutch group from Philips Eindhoven established oscillatory coupling for this system also for epitaxial growth with [100] orientation. Bill Egelhoff from NIST in Gaithersburg, well-known for very fine work in epitaxy, however, reproached vigorously. I still remember a MRS meeting in San Francisco where he defended his point. Then Professor Gradmann (UG in Figure 2) from Clausthal University in Germany stood up and said: "Bill I think I know what happened. Your samples are too good. You have very fine surfaces which nucleate antiphase domains that give large angle grain boundaries between the islands. Then you can get diffusion of Co along the boundaries and the formation of magnetic bridges. The resulting F-coupling swamps any possible AF coupling." Bill Egelhoff responded simply: "I like that". Indeed, somewhat later, Jürgen Kirschner (JK in Figure 2) and his group from Freie Universität in Berlin added further evidence to this case by showing that grain boundary formation comes from the usual competition between ABABA.. and ABCABC... type stacking for hcp type structures along [111].

Coming back to Bill Egelhoff, it has also to be mentioned that he used oxygen as a surfactant to set the record values for GMR of 25% in the simple trilayer structures of Co and Cu and also dual spinvalves with two Cu interlayers. In multilayered structures, of course, the effect is much stronger.

This example demonstrates again that coupling can change drastically with growth, which makes comparison with theory ambiguous. Then how do we know that current theories on this phenomenon as proposed for the first time by Yako Yafet are in essence correct? An important contribution to this comes from Bob Celotta's group at NIST. They grew trilayer-type samples, where one magnetic layer was replaced by a single crystal Fe whisker, which at the same time is used as a substrate. For the interlayer, various materials were chosen with an emphasis on good matching to Fe. Coupling is seen via magnetic domains, which are made visible using SEMPA. These indeed superb and celebrated experiments left no doubt that theories based on the RKKY interaction predict oscillatory coupling correctly. In real cases, however, the coupling could still be good for many surprises, as is clear from the anecdote above.

Another phenomenon which also finally turned out to be related to growth is that of 90° type coupling. This was proposed by A. Hubert and his group as the reason for special magnetic domain structures occurring in samples with wedge-shaped interlayers. Alex Hubert (AH) and his coworker Rudi Schäfer (RS) can also be seen in Figure 2.

For the size of the GMR effect, there is a large difference between trilayers and multilayers, as shown in Figure 9 of [1]. This has been known since GMR was announced publically for the first time at the ICMFS in Le Creusot in France, organized by Irena Puchalska (IR) and Horst Hoffmann (HH). I also met Albert Fert there for the first time (as he attended the conference but was not in the original group photo, I added his photo (AF) on the right-hand side of Figure 2). After we had compared our results and came to the conclusion that we had seen the same effect and thus confirmed it to each other, we were ready for a glass of red wine from Burgundy. We were not the only ones who enjoyed that conference. A group of talented musicians among the participants entertained us with piano concertos (Alex Malozemoff-AM and Jaques Miltat-JM), Ulrich Gradmann (UG) and Mrs. Yafet on the violin and finally Klaus Rohrmann (KR) with a solo on a water hose. The arrival of GMR was adequately celebrated indeed!

Figure 6. Bob Camley who together with Jozef Barnas invented the Camley- Barnas model describing GMR in Jülich.

When I returned home from the conference in Le Creusot, I was in a very fortunate situation that I had two excellent theoreticians as visiting scientists in Jülich. A picture of Jozef Barnas has already been shown in Figure 4, a passport photo of Bob Camley is shown in Figure 6 . I now had both previous collaborators on spinwaves together again in Jülich, but now the new exciting topic was the GMR effect. Before long, they had worked out a theory based on Boltzmann's diffusion equation for the evaluation of the GMR experiments which became known as the Camley- Barnas model.

Figure 7. My team in the time "beyond" the discovery of coupling and GMR.

The title of my Nobel lecture includes the term "beyond". The team with whom I obtained most of the results "beyond" can be seen in Figure 7. We conducted much work on Si interlayers, where we found very strong coupling but dissapointingly weak tunnel magnetoresistance. We also started on "current-induced magnetic switching" (CIMS) as invented by John Slonczewski and Luc Berger. John and his wife Ester stayed with me various times in Jülich. My successor as group leader, Daniel Bürgler, can be seen in the middle of Figure 7 (3rd from the left).

Figure 8. My hosts in Sendai. From left to right: T. Miyazaki, A. Yoshihara, H. Fujimori, K. Takanashi, Mrs. Yoshihara, Mrs. Abe.

In 1998, I was invited by Hiroyasu Fujimori (HF in Figure 2 and 3rd from left in Figure 8) to come to Sendai, Japan, for 6 months. During this stay, I also spent two months in Tsukuba as a guest of Yoshishige Suzuki. A group photo with my collaborators in Sendai is shown in Figure 8. Of these, Koki Takanashi in 1995 had spent a year in Jülich as a postdoc. Terunobu Miyazaki is one of the inventors of tunnel magnetoresistance (TMR) at room temperature and with Akira Yoshihara I worked on BLS from spinwaves.

I officially retired in 2004, but kept a desk and and some office space in the research centre within the Institute for Electronic Properties headed by Claus Schneider (CS in Figure 2). In this way, I was able to maintain my contacts with the scientific community, of which I wish to mention only as example a visit to Vladimir Ustinov (VU in Figure 2) in Ekaterinenburg. Now – as a result of being awarded this great Prize, I have also recieved a new contract from my employer in the form of a "Helmholtz professorship". As a result, I call myself jokingly a "re-entrant magnetician". I should explain to the non-expert that re-entrant magnetism is where magnetism disappeares as a function of some parameter (like pressure or temperature) at a certain value but comes back at another value. Obviously, the relevant parameter in my case is age!

I would like to add a remark on my religious believes. Brought up rather conservative catholique I see religions now more or less in the spirit of Lessing's (German dramatist 1729-1781) ring parabola which I would top by saying that – not only does nobody know which is the right ring (standing for religion) – but there indeed is no such thing as a right or false ring. Per se they are all equivalent. What really counts is how religions are practised, for example, with tolerance. And yet I believe that there is more than what we see, hear etc., or can detect with instruments. But it is a feeling borne out of many details of my personal experience and therefore impossible to share or communicate.

REFERENCES

[1] Nobel lecture of P. Grünberg, this volume.

FROM SPINWAVES TO GIANT MAGNETORESISTANCE (GMR) AND BEYOND

Nobel Lecture, December 8, 2007

by

PETER A. GRÜNBERG

Institut für Festkörperforschung, Forschungszentrum Jülich, Germany.

DISCOVERY OF LIGHTSCATTERING FROM DAMON ESHBACH SURFACE SPINWAVES

The "Institute for Magnetism" within the department for solid state physics at the research centre in Jülich, Germany, which I joined in 1972, was founded in 1971 by Professor W. Zinn. The main research topic was the exploration of the model magnetic semiconductors EuO and EuS with Curie temperatures T_c=60 K and T_c=17 K, respectively. As I had been working with light scattering (LS) techniques before I came to Jülich, I was very much interested in the observation of spinwaves in magnetic materials by means of LS.

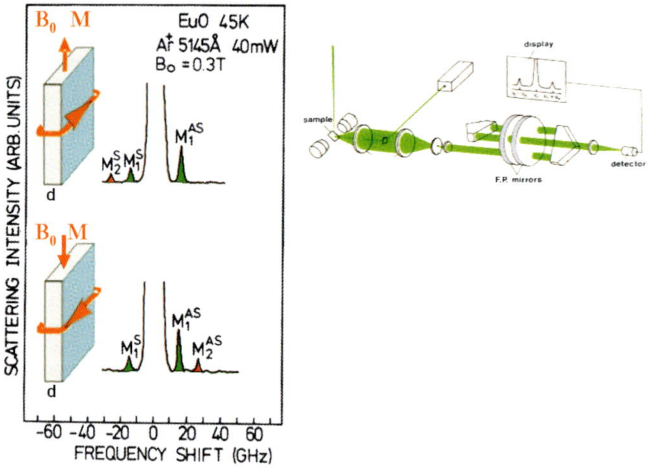

Figure 1. Right: Schematic of a Brillouin lightscattering (BLS) spectrometer used for the detection of spinwaves. Spectra of scattering from bulk spinwaves (marked with green) and DE surface spinwaves (red) are seen on the left. For the upper and lower part on the left, the field B_0 and magnetization M have been reversed. From [1].

LS can be performed with grating spectrometers, which is called Raman spectroscopy, and alternatively by Brillouin light scattering (BLS) spectroscopy. In the latter case, a Fabry Perot (FP) interferometer is used for the frequency analysis of the scattered light (see right-hand side of Figure 1.). The central part consists of two FP mirrors whose distance is scanned during operation. BLS spectroscopy is used when the frequency shift of the scattered light is small (below 100GHz), as expected for spinwaves in ferromagnets.

In the early 1970s, an interesting instrumental development took place in BLS, namely the invention of the multipass operation, and later, the combination of two multipass interferometers in tandem. The inventor was Dr. J.A. Sandercock in Zürich. Since we had the opportunity to install a new laboratory we decided in favour of BLS, initially using a single three-pass instrument as displayed on the right-hand side of Figure 1. With this, we started investigating spinwaves in EuO.

We indeed were able to find and identify the expected spinwaves as shown by the peaks in Figure 1 (marked green). Different intensities on the Stokes (S) and anti-Stokes (aS) side were known from other work to be due to the magnetooptic interaction of light with the spinwaves.

The peaks marked red remained a puzzle for some time until good luck came to help us. Good luck in this case was a breakdown of the system, a repair and unintentional interchange of the leads when reconnecting the magnet to the power supply. To our surprise S and aS side were now reversed. To understand what this means, one has to know that classically S and aS scattering is related to the propagation direction of the observed mode which is opposite for the two cases. This can be understood from the corresponding Doppler shift which is to higher freqencies when the wave travels towards the observer and down when away from him. The position of the observer here would be the same as of the viewer in Figure 1. The appearance of the red peak in the spectra on only either the S or the aS side can be explained by a unidirectional propagation of the corresponding spinwave along the surface of the sample. It can be reversed by reversing B_0 and M.

The unidirectional behaviour of the wave can be understood on the basis of symmetry. For this, one has to know that axial vectors which appear in nature, such as B and M on the left-hand side of Figure 1, reverse their sign under time inversion and so does the sense of the propagation of the surface wave as indicated. The upper and lower part of Figure 1 left-hand side therefore are linked by time inversion symmetry which is valid without damping. Hence, the unidirectional behaviour reflects the symmetry of the underlying system.

Finally, the observed wave could be identified as the Damon Eshbach (DE) surface mode known from theory and from microwave experiments.

From the magnetic parameters of EuO, one predicts in the present case that the penetration depth of the DE mode will be a few 100 Å. Sample thickness d is of the order of mm. Therefore, for the present purpose, EuO is opaque. In this case, the wave travelling on the back side of the sample in the opposite direction to the wave on the front side cannot been seen in this experiment. BLS is then either S *or* aS but not both at the same time. Due to all of these unique features, the results of Figure 1 have also been chosen as examples for current research in magnetism in a textbook on "Solid State Physics" (H. Ibach, H. Lüth, Springer 1995, p.186).

BLS FROM SPINWAVES IN SINGLE THIN MAGNETIC FILMS

We stick to the geometry of Figure 1 but with much smaller thickness d, of the order of 200 Å. As material, we chose Fe. Bulk modes now split into a bunch of standing modes and shift to higher frequencies. In the example in Figure 2, only the DE mode and the first standing mode are within the observable frequency range. The mode patterns (left) represent amplitude distributions, where the local precession is generally elliptical. DE modes with given propagation direction have finite amplitudes on both surfaces and therefore both directions are seen as S and aS scattering. As in Figure 1, the sample surface from which most of the scattering is observed is shown in turquoise.

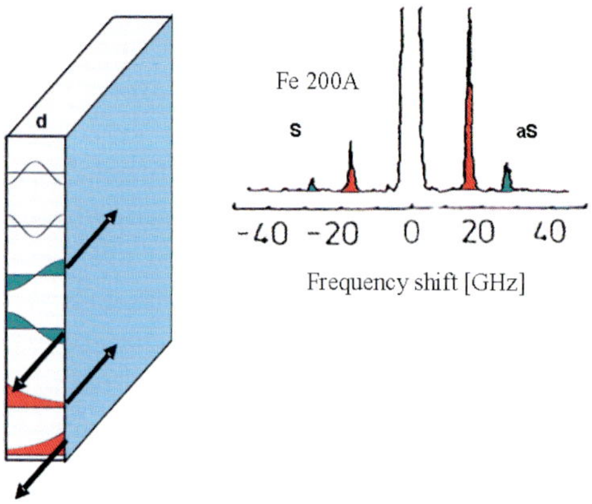

Figure 2. DE surface spinwave and 1st standing spinwave of a single thin film of Fe and its experimental observation by means of BLS. Opposite propagation directions (black arrows) correspond to S and aS scattering. Higher standing waves as indicated in the upper part on the left-hand side are beyond the observed frequency range [2].

DIPOLAR COUPLED DE MODES IN MAGNETIC DOUBLE LAYERS WITH PARALLEL AND ANTIPARALLEL MAGNETIC ALIGNMENT

Next, we consider magnetic double layers (sometimes also called trilayers) as illustrated on the left-hand side of Figure 3 [2]. There are four magnetic surfaces, each carrying a DE wave, which interact via the oscillating strayfields that they produce. The calculation proceeds by using the Landau Lifshitz equation after linearization and by applying the boundary conditions following from Maxwells equations. The middle section of the Figure displays the result for the mode frequencies as a function of the interlayer thickness d_0 multiplied by the value of the wavevector q of the wave. q is determined by the geometry of the BLS experiment, including the wavelength of the laser light used. A typical value is $q=1{,}73\cdot10^{-2}$ nm^{-1}. Hence $qd_0=1$ corresponds to $d_0=58nm$. This is the decay length of the DE waves away from the surface.

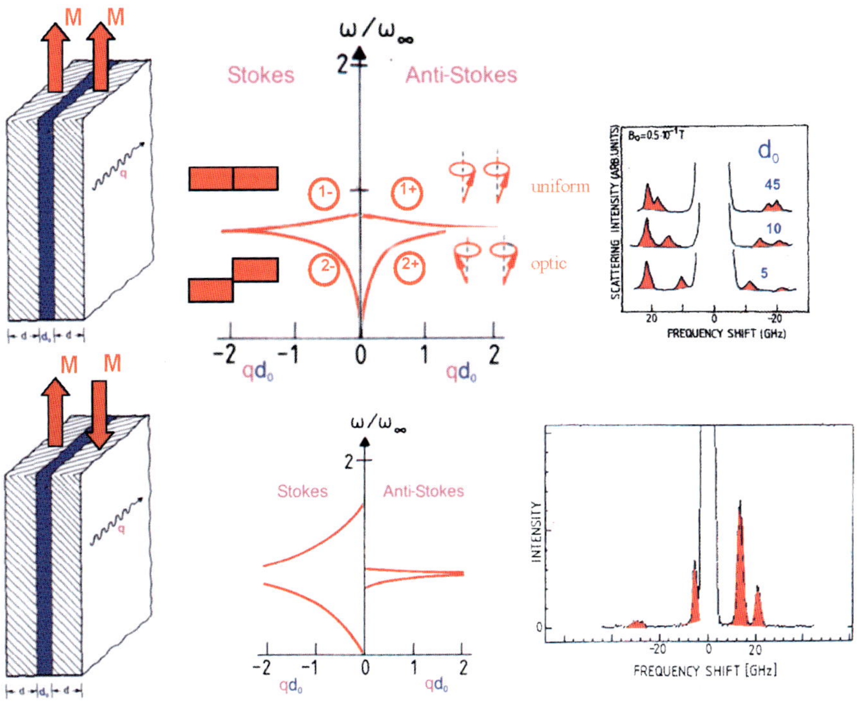

Figure 3. Coupled DE modes in magnetic double layers for parallel (upper part) and antiparallel (lower part) alignment of the magnetization M [2]. Calculation in the dipolar limit. For the spectrum in antiparallel alignment, a sample showing AF coupling (to be discussed later) was used. Peak positions are mainly due to the dipolar interaction with small corrections due to exchange coupling.

For the parallel alignment, the calculation predicts a symmetric graph, i.e. the same frequencies for both signs of q. For the antialignment, the predicted branches are not symmetric as also revealed by the experiment.

Figure 3 displays a few more details of the nature of these modes. The middle part reveals that the upper branches 1+ and 1- are characterized by an in-phase precession of the magnetizations M, whereas for the lower branches 2+ and 2- they are out of phase by 180°. In the middle part on the left-hand side, this aspect is again illustrated by showing the amplitude distribution inside a cross-section, where the DE attenuation has been neglected. Note the phase jump by 180° in the middle part of the structure.

DOUBLE LAYER WITH THE INCLUSION OF INTERLAYER EXCHANGE

Qualitatively, the behaviour of the dipolar modes due to the inclusion of ferromagnetic (F) interlayer exchange coupling (IEC) is easy to predict. This is because the result of full F coupling is a single film of which we already know the result as displayed in Figure 2. Why, in Figure 4, is it the lower dipolar branch from which the first standing mode evolves and not the upper one?

It becomes clear from the nature of the modes as displayed in the middle upper part of Figure 3 and has to do with the conservation of the symmetry character. When IEC is "turned on", the amplitude pattern with the 180° phase jump is changed into the first standing mode under conservation of parity. Note that a 180° phase jump would generate a very high exchange energy which is avoided by the formation of a node (as in Figure 2).

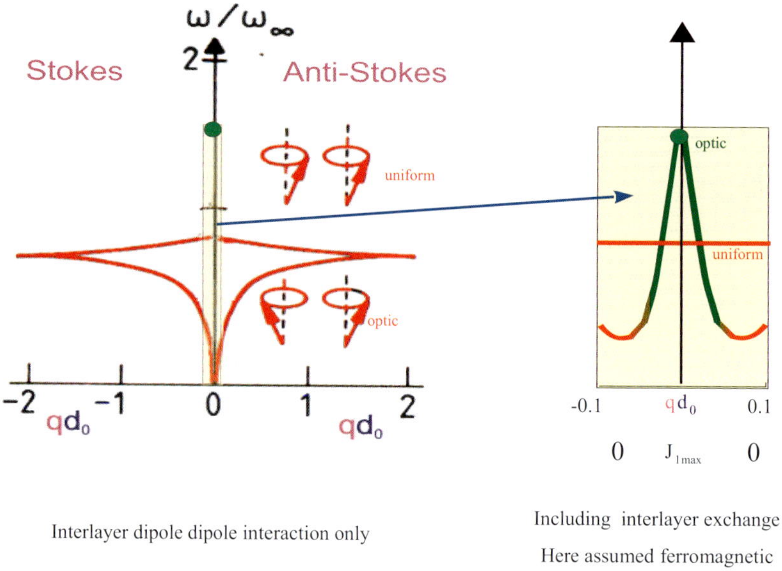

Figure 4. The expected effect of ferromagnetic interlayer exchange on the branches representing dipolar coupling (red). In the limit of full exchange for d_0=0, the lower branch is replaced by the green dot representing the first standing mode of the combined single film. The general behaviour for small d_0 is shown on the right-hand side. For the lower branch, there is a transition from dipolar dominated (red) to interlayer exchange dominated (green). The upper dipolar branch becomes the DE mode of the combined film. In the displayed region of small d_0, its frequency is practically constant (horizontal red line).

In Figure 3, we looked at the possibility of an antiparallel alignment and one way of achieving it would be antiferromagnetic (AF) interlayer exchange coupling. If we want to recognize this type of coupling from the behaviour of the spinwaves, what do we expect? At this point, it is important to mention that most of our experiments are performed in saturation. Hence, regardless of whether the coupling is F or AF, we would apply a high enough field for the BLS experiment to get parallel alignment (with an important exception as seen below). This has the advantage of a well-defined experimental situation.

The upper branch in the case of magnetic saturation is given by the uniform precession i.e. total moment and therefore independent of IEC.

For the lower branch, one can argue as follows. For magnetizations aligned in parallel, an external field as well as ferromagnetic IEC should increase the restoring force and hence also the spinwave frequency. Opposite to this, AF type IEC should weaken the effect of the external field and thus decrease the frequency. In the present case, the dipolar interaction works in a similar way. Hence, from these simple arguments one expects the frequency of the lower branch to be further decreased for antiferromagnetic IEC and to be increased as shown in Figure 4 for ferromagnetic IEC. On the other hand, if IEC is antiferromagnetic, then for small enough external fields, we should obtain antialignment. In this case, the branches presented in the lower part of Figure 3 could be used to identify the antialignment.

The concept just presented was later used and worked out in more detail by us [3] and by Cochran and Dutcher [4]. From the latter work, we have adopted the term "uniform mode" for the in-phase precession and "optic mode" for the 180° out of phase precession. The principle was also used by other groups for the evaluation of exchange coupling from microwave experiments.

Due to the insight just described, we became aware in the early 1980s that we had a powerful tool in our hands, which could be used to investigate a long-standing open question concerning coupling phenomena in layered magnetic structures. At that time, it was suggested that ferromagnetic coupling could be caused by pinholes and ferromagnetic bridges across the interlayer. As Néel showed, it could also be due to meandering interlayers. This effect was therefore called orange peel or Néel-type coupling. Antiferromagnetic-type coupling could possibly occur due to edge effects, such as flux closure. Real intrinsic types of interlayer coupling, similar to the RKKY interaction in dilute magnetic alloys, was not yet known.

The situation was further hampered by insufficient sample preparation. In our first attempt to explore coupling and its dependence on the interlayer material and thickness, we only found ferromagnetic-type coupling –presumably resulting from pinholes.

For various reasons, we now concentrated on Cr interlayers. First, Cr by itself is an antiferromagnet. Based on its magnetic structure for thin enough interlayers, one expects the coupling to oscillate between F and AF with a period of two monolayers (ML). Second, Cr matches Fe very nicely both in its crystallographic and thermodynamic properties, which is important for the growth. Fe/Cr/Fe structures therefore appeared to be most promising.

DISCOVERY OF ANTIFERROMAGNETIC (AF) COUPLING

In this situation, I was very fortunate to have the opportunity to take sabbatical leave and go to Argonne National Laboratories in the US. In the team around M. Brodsky in Materials Sciences, I found a group with expertise in preparing Fe/Cr/Fe samples on substrates of cleaved rocksalt. For comparison, we also prepared Fe/Au/Fe samples.

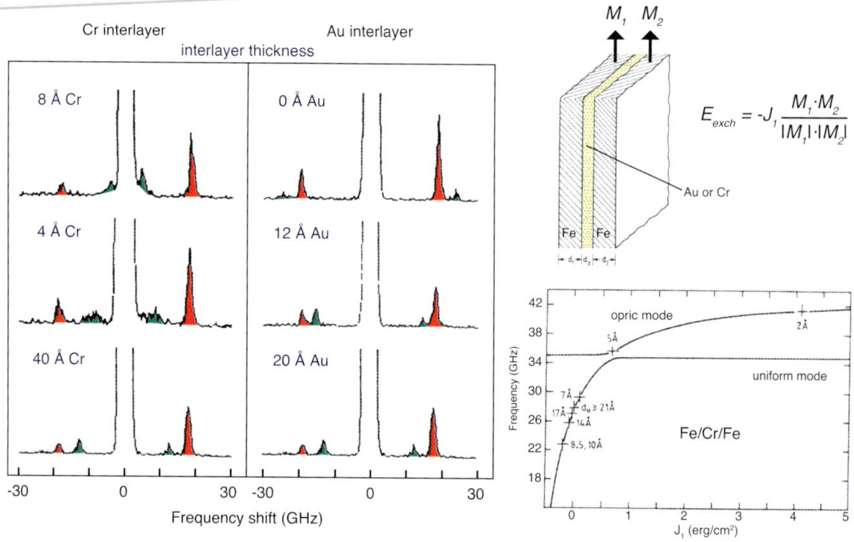

Figure 5. Left-hand side: Representative BLS spectra from spinwaves in Fe/X/Fe, where X=Cr and Au [5]. The plot on the right-hand side displays the mode frequencies as a function of the interlayer exchange parameter J_1 [3] where the theory is represented by the solid lines.

Spectra from both systems in strong enough fields for full saturation of the samples are displayed on the left-hand side of Figure 5. The spectrum with d_0=0 represents the fully F-coupled limit, as in Figure 3. The two spectra at the bottom represent pure dipolar coupling. From the Fe/Au/Fe data and comparison with Figure 4, we concluded that the coupling for the Au interlayers is ferromagnetic. However, by looking at the spectrum in the upper corner on the left-hand side for the Cr interlayers, the reader might share with me the sensation I had at that time when I viewed a BLS spectrum representing antiferromagnetic IEC for the first time.

In a phenomenological approach, IEC is treated via the associated aerial energy density E_{exch}, as quoted in the Figure. As already discussed, on physical grounds, the optic mode increases its frequency for F coupling (J_1>0) and decreases it for AF coupling (J_1<0).

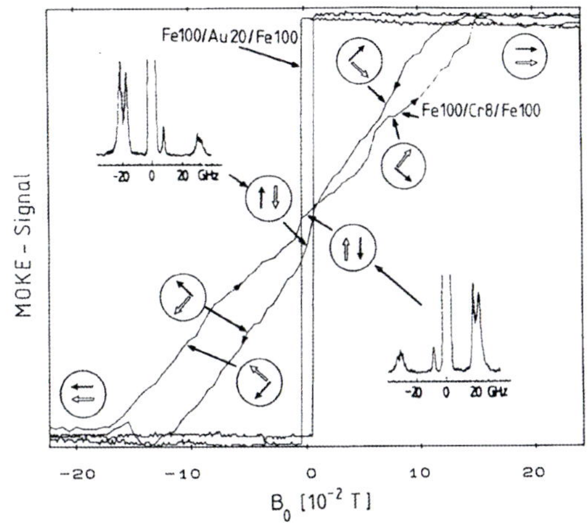

Figure 6. Remagnetization curves of samples with F-coupling (Fe/Au/Fe) and AF-coupling (Fe/Cr/Fe).

Figure 6 displays the effect of F and AF coupling on the remagnetization (so called hysteresis) curve. Plotted is the magneto-optic Kerr effect (MOKE) signal, which is roughly proportional to the magnetization M, vs. the external field B_0. There is a strong shearing for the Fe/Cr/Fe sample, which was not observed for the other (Fe/Au/Fe) sample, which displays the remagnetization of an uncoupled or F-coupled sample with small coercivity. F-coupled and uncoupled cannot be distinguished. Furthermore, one has to consider that shearing of hysteresis curves is not very unique. It is a phenomenon that occurs very often and can have many causes. The use of more sophisticated tools like BLS is not a luxury for a first observation. Note also that the information given by the MOKE curve is assisted by BLS spectra to make sure there is AF alignment at these points. However, once it is certain that shearing comes from AF coupling, one might as well employ this simpler method to obtain large amounts of data.

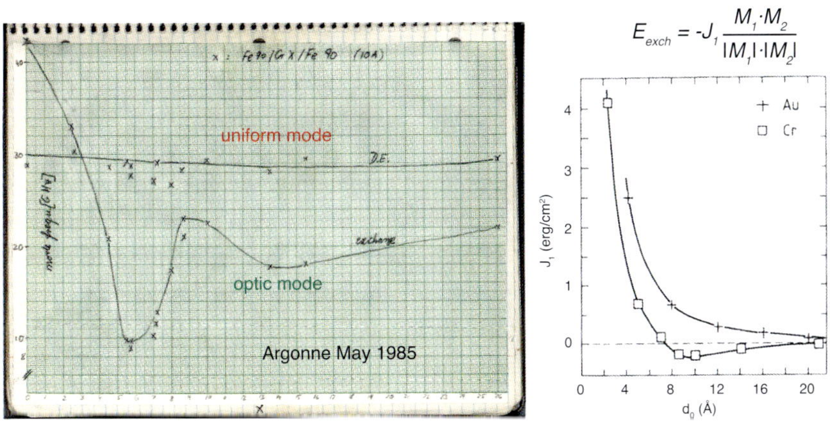

Figure 7. Left-hand side: First measurement of coupling for a wider range of Cr thicknesses, given in terms of spinwave frequencies. Note that the DE mode stays almost constant whereas the optic (or exchange-) mode shows two minima, indicative of AF coupling. Right-hand side: Evaluation of the frequencies for another sample not showing the second minimum, with respect to the coupling parameter J_1.

In Figure 7, we see the result of a more extended investigation of the coupling represented by mode frequencies as a function of the Cr interlayer thickness given in Ångströms. This is the first rather complete result obtained in May 1985, and is directly copied from the lab book. The minima of the lower branch indicate AF coupling in that range. The second more shallow minimum initially appeared to be not sufficiently reproducible and was not communicated in our first publication. Furthermore, we were not very happy with the NaCl substrates because of their sensitivity to humidity. Since meanwhile epipolished GaAs had become available, we switched to this substrate material. Finally, AF coupling could reproducibly be observed where the result of Figure 7 seemed generic instead of the expected period of 2ML. However, years later due to improved growth, this and other theoretically expected features caused by the noncommensurate spin density wave of Cr could be observed [6].

In 1986, when we published our results [5], there were two other publications on interlayer exchange coupling, namely of Dy and of Gd across Y interlayers observed by means of neutron scattering (cited in ref. 7). For Gd interlayers, it was reported to be oscillatory, as indicated by Figure 7 and for many other examples which followed.

DISCOVERY OF GIANT MAGNETORESISTANCE

Sir Nevil Mott proposed a two-current model for the description of the electrical resistivity of magnetic alloys. This model is based on the fact that the resistivity is due to electron scattering. In magnetic materials, this is dependent on spin orientation. Due to the quantum mechanical spatial quantization, this orientation is only parallel or antiparallel with respect to the local magnetization. As spin flip processes occur seldom, each of the two orientations defines a current.

With this picture, one expects that there would be a strong resistivity change if one could manage to change the direction of the local magnetization within the mean free path of the electrons or on an even shorter scale. In a case where the scattering rates are different, there is a better chance that the total scattering rate is increased. As mean free paths are of the order of 10 nm, the 1 nm thickness by which the magnetic layers are separated in the coupled structures perfectly fulfills this condition. We expect that if the relative orientation of the magnetizations changes from parallel to antiparallel in a structure such as in Figure 8, left-hand side, the resistivity for the current will increase.

In order to have more experimental possibilities available, we used samples grown epitaxially on (110) oriented GaAs. The film plane was parallel to a (110) atomic plane and had an easy (EA) and a hard (HA) axis. For the thickness d of the individual Fe films, we chose d =12 nm. The Cr thickness was d_0 =1 nm, which corresponds to the first maximum of AF coupling. As a reference sample, we also made a single Fe film with thickness d=25 nm in order to measure the anisotropic magnetoresistance (AMR) effect for comparison. Laterally, the samples had the shape of a long strip with contacts at both ends.

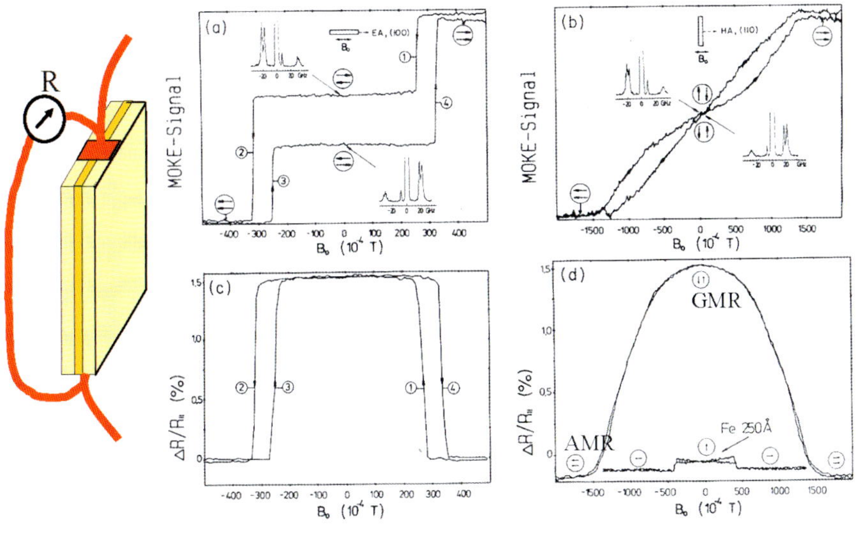

Figure 8. (a) and (b) MOKE curves and BLS spectra in order to identify magnetic alignment as shown by arrows. (c) and (d): relative resistance change corresponding to alignment [8].

In Figures 8(a) and 8(b), we can see the MOKE hysteresis loops from the double layers with AF coupling for B_0 along the EA and HA. The directions of the magnetization are indicated by the encircled pairs of arrows. This information is obtained from the MOKE intensities and the displayed LS spectra. As an example, let us discuss the hysteresis loop shown in Figure 8(a) in more detail. The field B_0 is applied along the EA, which is the long axis of the strip. It is clear that for a large enough B_0, the samples saturate in the field direction (parallel alignment). If we start with parallel alignment in the positive field direction and reduce B_0 then at a certain, but still positive value of B_0, the magnetization of one film reverses via domain-wall motion (point 1). Hence, in small fields we have antiparallel alignment. In a negative field, at point 2, the other film also reverses, and we have saturation. Points 3 and 4 mark the magnetization reversals when B_0 is scanned back.

Magnetoresistance (MR) traces are shown in Figure 8(c) and 8(d). Here, we have plotted the relative change of the resistivity $(R - R_{\uparrow\uparrow})/R_{\uparrow\uparrow}$ as we scan through the hysteresis loop. $R_{\uparrow\uparrow}$ is the resistivity for saturation along the EA. In Figure 8(d), we also show the MR of the single Fe film, taken for B_0 along the HA. Since for large B_0, the magnetization is along [110] and for small B_0, along [100], the maximum change of R seen in this trace displays the anisotropic MR= $(R_{\perp} - R_{\parallel})/R_{\parallel}$ of a 25-nm-thick Fe film. The value of - 0,13% is in reasonable agreement with the literature value of - 0,2%. For the single film and B_0 along the EA, there was no measurable MR effect. The reason for this is clear; the magnetization always stays aligned parallel to the EA and magnetization reversal takes place via domain-wall motion only. In Figure 8(d) for the Fe-Cr-Fe film, we have an MR effect due to both the anisotropic effect (negative values) and antiparallel alignment (positive values). The zero level of this plot is defined by $R=R_{||}$.

At the time of the discovery of GMR, it was well known that leading computer companies planned to develop AMR so it could be used for read heads in hard disk drives (HDDs). The comparison between AMR and the new effect (later, the term GMR was widely accepted) encouraged us to file a patent for using GMR in HDD.

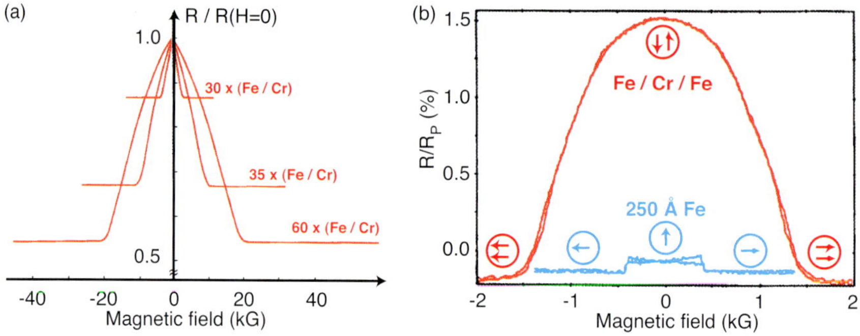

Figure 9. GMR effect (a) in a multilayer and (b) double layer of Fe interspaced by Cr. In (b) the AMR effect of a 250-Å-thick film of Fe is also shown for comparison.

In Figure 9 [9], we can see a comparison of the measurements using a multilayer as in Orsay and a double layer as in Jülich. The value for the multilayer of around 80% obtained in Orsay, which led to the term "giant" appears to have been surpassed by far by the TMR value of 500% for systems with MgO barriers. However, one has to be careful. There seems to be an important difference between dealing with insulators, semiconductors and metals. The magnetic semiconductor EuS, for example, changes its resistivity below its Curie point by many orders of magnitude. If we restrict the discussion to purely metallic systems, which has also some important implications for applications, the term "giant" is still justified.

The value 1.5% for the double layer displayed in Figure 9 (b) is even smaller but still large compared to the AMR effect of Fe (trace at the bottom). In Figure 8, the effect was measured using various configurations of external field relative to the axis of the crystal anisotropy to make sure that it really is due to the relative orientation of the magnetizations. The low values of the GMR effect as seen in Figure 8 and Figure 9 (b) is not a result of minor sample quality but is typical for the chosen film thicknesses.

GMR can be observed in the current in plane (CIP) and current perpendicular plane (CPP) geometry. The conventional geometry is CIP and is for most sensor applications (see below). The CPP configuration is expected to be of interest for future sensor designs.

As was already suggested in the first paper by Albert Fert and his group and later established by numerical evaluations of experiments [11], the microscopic explanation for the GMR effect is a dependence of the scattering rates on the orientation (parallel or antiparallel) of the electron spins with respect to the local magnetization.

We will now consider the situation in Figure 10 a. Two structures are shown – one with parallel and the other with antiparallel magnetization alignment. In an ideal situation, we assume that electrons with spin parallel to the local magnetization in their random walk are not scattered. In the figure, only one passage from the left to the right surface is displayed as representative for the entire motion. In this ideal picture, the electrons which are not scattered would cause a short circuit. This short circuit is removed when one magnetization is turned round and the electron with up spin enters this layer and finds its spin to be opposite to the local magnetization. The equivalent circuits for the various cases are shown at the bottom and display the increase in the resistivity due to the removal of the short circuit. In reality of course, both types of electrons are scattered. It is sufficient that one sort of spin is scattered more than the other for the resistivity to increase due to the antiparallel magnetization alignment. From symmetry for the CIP configuration, spin-dependent interface reflectivity does not contribute to the effect. The reason is that there is no change in momentum in the direction of the current due to translational symmetry during reflection at interfaces. This is different in the CPP geometry, where both spin-dependent scattering as well as spin-dependent reflectivity contribute.

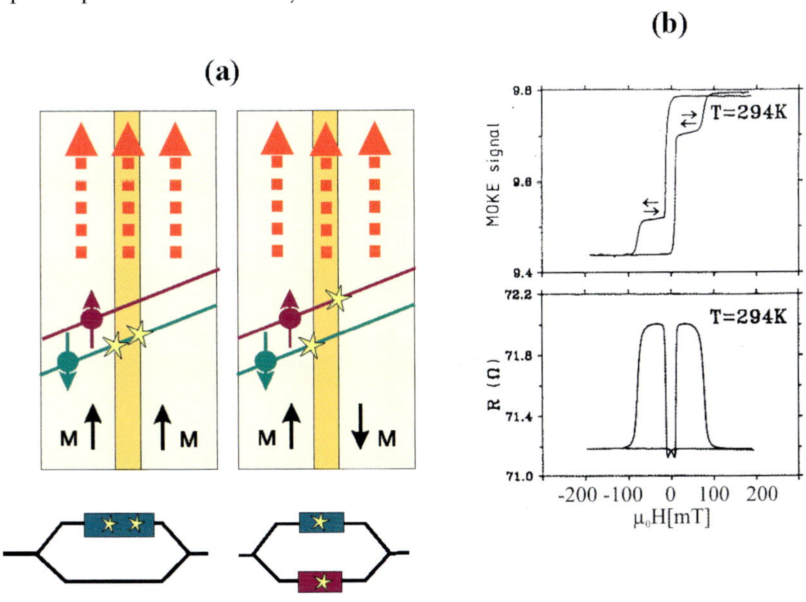

Figure 10. (a) Explanation of the CIP-GMR effect: spin-dependent electron scattering and redistribution of scattering events upon anti-alignment of magnetizations. The antiparallel alignment can be obtained by antiferromagnetic coupling, like in Figure 9, or due to hysteresis effects, like in Figure 10 (b).

The largest difference in resistivity occurs when an AF alignment is changed by an applied field into a F alignment. The AF alignment can be provided by

AF interlayer exchange as for the samples in Figure 9 or by different coercivities of successive magnetic layers, in particular by pinning the magnetization using an antiferromagnetic material in direct contact, known as "exchange biasing". If GMR is obtained via one of the latter methods and not via AF interlayer coupling, usually the term "spin valve system" is used in the literature, although there is no difference concerning the mechanism of the GMR effect. Such a case is displayed in Figure 10 b (see also the review article in ref. 9). The sample is a Co/Au/Co layered structure with one of the Co films deposited directly onto the GaAs substrate. Since this Co film is more strained than the one prepared on the Au interlayer, it has a higher coercivity. During magnetization reversal of the whole structure, it reverses later than the other, resulting in a small field range with antiparallel alignment. In Figure 10 (b) upper panel, this range is marked by antiparallel arrows. The curve in Figure 10 (b) lower panel shows the associated electrical resistance R(H) and its increase due to the antiparallel alignment. As will be seen in the section on Applications, the two different methods for obtaining antiparallel alignment lead to two different concepts for sensors based on GMR.

GMR: THE CAMLEY-BARNAS MODEL

At the time when the GMR effect was discovered, it was a coincidence that I had two visiting scientists in my team at Jülich, Jozef Barnas from Poznan in Poland and Bob Camley from Colorado Springs in the USA. Both are theoreticians whom I had been working with years before on spinwaves. Together, they worked out a method for the description of GMR by means of suitable parameters. Their method is based on Boltzmann's diffusion equation within the relaxation time approximation and soon became known as the Camley-Barnas model [10].

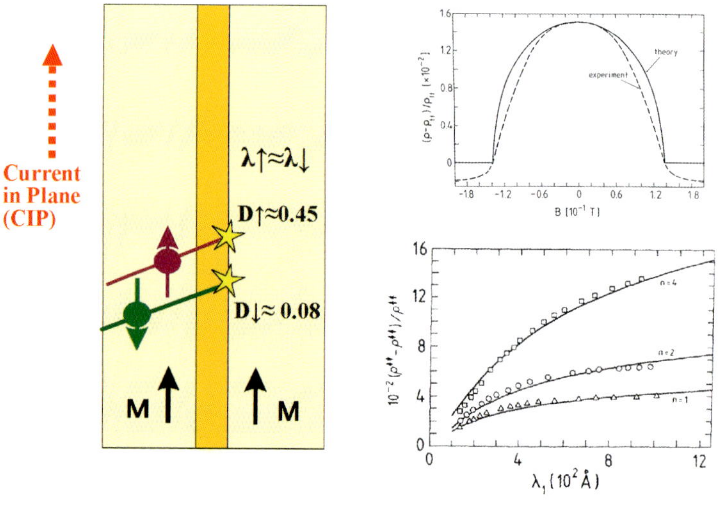

Figure 11. Comparison of theory and experiment within the Camley-Barnas model. Top right: resistance change vs. external field assuming AF coupling and cubic anisotropy. Bottom right: resistance change vs mean free path for multilayered structures (n=number of Cr layers). Left: assymetry of diffusive scattering parameters given by D↑ ≈0.45 and D↓≈ 0.08.

CIMS

Current-induced magnetic switching (CIMS) can be understood as an inverse GMR effect. In a trilayer structure, one film is prepared with a fixed magnetization M_{fixed} and the other with a magnetization which can be easily rotated (M_{free}). Using electron beam lithography, the samples are prepared in such a way that a strong current can be applied perpendicular to the sample plane. Then it can be shown that – no matter what the initial conditions are – when the current is such that the electron flow from the film with M_{fixed} to the film with M_{free}, M_{free} switches into the parallel orientation. When the current is in the opposite direction M_{free} switches into the antiparallel orientation. The effect was predicted by J. Slonczewski and by L. Berger and observed by Katine *et al.* for the first time. References to these works are given in [12], which also shows an example for "inverse" CIMS where the role of the two current directions is interchanged.

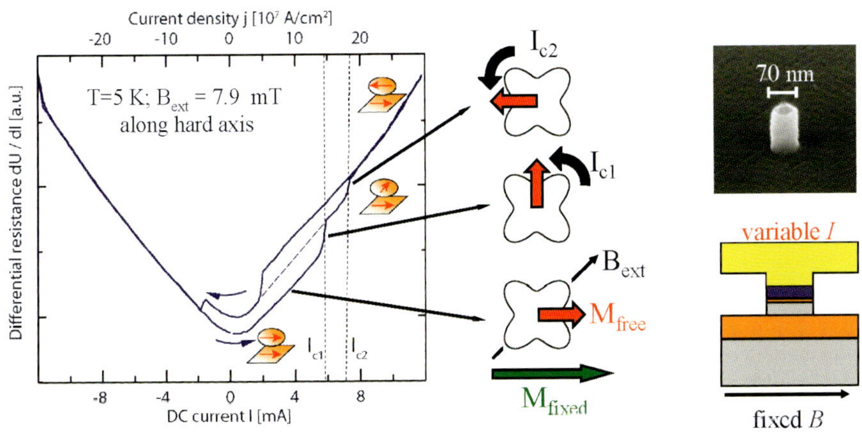

Figure 12. Two-step CIMS in Fe/Ag/Fe. Four energetically nearly identical states give rise to two-step switching [13].

Figure 12 displays an example for CIMS obtained recently in Jülich from epitaxial samples of Fe/Ag/Fe with fourfold cubic inplane anisotropy. The curves with the overall U shape display the resistivity of the sample in CPP geometry. The strong increase on both sides is due to the heating of sample by the strong current. The various steps in the curves are due to switching between the various configurations as indicated and the associated change in the CPP-GMR effect. The shamrock patterns in the middle part display the cubic anisotropy. Magnetizations always switch between easy axis indicated by minima in the polar plot.

APPLICATIONS

We now come to applications of the effects discussed. One can safely say that GMR has received much attention due to its impact on magnetic storage, in particular in hard disk drives (HDDs). Since the introduction of GMR-type sensors as reading elements, in around 1997, storage capacities have increased approximately 100 times. The introduction of GMR sensors into HDDs was particularly fast because magnetoresistive sensors based on the Anisotropic Magnetoresistivity Effect (AMR) had paved the way. Thanks to the fact that GMR is at least partly an interface effect (note that in Figure 10, the scattering events have been intentionally placed at the interfaces), one can make GMR sensors thinner than those based on AMR. A thinner layer is subject to less demagnetizing effects than a thicker one when one also wants to shrink the lateral dimensions in order to read a narrower track. In other fields where a small size is not as critical, AMR and GMR might still coexist for some time. It is expected that GMR will save costs in the long run, alone from the trivial fact that one can have more sensors on a wafer during production.

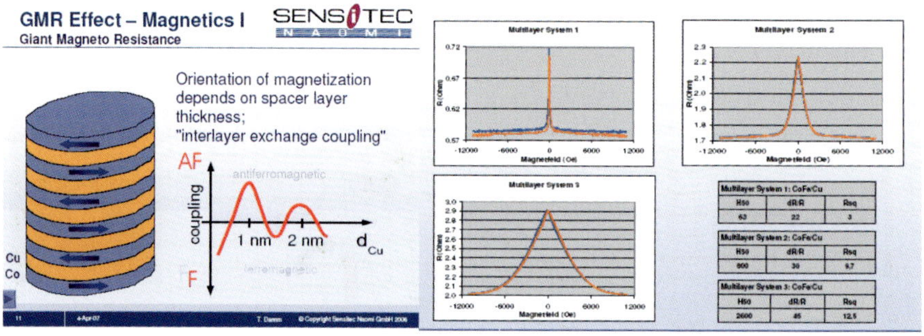

Figure 13. Working principle and data for GMR sensor with AF-coupled multilayer (courtesy of NAOMI- Sensitech, Germany).

There are two basic concepts for GMR-type sensors, which are connected with the two ways to obtain antiparallel alignment discussed in Figure 9 and Figure 10 b. For the sensor material in Figure 13, a Co/Cu multilayer was used, where the Cu thickness was chosen to provide antiferromagnetic coupling. Its strength and the associated properties can be chosen from the empirical coupling curve, which is also displayed. Three different panels and a table are displayed which present these properties, in particular saturation field and sensitivity. This type of sensor has the advantage of a large signal because multilayers can be used, but the disadvantage that it is unipolar (positive and negative fields yield the same signal as also displayed by the curves). They cannot be used for magnetic recording where the polarity of the field contains the essential information. However, there are many other applications in position and motion sensing of objects which are magnetic or marked by attaching a little permanent magnet to them. These range from tooth detection of rotating gear wheels to the detection of cars or even airplanes for traffic control. In future car parks, each lot will be equipped with a sensor and will give information regarding free lots to the customer via a panel at the entrance.

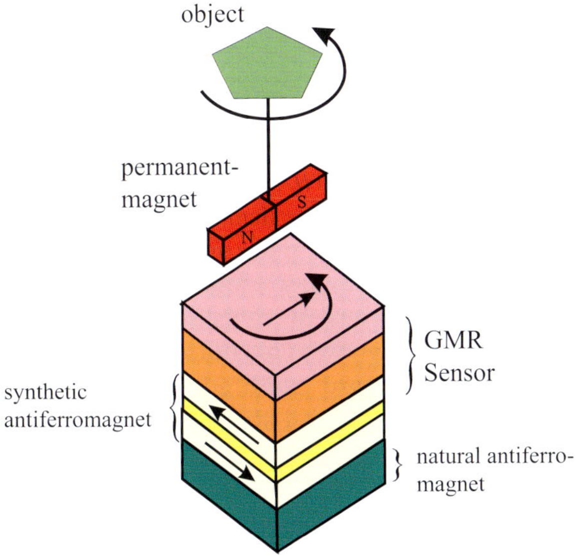

Figure 14. Working principle of a GMR- spinvalve-type sensor, used here for measurement of rotational angles.

The other type of GMR sensor also available on the market is of the spin-valve type (see above). It consists of a fixed layer (see Figure 14) and a free layer which can follow the applied field. There are various ways to stabilize the fixed layer, like the use of "synthetic antiferromagnets" with compensated overall magnetization such that they do not react to the applied field. They can be stabilized further by using natural antiferromagnets and the well-known effect of "exchange anisotropy". The disadvantage here is that multilayers yield no improvement. GMR signals are hardly more than 10 %. However the real advantage is that they are bipolar and thus can be used in magnetic recording and in electronic compasses.

Applications of antiferromagnetic coupling have already been indicated (synthetic antiferromagnets). This would be for use in sensors. However, magnetic storage media of HDDs have also benefited from the invention of AFC (antiferromagnetically coupled) media with improved stability of stored information.

REFERENCES

[1] **P. Grünberg and F. Metawe,** Phys. Rev. Lett. 39, 1561 (1977), *"Light Scattering from Bulk and Surface Spinwaves in EuO".*

[2] **P. Grünberg,** Appl. Phys. (1989) Vol. 66: "Light Scattering in Solids V", ed. by M. Cardona and G. Güntherodt, *"Light Scattering from Spinwaves in Layered Magnetic Structures" (review).*

[3] **J. Barnas, P. Grünberg,** J. Magn. Magn. Mat. 82, 186 (1989), *"Spin waves in exchange coupled epitaxial double layers"*

[4] **J. F. Cochran and J. R. Dutcher,** J. Appl. Phys. 64(1988)6092, *"Calculation of Brillouin light scattering intensities from pairs of exchange coupled films".*

[5] **P. Grünberg, R. Schreiber, Y. Pang, M. B. Brodsky, H. Sowers,** Phys. Rev. Lett. 57, 2442 (1986), *"Layered Magnetic Structures: Evidence for antiferromagnetic coupling of Fe-layers across Cr-interlayer.*

[6] **J. Unguris, R. J. Celotta, and D. T. Pierce,** Phys. Rev. Lett. 67, 140 (1991), *Observation of two different oscillation periods in the exchange coupling of Fe/Cr/Fe(100).*

[7] **D. E. Bürgler, P. Grünberg, S. O. Demokritov, M. T. Johnson,** Handbook of Magnetic Materials 13, Elsevier, ed. by K.H.J. Buschow, *"Interlayer Exchange Coupling in Layered Magnetic Structures"(review).*

[8] **G. Binasch, P. Grünberg, F. Saurenbach, W. Zinn,** Phys. Rev. B39, 4828 (1989), *"Enhanced magnetoresistance in layered magnetic structures with antiferromagnetic interlayer exchange".*

[9] **A. Fert, P. Grünberg, A. Barthelemy, F. Petroff, W. Zinn,** JMMM 140-144, 1 (1995), *"Layered magnetic structures: interlayer exchange coupling and giant magnetoresistance".*

[10] **R. E. Camley and J. Barnas,** Phys.Rev. Lett.63, 664, (1989), *"Theory of Giant Magneto-resistance Effects in Magnetic Layered Structures with Antiferromagnetic Coupling".*

[11] **J. Barnas, A. Fuß, R. E. Camley, P. Grünberg, W. Zinn,** Phys. Rev. B42, 8110 (1990), *"Novel magnetoresistance effect in layered magnetic structures: theory and experiment".*

[12] **P. Grünberg, D. E. Bürgler, H. Dassow, A. D. Rata, C. M. Schneider,** Acta Materialia 55 (2007) 1171–1182, *"Spin-transfer phenomena in layered magnetic structures:physical phenomena and materials aspects".*

[13] **R. Lehndorff, M. Buchmeier, D. E. Bürgler, A. Kakay, R. Hertel, C. M. Schneider,** *"Asymmetric spintransfer torque in single-crystalline Fe/Ag/Fe nanopillars"* Phys. Rev. B76, 214420 (2007).

Portrait photo of Peter A. Grünberg by photographer Ulla Montan.

Physics 2008

Yoichiro Nambu

"for the discovery of the mechanism of spontaneous broken symmetry in subatomic physics"

and the other half jointly to

Makoto Kobayashi and Toshihide Maskawa

"for the discovery of the origin of the broken symmetry which predicts the existence of at least three families of quarks in nature"

THE NOBEL PRIZE IN PHYSICS

Speech by Professor Lars Brink of the Royal Swedish Academy of Sciences.
Translation of the Swedish text.

Your Majesties, Your Royal Highnesses, Ladies and Gentlemen,

"The Earth is round." This simple sentence contains so much. It shows how we human beings see the objects around us in symmetric shapes. We have done so for as long as we can remember. Long ago the ancient Greeks classified geometric objects and introduced the concepts we use today. The sentence also shows the importance of symmetries in determining the laws of physics. These laws do not allow the earth to be flat or square. They have a built-in symmetry. But the earth is not exactly round. Its radius at the equator is slightly bigger than at its poles. There are mountains and valleys. So physicists would say that symmetry is "weakly broken". There are parts of the laws that determine the shape of the earth that break symmetry.

When we look in a mirror we accept the image we see as a faithful image of ourselves. But is it? Are we symmetrical if we draw a line from our head downward, dividing the body into a left and right part? No, not completely. Indeed, it is often said that this slight lack of symmetry makes a face look more interesting. Pablo Picasso experimented with this in his paintings of Dora Maar, where the two halves of her face represented different moods. But a real human face is more mirror symmetrical than that.

We often find situations where symmetry must be broken. If we find ourselves at the top of a completely symmetric mountain, almost like the Matterhorn, we are in an unstable situation. If we do not hold onto something, we will fall in one direction and slide down to the foot of the mountain. We do not know beforehand in which direction we will slide. Symmetry, which says that all directions down the mountain are equally probable, has been broken spontaneously. The laws are still symmetrical, but our position somewhere at the foot of the mountain, the "ground state" as physicists say, breaks this symmetry. We do not see the symmetry, but it is there.

How is this reflected in the fundamental laws of nature, the ones that govern physics at the smallest distances? When elementary particles and the forces that act between them were studied in experiments from the 1950s onward, it was found that everything that the underlying symmetries allow can happen. Symmetries must therefore impose enormously strong restrictions for the laws to be meaningful. How can this lead to our universe, with its four kinds of fundamental forces and great variety of elementary particles? The decisive idea came from Yoichiro Nambu, who in 1960 showed that the fundamental laws of nature can exhibit spontaneous broken symmetries, as illustrated by the person falling off the Matterhorn. There can be strong restrictions from large symmetries that are not seen directly in physical experiments, at least at the energies that can be produced in today's large particle accelerators. The symmetries are there as a consequence of the fundamental laws of physics, but for a system in its ground state there are no direct signs of the restrictions that symmetry should impose. This means, for instance, that the electromagnetic force and the weak nuclear force do not appear related, even though the underlying theory says so. This is just one example of Nambu's ideas that permeate all of modern physics and have been tested in numerous experiments.

Mirror symmetry is important for fundamental laws, too. It caused a major sensation when Lee and Yang pointed out in 1956 that this is not necessarily true in radioactive decays, a fact that was very quickly confirmed. Nevertheless, it was thought that mirror symmetry would indeed apply to all particle reactions if it were combined with the operation of changing a particle to its antiparticle. However, in 1964 Cronin and Fitch discovered, very surprisingly, that this combined symmetry could also be very weakly broken in certain special processes. During the late 1960s and 1970s the Standard Model that is now used to describe elementary particle physics was developed, partly based on Nambu's ideas. However, it could not explain this weak symmetry breaking, so the question arose of how one could extend the model without destroying the results that had so successfully been compared to experiments. In 1972 Makoto Kobayashi and Toshihide Maskawa examined the possibility of introducing more fundamental particles – quarks – and found that if there are six different quarks, the theory can indeed break the symmetry. It was a daring assumption, since at that time the everyday world seemingly needed only two different kinds of quarks, whereas three had already been discovered. Kobayashi and Maskawa suggested that there should be three more. These were eventually discovered, the last one in 1994. During the past decade, elementary particle physicists have measured Kobayashi and Maskawa's theory with great precision and found that it really does fit the data. Nature is endowed with six different quarks, at least at the energies that have so far been accessible. This also leads to an imbalance between matter and antimatter, a fact that Andrei Sacharov pointed out early on. Thanks to this imbalance, we are here today. Symmetry breaking allows for a world of matter.

小林先生、益川先生、

あなた方お二人、および南部先生は、 素粒子物理学理論に取って不可欠である対称性の破れに関する画期的な業績によって２００８年度のノーベル物理学賞を受賞されました。ここにスゥェーデン王立科学アカデミーを代表して心からお喜び申し上げます。それでは、国王陛下よりノーベル賞の授与がありますので、前にお進みください。

'Professor Kobayashi, Professor Maskawa,

Together with Professor Nambu you have been awarded the 2008 Nobel Prize in Physics for your seminal works on broken symmetries, which have been instrumental for the modern theories of elementary particles. It is an honour for me to convey the warmest congratulations of the Royal Swedish Academy of Sciences. I now ask you to step forward to receive your Nobel Prizes from the hands of His Majesty the King.'

Yoichiro Nambu

YOICHIRO NAMBU

I was born in 1921 in Tokyo and grew up in the provincial city of Fukui. I studied physics at the Imperial University of Tokyo from 1940 to 1942, graduating at the level of M.S. Then I was drafted into an army radar laboratory. After the end of the war, in 1946, I returned to the University of Tokyo as a kind of research associate. I received a doctorate in 1952. In 1950 I became professor at a newly created Osaka City University, a position I held until 1956. But from 1952 to 1954 I stayed at the Institute for Advanced Study in Princeton, USA, as a member, and from 1954 to 1956 at the University of Chicago as a research associate. I was made associate professor at Chicago in 1956, professor in 1958, and Distinguished Service Professor in 1971. From 1973 to 1976 I served as chairman of the department of physics. In 1976 I became Henry Judson Distinguished Professor, from which position I retired in 1991 and became Emeritus.

I married Chieko Hida in 1945 and have a son, Jun-ichi. I have been a citizen of the United States of America since 1970. I hold honorary degrees from Osaka City University (1980), Northwestern University (1985), and Osaka University (1997). I have been a member of the United States National Academy of Sciences and the American Academy of Arts and Sciences since 1971, and an honorary member of the Japan Academy since 1984.

A partial list of the prizes I received in the past is:
- Dannie Heineman Prize, American Physical Society (1970)
- Order of Culture, Government of Japan (1978)
- United States National Medal of Science (1982)
- Max Planck Medal, German Physical Society (1985)
- Dirac Medal, International Centre for Theoretical Physics, Trieste (1986)
- Sakurai Prize, American Physical Society (1994)
- Wolf Prize, Government of Israel (1995)
- Gian Carlo Wick Medal, World Federation of Scientists, Lausanne (1996)
- Bogoliubov Prize, Joint Institute for Nuclear Research, Dubna (2003)
- Benjamin Franklin Medal, Franklin Inst., Philadelphia (2005)
- Pomeranchuk Prize, Inst. Theoret. and Exper. Physics, Moscow (2007)

My interests in physics have been mainly on the theoretical side. The University of Tokyo was good in condensed matter physics, but I was more attracted to nuclear and particle physics where names like Nishina, Tomonaga, and Yukawa in other institutions were making great contributions. As a student I was exposed to cosmic ray and particle physics by attending seminars held by Nishina and Tomonaga at their laboratory nearby. I started my research career at the time when Tomonaga was developing his theory of renormalization, for which he would receive the Nobel Prize. I was able to approach his group and start working on his theory and other topics in particle physics. At his recommendation I obtained the position at Osaka City University and later was invited to the Institute for Advanced Study. I owe my move to Chicago to M. L. Goldberger.

My work on spontaneous symmetry breaking (SSB), for which I am receiving the Nobel Prize, started around 1959. It is a result of my experience in both condensed matter and particle physics. The BCS theory of superconductivity in 1957 led me to the idea of SSB as a general phenomenon in physics. My work on its specific application to particle physics as a mechanism for generation of the nucleon mass and the pion was first published in 1960. Since then I have pursued the subject in various areas.

SPONTANEOUS SYMMETRY BREAKING IN PARTICLE PHYSICS: A CASE OF CROSS FERTILIZATION

Nobel Lecture, December 8, 2008

by

YOICHIRO NAMBU*

Department of Physics, Enrico Fermi Institute, University of Chicago, 5720 Ellis Avenue, Chicago, USA.

I will begin with a short story about my background. I studied physics at the University of Tokyo. I was attracted to particle physics because of three famous names, Nishina, Tomonaga and Yukawa, who were the founders of particle physics in Japan. But these people were at different institutions than mine. On the other hand, condensed matter physics was pretty good at Tokyo. I got into particle physics only when I came back to Tokyo after the war. In hindsight, though, I must say that my early exposure to condensed matter physics has been quite beneficial to me

Particle physics is an outgrowth of nuclear physics, which began in the early 1930s with the discovery of the neutron by Chadwick, the invention of the cyclotron by Lawrence, and the 'invention' of meson theory by Yukawa [1]. The appearance of an ever increasing array of new particles in the subsequent decades and advances in quantum field theory gradually led to our understanding of the basic laws of nature, culminating in the present standard model.

When we faced those new particles, our first attempts were to make sense out of them by finding some regularities in their properties. Researchers invoked the symmetry principle to classify them. A symmetry in physics leads to a conservation law. Some conservation laws are exact, like energy and electric charge, but these attempts were based on approximate similarities of masses and interactions.

Nevertheless, seeing similarities is a natural and very useful trait of the human mind, The near equality of proton and neutron masses and their interactions led to the concept of isospin SU(2) symmetry [2]. On the other hand, one could also go in the opposite direction, and elevate a symmetry to a more elaborate gauged symmetry. Then symmetry will determine dynamics as well, a most attractive possibility. Thus the beautiful properties of electro-

* The Nobel Lecture of Yoichiro Nambu was presented by Giovanni Jona-Lasinio, La Sapienza, University of Rome, Italy.

magnetism were extended to the SU(2) non-Abelian gauge field [3]. But strong interactions are short range. Giving a mass to a gauge field destroys gauge invariance.

Spontaneous symmetry breaking (SSB), which is the main subject of my talk, is a phenomenon where a symmetry in the basic laws of physics appears to be broken. In fact, it is a very familiar one in our daily life, although the name SSB is not [4]. For example, consider an elastic straight rod standing vertically. It has a rotational symmetry; it looks the same from any horizontal direction. But if one applies increasing pressure to squeeze it, it will bend in some direction, and the symmetry is lost. The bending can occur in principle in any direction since all directions are equivalent. But you do not see it unless you repeat the experiment many times. This is SSB.

SSB in quantum mechanics occurs typically in a uniform medium consisting of a large number of elements. It is a dynamical effect. A symmetry allows some freedom of action to each of them but the interaction among them forces them, figuratively speaking, to line up like a crowd of people looking into the same direction. Then it is not easy to change the direction wholesale even if it is allowed by the symmetry hence does not take energy, because the action is not a local operator. So the symmetry appears to be lost. It is still possible to recover the lost symmetry by a global operation, but it would amount to a kind of phase transition. Some of the examples are:

physical system	broken symmetry
ferromagnets	rotational invariance (with respect to spin)
crystals	translational and rotational invariance (modulo discrete values)
superconductors	local gauge invariance (particle number)

SSB in a medium then has the following characteristic properties:

1. The ground state has huge degeneracy. A symmetry operation takes one ground state to another.
2. Only one of the ground states and a complete set of excited states built on it are realized in a give situation.
3. SSB is in general lost at sufficiently high temperatures.

In relativistic quantum field theory, this phenomenon becomes also possible for the entire space-time, for the "vacuum" is not void, but has many intrinsic degrees of freedom. In this context, it may play an important role in cosmology. As the universe expands and cools down, it may undergo one or more SSB phase transitions from states of higher symmetries to lower ones, which change the governing laws of physics.

I will now recall the chain of events which led me to the idea of SSB and its application to particle physics. One day in 1956 R. Schrieffer gave us a seminar on what would come to be called the BCS theory [5] of superconductivity. I was impressed by the boldness of their ansatz for the state vector, but at the same time I became worried about the fact that it did not appear to respect gauge invariance. Soon thereafter Bogoliubov [6] and Valatin [7] independently introduced the concept of quasiparticles as fermionic excitations in the BCS medium. The quasiparticles did not carry a definite charge as they were a superposition of electron and hole, with their proportion depending on the momentum. How can one then trust the BCS theory for discussing the electromagnetic properties like the Meissner effect? It actually took two years for me to resolve the problem to my satisfaction. There were a number of people who also addressed the same problem, but I wanted to understand it in my own way. Essentially it is the presence of a massless collective mode, now known by the generic name of Nambu-Goldstone (NG) boson, that saves charge conservation or gauge invariance.

The Bogoliubov-Valatin (BV) quasiparticles are described by the equations [8]

(1)
$$\begin{aligned} E\psi_{p,+} &= \varepsilon_p \psi_{p,+} + \Delta\psi^{\dagger}_{-p,-}, \\ E\psi^{\dagger}_{-p,-} &= -\varepsilon_p \psi^{\dagger}_{-p,-} + \Delta\psi_{p,+}, \\ E &= \sqrt{\varepsilon_p^2 + \Delta^2} \end{aligned}$$

Here, $\psi_{p,+}$ and $\psi^{\dagger}_{-p,-}$ are the wave functions for an electron and a hole of momentum p and spin + or –, ε_p is the kinetic energy relative to the Fermi energy, and 2Δ is the energy gap. In terms of spinlike matrices τ_i the corresponding Hamiltonian and the charge-current are

(2)
$$\begin{aligned} H_0 &= \varepsilon_p \Psi^{\dagger}\tau_3\Psi + \Delta\Psi^{\dagger}\tau_1\Psi, \\ \rho_0 &= \Psi^{\dagger}\tau_3\Psi, \quad j_0 = \Psi^{\dagger}(p/m)\Psi \end{aligned}$$

The BV ground state is $\psi^{\dagger}_p|0\rangle = 0$ for all p. The charge does not commute with H_0, and the continuity equation does not hold, which is the problem. But it has turned out that the same interaction that led to the BCS-BV ground state also leads to collective excitations f, which contributes to the charge-current and restores the continuity equation. The correct expression is

(3)
$$\begin{aligned} \rho &\simeq \rho_0 + \frac{1}{\alpha^2}\partial_t f, \\ j &\simeq j_0 - \nabla f, \\ \left(\nabla^2 - \frac{1}{\alpha^2}\partial_t^2\right) f &\simeq -2\Delta\Psi^{\dagger}\tau_2\Psi \end{aligned}$$

where f represents the NG mode. Physically it corresponds to excitations that tend to restore the lost symmetry. Its energy goes to zero in the long wavelength limit as it corresponds to the global symmetry operation. It also happens that the above NG mode mixes with the Coulomb interaction among the electrons because of their common long-range nature, and turns into the well-known plasmons with

$$\omega^2 = en^2/m \tag{4}$$

where e, n, m, are respectively the charge, density, and mass of the electron. The formal similarity of the BV equation to the Dirac equation naturally led me to transport the BCS theory to particle physics [9]. The gap Δ goes over to the mass M, which breaks chirality ~ γ_5 rather than the ordinary charge ~ 1: The axial current is the analog of the electromagnetic vector current in the BCS theory. If chirality is a broken symmetry, the matrix elements of the axial current between nucleon states of four-momentum p and p' should have the form

$$\Gamma_{\mu 5}(p',p) = \left(\gamma_\mu \gamma_5 - 2M\lambda_5\, q_\mu / q^2\right) F\left(q^2\right),\ q_\mu = p'_\mu - p_\mu \tag{5}$$

So chiral symmetry is compatible with a finite nucleon mass M provided that there exists a massless pseudoscalar NG boson. In reality, there is a pseudoscalar pion, and the vector and axial vector interactions that appear in weak decays of the nucleons and the pion had the properties

$$g_V \simeq g_A \qquad g_\pi \simeq \sqrt{2}M\, g_A / G \tag{6}$$

where g_V, and g_A are vector and axial vector couplings of the nucleon, g_π is axial coupling for the pion, G is the pion-nucleon coupling, and M is the nucleon mass. The second relation was known as the Goldberger-Treiman relation [10], and it implies that the matrix element of the axial vector part of nucleon decay is

$$\Gamma_{\mu A} \simeq \gamma_\mu \gamma_5 - 2M\gamma_5\, q_\mu / \left(q^2 - m_\pi^2\right), \tag{7}$$

which differs from Eq. (5) by the presence of pion mass. In view of the smallness of m_π compared to M, I made the hypothesis that the axial current is an approximately conserved quantity, the nucleon mass is generated by an SSB of chirality, and the pion is the corresponding NG boson which should become massless in the limit of exact conservation (Proton and neutron masses should also become equal.)

The model system [11] I worked out subsequently with Jona-Lasinio is a concrete realization of the proposed SSB. It has the form similar to the BCS model

(8) $$L = -\bar{\psi}\gamma^{\mu}\partial_{\mu}\psi + g\left[(\bar{\psi}\psi)^2 - (\bar{\psi}\gamma_5\psi)^2\right]$$

which is invariant against the particle number and chiral transformations

(9) $$\begin{aligned} \psi &\to \exp(i\alpha)\psi, \quad \bar{\psi} \to \bar{\psi}\exp(-i\alpha) \\ \psi &\to \exp(i\gamma_5\alpha)\psi, \quad \bar{\psi} \to \bar{\psi}\exp(i\gamma_5\alpha) \end{aligned}$$

After SSB, the "nucleon" acquires a mass $M \sim 2g\langle\bar{\psi}\psi\rangle$. Although the model is non-renormalizable, it is easy to demonstrate the SSB mechanism. The generated mass M is determined by the "gap equation"

(10) $$\frac{2\pi^2}{g\Lambda^2} = 1 - \frac{M^2}{\Lambda^2}\ln\left(1 + \frac{\Lambda^2}{M^2}\right)$$

where Λ is a cutoff. Bound states of nucleon-antinucleon (meson) and nucleon-nucleon (dibaryon) pairs of spin 0 and 1 were also found. In particular, the masses of $0^-, (\sim\bar{\psi}\gamma_5\psi)$ and $0^+, (\sim\bar{\psi}\psi)$ mesons are found to be 0 and $2M$ respectively. A more realistic two-flavor model was also considered by generalizing Eq. (8) to

(11) $$L = -\bar{\psi}\gamma^{\mu}\partial_{\mu}\psi + g\left[(\bar{\psi}\psi)^2 - \sum_i(\bar{\psi}\gamma_5\tau_i\psi)(\bar{\psi}\gamma_5\tau_i\psi)\right]$$

with a similar gap equation. We get an isovector 0^- pion and an isoscalar 0^+. The actual pion mass was generated by a small explicit bare mass in the Lagrangian of the order of 5 MeV. This also induced a change of axial coupling constant g_A in the right direction.

Other examples of the BCS-type SSB are ^{3}He superfluidity and nucleon pairing in nuclei [12]. In general there exist simple mass relations among the fermion and the bosons in BCS-type theories [13]. The BCS theory also accounts for the generation of the London mass for the electromagnetic field. This problem is made simple in terms of the Higgs scalar field [14]. The relativistic analog of the London relation is, in momentum space,

(12) $$\begin{aligned} j_{\mu}(q) &= K_{\mu\nu}(q)\Lambda^{\nu}, \\ K_{\mu\nu} &= \left(\delta_{\mu\nu} - q_{\mu}q_{\nu}/q^2\right)K(q^2), \\ K(q^2) &\simeq q^2/(q^2 - m^2) \end{aligned}$$

The third relation shows the massless NG boson turning into a massive "plasmon", a process corresponding to Eq. (4). This was successfully applied to weak gauge field in the Weinberg-Salam (WS) theory [15] of electroweak unification. The fermion masses are also generated and break chiral invariance. These so-called current masses for the up and down quarks play the role of the bare mass in the NJL model.

In the current standard model of particle physics, the NJL model may be regarded as an elective theory for the QCD with respect to generation of the so-called constituent masses. One is interested in the low energy degrees of freedom on a scale smaller than some cut-off Λ~ 1 Gev. The short distance dynamics above Λ as well as the confinement may be treated as a perturbation. The problem has been extensively studied by many people. The Lagrangian adopted by Hatsuda and Kunihiro [16] is

$$L = L_{QCD} + L_{NJL} + L_{KMT} + \delta L \tag{13}$$

L_{NJL} is for the quarks, and contains "current mass" terms. L_{KMT} refers to the Kobayashi-Maskawa-t'Hooft chiral anomaly

$$L_{KMT} = g_D \left(\det\left[\bar{q}_i\left(1-\gamma_5\right)q_j\right] + h.c.\right) \tag{14}$$

Both of them contributes to the explicit breaking of chiral invariance

(δL contains the effects of confinement and one gluon exchange.) The WS theory resembles the Ginzburg-Landau [17] description of superconductivity, which was shown to follow from the BCS theory by Gor'kov [18]. In the same way the NJL model goes over to the model of Gell-Mann and Levy [19]. If this analogy turns out real, the Higgs field might be an effective description of underlying dynamics.

Finally I will end this lecture with a comment on the mass hierarchy problem. Hierarchical structure is an outstanding feature of the universe. The masses of known fundamental fermions also make a hierarchy stretching 11 orders of magnitude. Mass is not quantized in a simple regular manner like charge and spin. Mass is a dynamical quantity since it receives contributions from interactions. But we do not see yet a pattern like those in the hydrogen atom, which led to quantum mechanics, or the Regge trajectories, which led to the dual string picture.

The BCS mechanism seems relevant to the problem, as was remarked earlier. It generates a mass gap for fermions, plus the Goldstone and Higgs modes as low-lying bosons. The bosons may act in turn as an agent for further SSB, leading to the possibility of hierarchical SSB or "tumbling" [20]. In fact we already have examples of it [21]:

1. The chain atoms – crystal – phonon – superconductivity. The NG mode for crystal formation is the phonon, which induces the Cooper pairing of electrons to cause superconductivity
2. The chain QCD – chiral SSB of quarks and baryons – π, σ and other mesons – nuclei formation and nucleon pairing – nuclear collective modes. No further elaboration would be required

I am greatly thankful to G. Jona-Lasinio for his help in the planning of this lecture.

REFERENCES

[1] See, for example, Y. Nambu, *J. Phys. Soc. Japan*, 76 (2007), 111002.
[2] W. Heisenberg, *Z. Phys.* 77 (1932) 1.
[3] C. N. Yang and R. L. Mills, *Phys. Rev.* 96 (1954), 191.
[4] The name is due to M. Baker and S. Glashow, *Phys. Rev.* 128 (1962) 2462.
[5] J. Bardeen, L. N. Cooper, and J. R. Schrieffer, *Phys. Rev.* 108 (1957), 1175.
[6] N. N. Bogoliubov, *J. Exptl. Theoret. Phys.* (U.S.S.R.) 34 (1958), 58, 33. This is a fermionic version of transformation, which he first introduced in a description of superfluidity. *J. Phys.* (1947) Vol. 11, No.1, 23–32.
[7] G. Valatin, *Nuovo cimento* 7 (1958), 843.
[8] Y. Nambu, *Phys. Rev.* 117 (1960), 648.
[9] Y. Nambu, *Phys. Rev. Lett.* 4 (1960), 380.
[10] M. L. Goldberger and S. Treiman, *Phys. Rev.* 110 (1958), 117.
[11] Y. Nambu and G. Jona-Lasinio, *Phys. Rev.* 122 (1961), 345; *Phys. Rev.* 124 (1961), 246. The first presentation of the model was made at Midwest Conference on Theoretical Physics: "A superconductor model of elementary particles and its consequences", in the *Proceedings*, eds. F. J. Biedenhahn *et al.*, Purdue University, 1960, p. 1.
[12] A. Arima and F. Iachello, *Phys. Rev. Lett.* 35 (1975), 1069; *Ann. Phys.* 99 (1976), 253.
[13] E.g., Y. Nambu, *Physica* 15D (1985), 147; *Phys. Lett.* 201 (1988), 1.
[14] P. W. Anderson, *Phys. Rev.* 130 (1963), 439; F. Englert and R. Brout, *Phys Rev. Lett.* 13 (1964), 321; P. W. Higgs, *Phys. Rev. Lett.* 13 (1964), 508.
[15] S. Weinberg, *Phys. Rev. Lett*, 19 (1967), 1264; A. Salam, in *Elementary Particles*, Almqvist Förlag AB (1968), 167.
[16] T. Hatsuda, T. Kunihiro, *Phys. Rep.* 247, 221 (1994).
[17] V. I. Ginzburg and L. D. Landau, *J. Exptl. Theoret. Phys.* (U.S.S.R.) 20 (1950), 1064.
[18] L. P. Gor'kov, *J. Exptl. Theoret. Phys.* (U.S.S.R.) 36 (1959), 1918.
[19] M. Gell-Mann and M. Lévy, *Nuovo cimento* 16 (1960), 705.
[20] S. Dimopoulos, S. Raby, and L. Susskind, *Nucl. Phys.* B169 (1980), 373.
[21] Y. Nambu, "Masses as a problem and as a clue", presented at the International Centre for Theoretical Physics, Trieste, (2004).

Portrait photo of Yoichiro Nambu by courtesy of University of Chicago.

小林 誠

MAKOTO KOBAYASHI

I was born in Nagoya, Japan on April 7, 1944. As it was in the middle of the Second World War, I was evacuated to Kawagoe Village in Mie Prefecture the following year to escape the aerial bombardment over Nagoya. Soon after the war ended, my father passed away. As I was only two years old at the time, I have no memory of him. My father, Hisashi, was a physician. At the end of his career, he served as the director of the central public health centre in Nagoya.

After my father's death, we moved back to Nagoya to find our house rendered to ashes by the bombing. So we stayed at my mother's family house. My mother, Ai, came from a family surnamed Kaifu. At that time, my mother's parents and their elder son's family were living in the house. The latter's elder son, Toshiki Kaifu, who is my cousin, later became Japan's prime minister. I don't have any siblings. I have another cousin named Norio Kaifu, who is an astronomer. He served as the director of the National Astronomical Observatory of Japan.

In 1975, I married Sachiko Enomoto. My son, Junichiro, was born in 1977. Sachiko died of cancer at age 39. Junichiro majored and obtained a master's degree in urban planning. He currently works at a private consultant firm. In 1990, I remarried. My wife Emiko Nakayama's father, Tadasi, was a mathematician known for his research on Frobenius algebras. As he had passed away when Emiko was a child, I never had a chance to meet him. Emiko gave birth to our daughter Yuka.

I went to elementary and middle school in regular public schools. Nothing particularly unique happened during those school years. In high school, I played tennis every day. Though I never really got that good at it, I continued to enjoy playing tennis in my adult years. At around that time, I read *The Evolution of Physics* by Albert Einstein and Leopold Infeld. I don't remember to what extent I actually understood it, but the book sparked my interest in physics.

I entered the Physics Department of Nagoya University. Besides it being my local university, what also attracted me to Nagoya University was the presence of Shoichi Sakata on its faculty. The model of elementary particles bearing his name was famous, so even high school students knew about it.

As a graduate student, I began my research in particle physics as a member of Prof. Sakata's lab. A free atmosphere abounded in the lab; treated impartially, the graduate students were allowed to participate in discussions among the researchers. I believe that I learned a lot from that experience.

Regrettably, Prof. Sakata passed away while I was still a graduate student. During my graduate student days, I engaged in many discussions with Prof. Yoshio Ohnuki. I also met Toshihide Maskawa at Nagoya University. As I recall, our first contact was when I was an undergraduate student: As a graduate student he was helping us with our studies. I began conducting joint research with Maskawa after entering graduate school. Our research theme at that time was on chiral symmetry. We were trying to approach the subject from a quark-model perspective.

In March 1972, I received my doctoral degree in physics from Nagoya University. At that time, it was not easy for post-doctoral researchers to find positions. Fortunately, however, I was hired as a research associate in the Physics Department of Kyoto University. I moved to Kyoto in April. I resumed joint research with Maskawa, who had transferred to Kyoto University a little before me. We worked on CP violation, the research for which we later received the Nobel Prize. Engaging each other in discussion, we advanced our research quickly, completing it in a relatively short period of time. By the end of August, we had finished writing our paper.

With the discovery of the J/ψ-particle in 1974, there was, as in other countries, quite a commotion in Japan. Many theories were ventured as to the character of the J/ψ-particle; ultimately, however, it was determined to be a charmonium, which is a bound state of the c quark and its anti-particle. Before that, the hint of a fourth quark was obtained by Kiyoshi Niu in his experiment exposing emulsion chambers to cosmic rays. In this connection, a few Japanese groups, including mine, were investigating a four-quark model but were not able to predict the long lifetime of the charmonium state.

In 1975, the tau lepton was discovered. Since this indicated the existence of third-generation quarks, the six-quark model we proposed began to attract attention.

Although I did not contribute directly to the development of the six-quark model, I did write a paper with Katsuhiko Sato that was somewhat related. We tried to elucidate the limitations on the mass and lifetime of neutrinos in using cosmological arguments when similar flavour mixing exists within a leptonic sector.

During that period, KEK (the National Laboratory for High Energy Physics, now the High Energy Accelerator Research Organisation) had started operating its proton synchrotron accelerator, and discussions were underway on planning the following TRISTAN project. My first relationship with KEK was participation in these discussions. I was then hired as an associate professor in KEK's Theory Division and moved to Tsukuba in 1979. At that time, the Theory Division was headed by Hirotaka Sugawara. Motohiko Yoshimura came to the Division about the same time I did. That year, I received the Nishina Memorial Prize.

Upon arriving at KEK, I became engaged in drafting the proposal for the TRISTAN project. It was originally intended to be an electron-positron-proton collider. However, the one approved for construction in 1981 was an electron-positron collider. Operation started in 1987. With our dream to discover the top quark unfulfilled, TRISTAN was shut down in 1995.

During that period, I spent three months at CERN (the European Organisation for Nuclear Research) from November 1982. While I was there, the W particle was discovered. It was a very exciting experience for me. However, just before my arrival at CERN, J.J. Sakurai died while working there as a visiting researcher. In 1985, I was awarded the J.J. Sakurai Prize, instituted in that year by the American Physical Society. The same year I received the Japan Academy Prize.

In 1989, I was appointed head of KEK's Physics Division II, where I assumed responsibility for an experimental research group. We set about in earnest to prepare plans for a post-TRISTAN accelerator: It would be the B-factory accelerator constructed inside the TRISTAN tunnel and operated with an aim to proving CP violation in a B-meson system. Construction was approved and started in 1994. Experiments using the B-factory started in 1999, and the initial results were obtained in 2000.

In 2003, I was appointed director of KEK's Institute of Particle and Nuclear Studies, in which position I was directly responsible for the Institute's experimental research activities including those conducted using the B-factory accelerator. At the time, Yoji Totsuka was appointed as KEK's director general. Later, I came to work together with Totsuka again at the Japan Society for the Promotion of Science (JSPS), but he regrettably passed away in July 2008.

During my tenure as director of the Institute, KEK was converted from a government organisation to an independent research corporation. I became very busy in carrying out this reorganisation. Fortunately, we are able to steadily improve the performance of the B-factory accelerator during that period. Its experimental results showed our 6-quark theory to be virtually accurate.

I retired from KEK when my tenure as the Institute's director ended in 2006. For a while, I was able to enjoy a life of relative freedom. In the meantime, I was invited to become an IIAS (International Institute for Advanced Studies) Fellow. In that capacity, I travel on occasion to the Kansai district to hold discussions and write papers with Taichiro Kugo, with whom I have conducted joint research in the past.

In October 2007, I became an executive director of JSPS, where I enjoy many opportunities to meet researchers from a wide spectrum of fields.

In recent years, I received the Person of Cultural Merit award in 2001 and the Order of Culture award in 2008, both from the Japanese government, and the High Energy Particle Physics Prize in 2007 from the European Physical Society.

CP VIOLATION AND FLAVOUR MIXING

Nobel Lecture, December 8, 2008

by

MAKOTO KOBAYASHI

KEK, Oho, Tsukuba-shi, 305-0801, and JSPS, Ichibancho, Chiyodaku-ku, Tokyo, 102-8471, Japan.

1. INTRODUCTION

We know that ordinary matter is made of atoms. An atom consists of the atomic nucleus and electrons. The atomic nucleus is made of a number of protons and neutrons. A proton and a neutron are further made of two kinds of quarks, u and d. Therefore the fundamental building blocks of ordinary matter are the electron and two kinds of quarks, u and d.

The standard model, which gives a comprehensive description of current understanding of the elementary particle phenomena, however, tells us that the number of species of quarks is six. The additional quarks are called s, c, b and t. The reason why we do not find them in ordinary matter is that they are unstable in the usual environment. Similarly, the electron belongs to a family of six members called leptons. Three types of neutrinos are included among these six.

Another important ingredient of the standard model is fundamental interactions. Three kinds of interactions act on the quarks and leptons. The strong interaction is described by Quantum Chromodynamics (QCD) and the electro-magnetic and weak interactions by the Weinberg-Salam-Glashow theory in a unified manner. All of them belong to a special type of field theory called gauge theory.

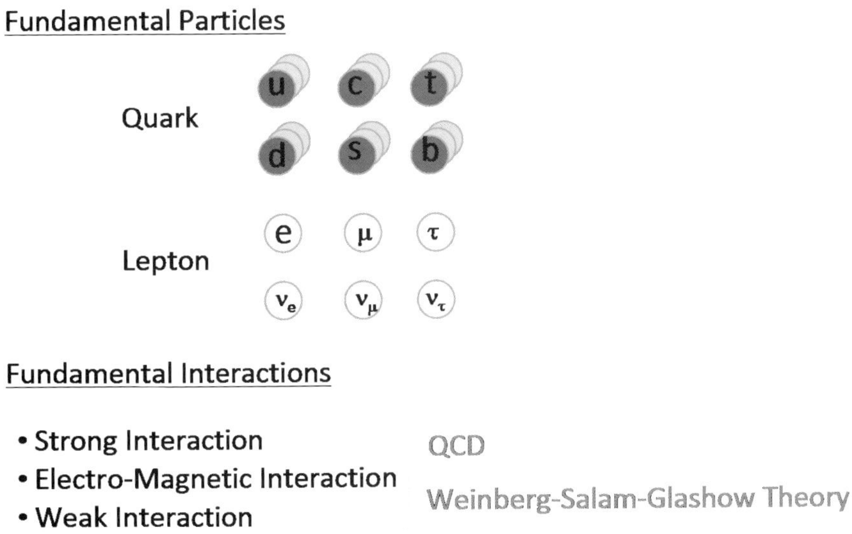

Figure 1. The Standard Model.

The standard model was established in the 1970s. It was triggered by the development of studies of gauge theories. In particular, it was proved that generalised gauge theory is renormalisable [1]. This opened the possibility that all the interactions of an elementary particle can be described by the quantum field theory without the difficulty of divergence. Before this time, such description was possible only for electro-magnetic interaction.

The discovery of the new flavours made in 1970s played an important role in the establishment of the standard model. In particular, the τ-lepton and c- and b-quarks were found in the 1970s. When we proposed the six quark model to explain CP violation with Dr. Toshihide Maskawa in 1973 [2], only three quarks were widely accepted, and a slight hint of the fourth quark was there, but no one thought there would be six quarks.

In the following, I will describe the development of the studies on CP violation and the quark and lepton flavours, putting some emphasis on contributions from Japan. The next section will be devoted to the pioneering works of the Sakata School, from which I learned many things. The work on CP violation will be discussed in Section 3. I will explain what we thought and what we found at that time. Section 4 will describe subsequent development related to our work. Experimental verification of the proposed model has been done by using accelerators called B-factories. A brief outline of those experiments will be given. Finally in Section 5, I will look briefly at flavour mixing in the lepton sector, because this is a phenomenon parallel to the flavour mixing in the quark sector, and Japan has made unique and important contributions in this field.

2. SAKATA SCHOOL

Both Dr. Maskawa and I graduated from and obtained our PhD's from Nagoya University. When I entered the graduate program, the theoretical particle physics group of Nagoya University was known for unique research activity and was led by Professor Shoichi Sakata.

Figure 2. Shoichi Sakata 1911–1970 (Courtesy of Sakata Memorial Archival Library).

In the early 1950s, a number of strange particles were discovered, with the first evidence having been found in the cosmic ray events of 1947. In the current terminology, strange particles contain an s-quark or anti-s-quark as a constituent, while non-strange particles do not. But what we are about to consider is the era before the quark model appeared.

In 1956, Sakata [3] proposed a model which is known as the Sakata model. In this model, all hadrons, strange and non-strange, are supposed to be composite states of the triplet of baryons, the proton(p), the neutron(n), and the lambda(Λ). In other words, three baryons, p, n, and Λ are the fundamental building blocks of the hadrons in the model. Eventually, the Sakata model was replaced by the quark model, where the triplet of quarks, u, d, and s replace p, n, and Λ. But the root of the idea of fundamental triplet is in the Sakata model.

In the following, we focus on the weak interactions in the Sakata model. Usual beta-decays of the atomic nucleus are caused by the transition of a neutron into a proton. Similarly we can consider the transition of a lambda into a proton. In the Sakata model, all the weak interaction of the hadrons can be explained by these two kinds of transitions among the fundamental triplet.

This pattern of the weak interaction is quite similar to the weak interaction of the leptons;

$$\begin{array}{ccc} & p & \\ \swarrow\nearrow & & \nwarrow\searrow \\ n & & \Lambda \end{array} \qquad \Leftrightarrow \qquad \begin{array}{ccc} & \nu & \\ \swarrow\nearrow & & \nwarrow\searrow \\ e & & \mu \end{array}$$

It should be noted that at that time, the neutrino was thought of as a single species. This similarity of the weak interaction between the baryons and the leptons was pointed out by Gamba, Marshak and Okubo [4].

In 1960, Maki, Nakagawa, Ohnuki and Sakata [5] developed the idea of baryon-lepton or B-L symmetry further and proposed the so-called Nagoya model. They considered that the triplet baryons, p, n, and Λ are composite states of a hypothetical object called B-matter and the neutrino, the electron, and the muon, respectively;

$$p = \left\langle B^{+}\nu \right\rangle, \quad n = \left\langle B^{+}e \right\rangle, \quad \Lambda = \left\langle B^{+}\mu \right\rangle$$

where B-matter is denoted as B^+.

Although the composite picture of the Nagoya model did not lead to remarkable progress, some ideas in the Nagoya model developed in an interesting way. In 1962, it was discovered that there exist two kinds of neutrinos, corresponding to the electron and the muon, respectively. When the results of this discovery at BNL [6] were to come out, two interesting papers were published, one written by Maki, Nakagawa and Sakata [7] and the other by Katayama, Matsumoto, Tanaka and Yamada [8]. Both papers discussed the modification of the Nagoya model to accommodate two neutrinos in the model.

In the course of the argument to associate leptons and baryons, Maki *et al.* discussed the masses of neutrinos and derived the relation describing the mixing of the neutrino states;

$$\nu_1 = \cos\theta\, \nu_e + \sin\theta\, \nu_\mu\,,$$
$$\nu_2 = -\sin\theta\, \nu_e + \cos\theta\, \nu_\mu\,,$$

where ν_1 and ν_2 are the mass eigenstates of neutrinos, and they assumed that the proton is the composite state of the B-matter and ν_1. Although the last assumption is not compatible with the current experimental evidence, it is remarkable that they did present the correct formulation of lepton flavour mixing. To recognise their contribution, the lepton flavour mixing matrix is called the MNS matrix today.

Lepton flavour mixing gives rise to the phenomenon called neutrino oscillation. Many years later, neutrino oscillation was discovered in an unexpected manner. We will come back to this point later.

Another important outcome of this argument is the possible existence of the fourth fundamental particle associated with ν_2. This was discussed by Katayama *et al.* in some detail. At the time, the fundamental particles were still considered baryons but the structure of the weak interaction discussed here is the same as that of the Glashow-Illiopoulos-Miani [9] scheme.

These works were revived in 1971, when Niu and his collaborators found new kind of events in emulsion chambers exposed to cosmic rays [10]. One of the events they found is shown in Figure 3. In this event, we see kinks on two tracks, which indicate the decay of new particles produced in pairs. The estimated mass of the new particle was 2~3 GeV and the life was a few times 10^{-14} sec. under some reasonable assumptions.

When this result came to his attention, Shuzo Ogawa, a member of the Sakata group, immediately pointed out that this new particle might be related to the fourth element expected in the extended version of the Nagoya model [11]. By that time, the Sakata model had already been replaced by the quark model, so that what he meant was that those new particles might be charmed particles, in the current terminology. Following this suggestion, several Japanese groups, including mine, began to investigate the four-quark model [12]. At that time, I was a graduate student at Nagoya University.

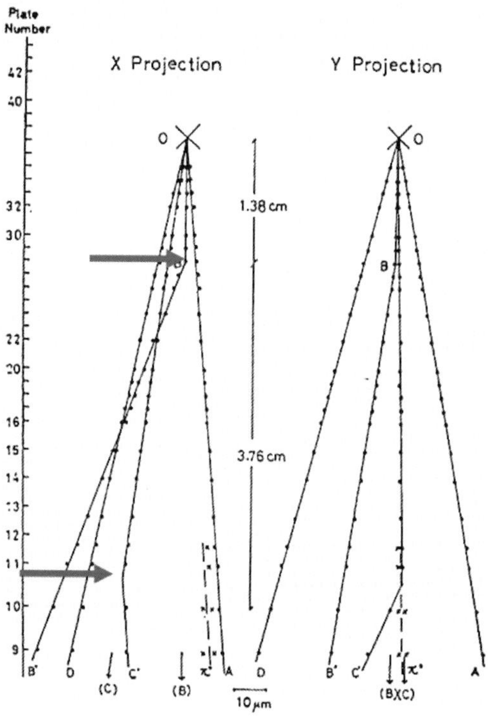

Figure 3. A cosmic ray event [10].

So far I have explained about the unique activities of the Sakata School. I mentioned the four-quark model in some detail. But I do not mean to imply that the six-quark model we proposed is a simple extension of the four-quark model. What was most important for me was the atmosphere of the particle physics group of Nagoya University. Although most of the work we discussed in this section had been done by Sakata and his group before I entered the graduate course, the spirit created by this work was still there. I learned the importance of capturing the entire picture, which is necessary for this kind of work.

3. SIX-QUARK MODEL

In 1971, the renormalisability of the non-Abelian gauge theory was proved [1]. This enabled a description of the weak interactions with the quantum field theory in a consistent manner, and the Weinberg-Salam-Glashow [13] theory began to attract attentions. In 1972, I obtained my PhD from Nagoya University and moved to Kyoto University. Then my work on CP violation started.

CP violation was first found in 1964 by Cronin, Fitch *et al.* [14] in the decay of the neutral K-meson. CP violation means violation of symmetry between particles and anti-particles. The discovery of CP violation implies that there is an essential difference between particles and anti-particles.

We thought that if the gauge theory can describe the interactions of particles consistently, CP violating interaction should also be included in it. It was rather straightforward to solve the problem. We simply investigated conditions for CP violation in the renormalisable gauge theory. What we found then is summarised as follows [2].

At that time only three quarks were widely accepted, but the three-quark model had some flaws in the gauge theory. Therefore, from a theoretical point of view, the four-quark model of the GIM type was considered preferable. However, it is impossible to accommodate CP violation in a model of the GIM type. We found that even if we relax the conditions for the GIM type, we cannot make any realistic model of CP violation with four quarks. This implies that there must be some unknown particles besides the fourth quark. I thought that this was quite strong and an important conclusion of our argument.

Then we considered a few possible mechanisms of CP violation by introducing new particles. We proposed the six-quark model as one such possible mechanism.

Below we will discuss the quark flavour mixing in some detail, in order to understand why four quarks are not enough and six quarks are needed to accommodate CP violation.

In the frame work of the gauge theory, flavour mixing arises from a mismatch between gauge symmetry and particle states. Gauge symmetry lumps a certain number of particles into a group called a multiplet. However, each multiplet member is not necessarily identical to a single species of particles, but sometimes it is a superposition of particles. The flavour mixing is nothing but this superposition. In the present case, the relevant gauge group is SU(2) of the Weinberg-Salam-Glashow theory and the multiplet is a doublet.

Assuming that four quarks consist of two doublets of the SU(2) group, we can denote the most general form of them as

$$\begin{pmatrix} u \\ d' \end{pmatrix}, \begin{pmatrix} c \\ s' \end{pmatrix},$$

where d' and s' are the superposition of real quark states d and s, described in a matrix form as

$$\begin{pmatrix} d' \\ s' \end{pmatrix} = \begin{pmatrix} V_{ud} & V_{us} \\ V_{cd} & V_{cs} \end{pmatrix} \begin{pmatrix} d \\ s \end{pmatrix},$$

where the matrix describing the mixing should be what is called a unitary matrix in mathematics.

The next problem is what the condition for CP violation is. In quantum field theory, CP violation is related to complex coupling constants. To be more concrete in the present formulation, CP violation will occur if irreducible complex numbers appear in the elements of the mixing matrix. The matrix elements of a unitary matrix are complex numbers in general, but some of them can be made real by adjusting the phase factor of the particle state without changing the physics results. In such case, those complex numbers are called reducible, and otherwise, irreducible. Therefore, one condition of CP violation is that there remain complex numbers which cannot be removed by the phase adjustment of the particle states.

In the four-quark model, adjustment factors are described by two diagonal matrices whose elements are mere phase factors. It is easy to see that, if we choose them properly, we can make any 2x2 unitary matrices into a real matrix:

$$\begin{pmatrix} e^{i\phi_u} & 0 \\ 0 & e^{i\phi_c} \end{pmatrix} \begin{pmatrix} V_{ud} & V_{us} \\ V_{cd} & V_{cs} \end{pmatrix} \begin{pmatrix} e^{-i\phi_d} & 0 \\ 0 & e^{-i\phi_s} \end{pmatrix} = \begin{pmatrix} \cos\theta & \sin\theta \\ -\sin\theta & \cos\theta \end{pmatrix}$$

Therefore, in this case, we cannot accommodate CP violation.

How does this argument change in the six-quark model? In this case, we can express the flavour mixing as follows:

$$\begin{pmatrix} u \\ d' \end{pmatrix}, \begin{pmatrix} c \\ s' \end{pmatrix}, \begin{pmatrix} t \\ b' \end{pmatrix},$$

$$\begin{pmatrix} d' \\ s' \\ b' \end{pmatrix} = \begin{pmatrix} V_{ud} & V_{us} & V_{ub} \\ V_{cd} & V_{cs} & V_{cb} \\ V_{td} & V_{ts} & V_{tb} \end{pmatrix} \begin{pmatrix} d \\ s \\ b \end{pmatrix}.$$

This time the mixing matrix is a 3x3 unitary matrix. In this case, however, we cannot remove all the phase factors of the matrix elements by adjusting the phases of the quark states. The best we can do by adjusting the phases is to express them by a certain standard form with four parameters. A popular parameterisation is

$$V = \begin{pmatrix} c_{12}c_{13} & s_{12}c_{13} & s_{13}e^{-i\delta} \\ -s_{12}c_{23} - c_{12}s_{23}s_{13}e^{i\delta} & c_{12}c_{23} - s_{12}s_{23}s_{13}e^{i\delta} & s_{23}c_{13} \\ s_{12}s_{23} - c_{12}c_{23}s_{13}e^{i\delta} & -c_{12}s_{23} - s_{12}c_{23}s_{13}e^{i\delta} & c_{23}c_{13} \end{pmatrix},$$

where $c_{ij} = \cos\theta_{ij}$ and $s_{ij} = \sin\theta_{ij}$ with i,j = 1, 2, 3. We note that, unless $\delta = 0$, the imaginary part remains in the matrix elements and therefore CP symmetry is violated.

Taking into account the hierarchy of the actual values of the parameters, the following approximate parameterisation is frequently used in phenomenological analyses.

$$V \approx \begin{pmatrix} 1-\lambda^2/2 & \lambda & A\lambda^3(\rho - i\eta) \\ -\lambda & 1-\lambda^2/2 & A\lambda^2 \\ A\lambda^3(1-\rho-i\eta) & -A\lambda^2 & 1 \end{pmatrix} + O(\lambda^4)$$

In this parameterisation, if η is not zero, the system is violating CP symmetry.

We thought that this mechanism of CP violation is very interesting and elegant, but we had no further reason to single out the six-quark model from the other possibilities. The model was not so special, because if the system has sufficiently many particles, it is not difficult to violate CP symmetry. However, the subsequent experimental development pushed up the six-quark model to a special position.

In 1974, the J/ψ particle was discovered [15], and soon it turned out that it is the bound state of the fourth quark c and its anti-particle. The discovery had a great impact on particle physics, but it had little effect on the six-quark model.

In 1975, the τ-lepton was discovered [16]. This discovery had a significant effect on our model. The τ-lepton is the fifth member of leptons. Although it is a lepton, it suggested the existence of a third family in the quark sector, too. That was when people began to pay attention to our model. Early works which discussed the six quark model include Ref. [17] and Ref. [18].

In 1977, the Upsilon particle was discovered [19], and it turned out that it is a bound state of the fifth quark b and anti-b. The discovery of the last quark, t, occurred as recently as in 1995 [20], but before that time the six-quark model became a standard one.

Meanwhile, it was pointed out that we could expect large CP asymmetry in the B-meson system [21]. This opened the possibility to test the model with B-factories. B-meson implies a meson containing b- or anti-b as a constituent, and B-factory means an accelerator, which produces a lot of B-mesons like a factory.

4. EXPERIMENTAL VERIFICATION AT B-FACTORIES

In order to verify the six-quark model experimentally, two B-factories, KEKB at KEK in Japan and PEPII at SLAC in the US, were built. Those accelerators are unusual ones. Colliding electrons and positrons have different energies, so that the B-mesons produced are boosted. Both experimental groups, Belle(KEKB) and BaBar(PEPII), are large international teams organised with participation from many countries.

They were approved and started experiments almost at the same time. PEPII/BaBar ceased operation this year, while KEKB/Belle is still running. They achieved luminosities more than 10^{34} $cm^{-2}s^{-1}$, which are record high. Luminosity is a key parameter representing the performance of the colliding accelerator.

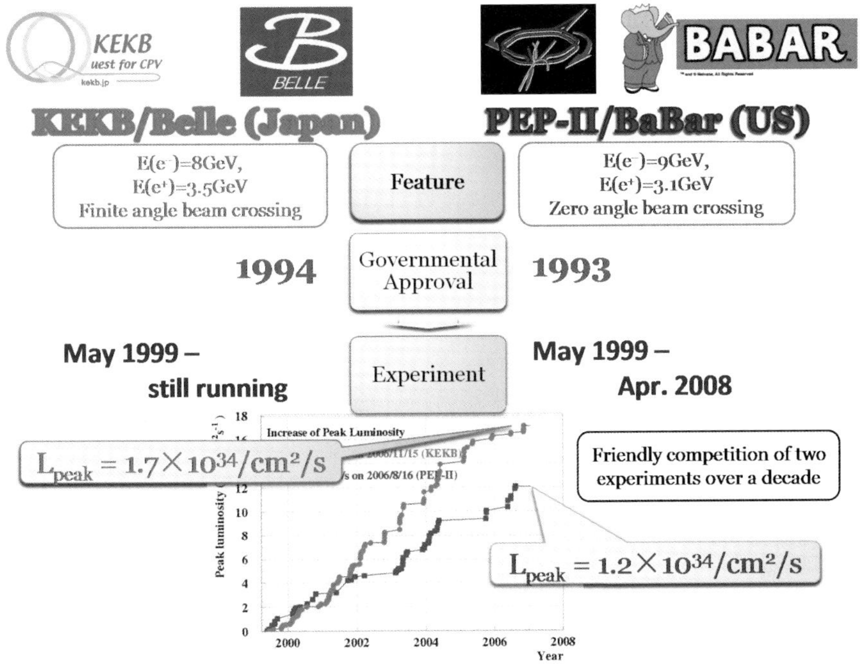

Figure 4. KEKB/Belle and PEP-II/BaBar.

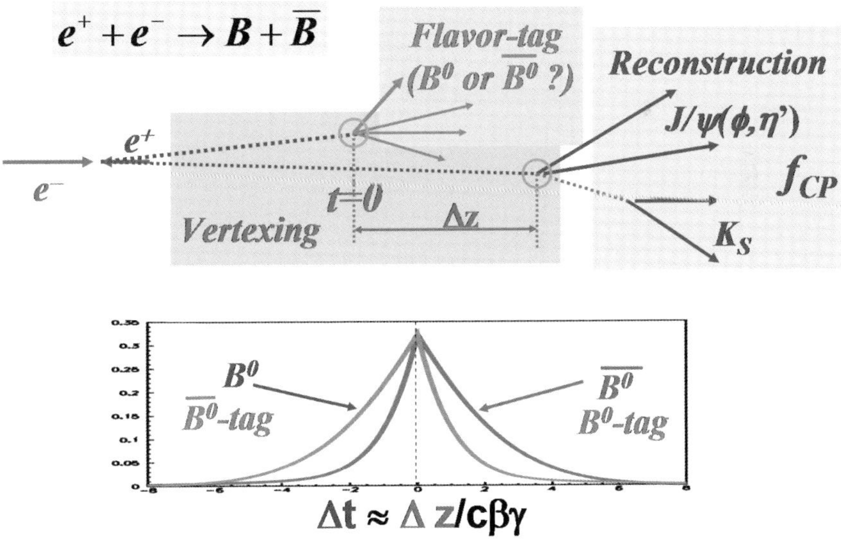

Figure 5. A typical method of measuring CP violation in the B-meson decay.

One of the typical methods of measuring CP violation in the B-factory experiment is shown in Figure 5. The six-quark model predicts fairly large asymmetry between B-meson and anti-B-meson in the decay time distribution of, for example, $B(\bar{B}) \rightarrow J/\psi + K_S$. Thanks to the boost of the produced B-mesons, we can find the decay time distribution of B or anti-B by measuring its decay position. This requires, however, measuring the decay position to the accuracy of 10 microns, so that a sophisticated device called a vertex detector is installed.

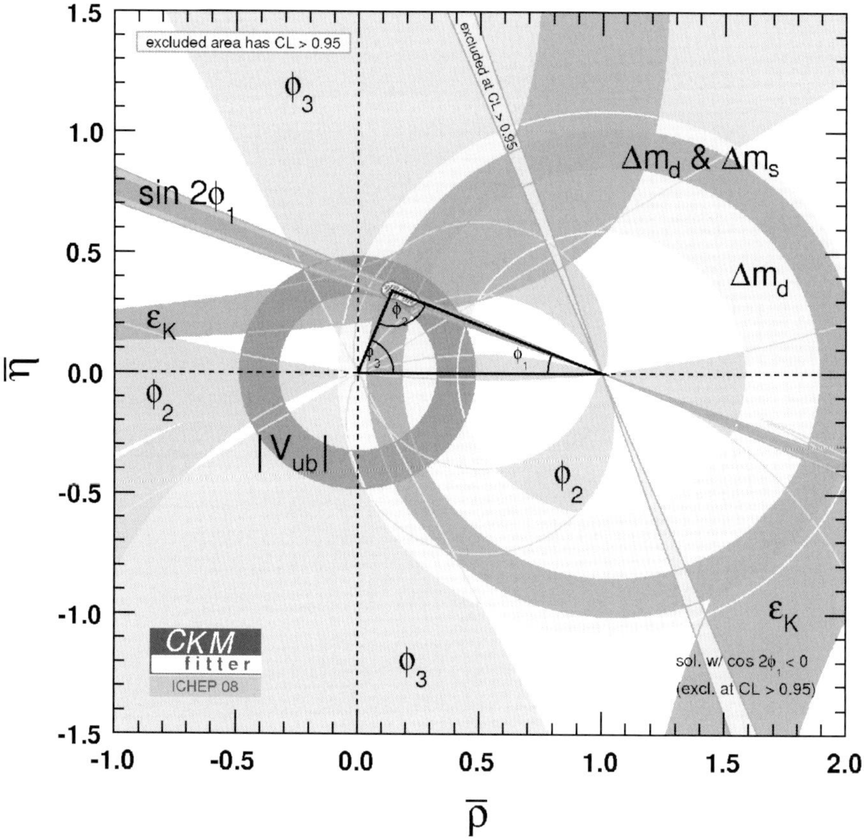

Figure 6. The results of the B-factory experiments [23].

The most important results of the experiments are well described by Figure 6 [22]. The coloured circles and cones show the experimental constraints on the mixing parameters ρ and η. All the constraints overlap on a narrow region coloured by red. This means the six-quark model can explain all those results by choosing the parameters in this region.

In the light of the B-factory results, the present status of CP violation may be summarised as follows.

- B-factory results show that quark mixing of the six-quark model is the dominant source of the observed CP violation.
- B-factory results, however, allow small room for additional source from new physics beyond the standard model.
- And matter dominance of the Universe seems to require new sources of CP violation, because it appears that CP violation of the six-quark model is too small to explain matter dominance.

It has been proposed that the last point may be related to lepton flavour mixing, which is the counterpart of quark mixing. In regard to lepton flavour mixing, very important contributions have been made in Japan, which will be discussed in the next section.

5. LEPTON FLAVOUR MIXING

The most important achievement is the discovery of neutrino oscillation at Super Kamiokande, which is a huge water tank detector built in the Kamioka mine in central Japan [24].

They were observing neutrinos produced by cosmic rays in the atmosphere surrounding the earth. Since neutrinos penetrate the earth, those neutrinos come to the detector also from below. The neutrino oscillation implies the species of neutrino changes during its flight. So if the neutrino oscillation takes place while neutrinos are travelling the distance from the other side of the earth, the observed number of the particular kind of neutrino will be reduced.

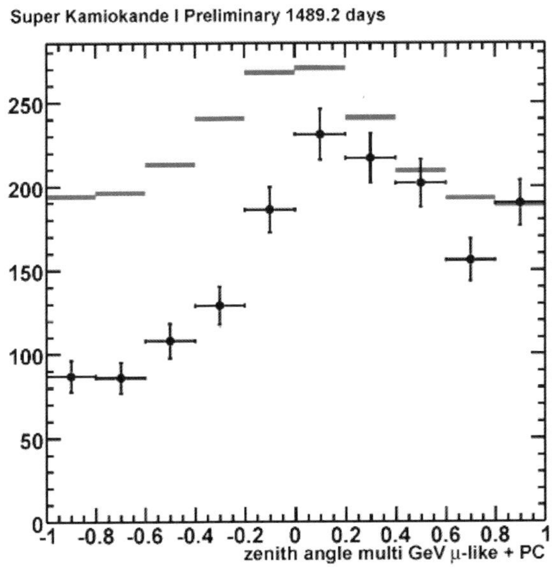

Figure 7. The results of the observation of atmospheric neutrinos [J. Raaf, Neutrino 2008].

Figure 7 shows the result of their observation. The red bars indicate the expectations for the non-oscillation case, and the crosses are real data. The results show a clear deficit of the observed neutrinos and are completely consistent with neutrino oscillation.

Figure 8. Yoji Totsuka 1942-2008 (Courtesy of KEK).

This great discovery was led by Yoji Totsuka. To our deep regret, he passed away in this last July.

The neutrino oscillation was further confirmed by two experiments using man-made neutrinos. One is the K2K experiment [25]. In this experiment, neutrinos were produced by the proton synchrotron in the KEK laboratory and those neutrinos were observed by the Super-Kamiokande. Figure 9 shows the observed neutrino spectrum. Data show a clear oscillation pattern.

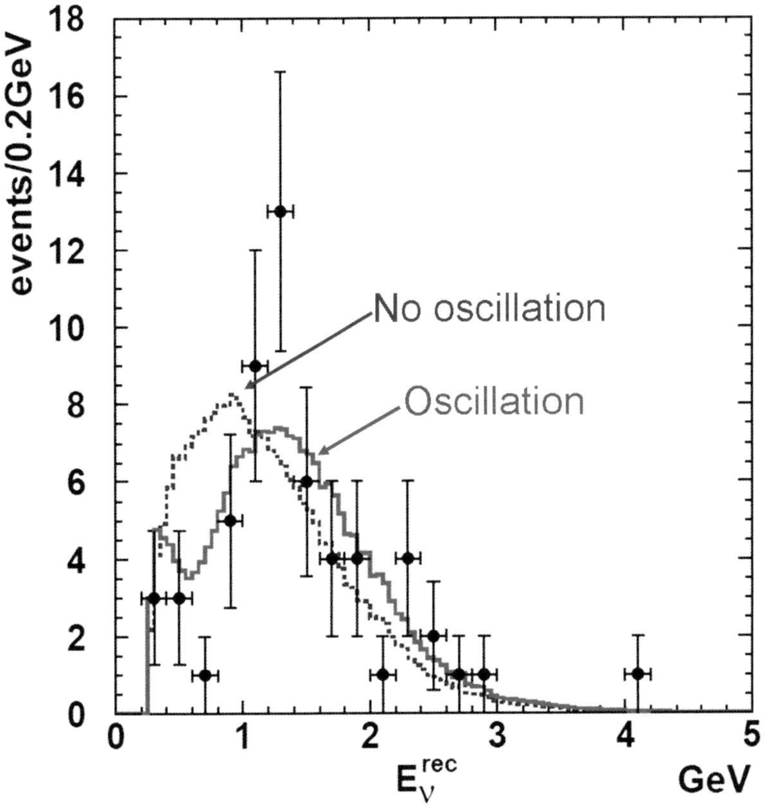

Figure 9. The result of the K2K experiment [25].

The other experiment is the KamLAND experiment [26]. The KamLAND detector uses liquid scintillator instead of water, and it is also located in the Kamioka mine. They observed neutrinos produced in the nuclear reactors in the surrounding area. Data show a clean agreement with the oscillation. (Figure 10).

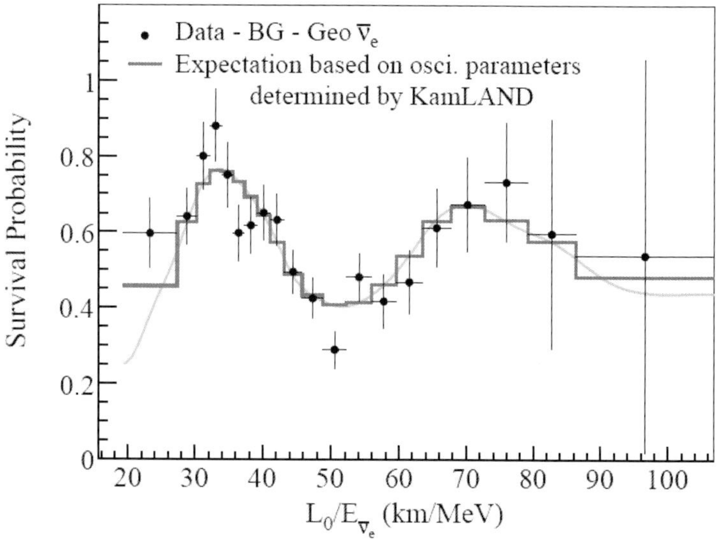

Figure 10. The result of the KamLAND experiment [26].

While we have seen past and present experiments, the T2K experiment is an upcoming future experiment. Neutrinos will be produced by the newly built accelerator J-PARC located in Tokai, 60 km north-east of KEK, and sent to Super Kamiokande. The distance to the detector is more or less the same as the K2K, but the intensity will be much higher. The T2K experiment aims the ν_e appearance measurement, which means the observation of ν_μ to ν_e oscillation. This measurement has crucial importance for estimating the possible size of CP violation in the lepton sector, which may have some implication for the matter dominance of the universe.

In summary, I think that Japan has made important contributions to flavour physics. This includes the early activities of the Sakata group on both hadron and lepton flavours, and the experimental studies of B-meson system at KEK B-factory and the observations of neutrino oscillation at Super-Kamiokande and KamLAND. I am very glad that I was able to be an eyewitness to many of these developments. In particular, it was unforgettable that I could work together with my colleagues on the B-factory experiments. And above all, I am very happy that I could contribute to these developments through my work with Dr. Maskawa.

REFERENCES

1. G. 't Hooft, "Renormalizable Lagrangians for Massive Yang-Mills Fields", *Nucl.Phys.* **B 3** (1971) 167; G. 't Hooft and M. J. G. Veltman, "Regularization and Renormalization of Gauge Fields", *Nucl. Phys.* **B 44** (1972) 189.
2. M. Kobayashi and T. Maskawa, "CP Violation in the Renormalizable Theory of Weak Interactions", *Progr. Theor. Phys.* **49** (1973) 652.
3. S. Sakata, "On a Composite Model for the New Particles", *Prog. Theor. Phys.* **16**(1956) 686.
4. A. Gamba, R. E. Marshak and S. Okubo, "Proceedings of the Ninth Conference on High Energy Nuclear Physics", *Proc. Nat. Acad. Sci.* **45** (1959) 881.
5. Z. Maki, M. Nakagawa, Y. Ohnuki and S. Sakata, "A Unified Model for Elementary Particles", *Progr. Theor. Phys.* **23** (1960) 1174.
6. G. Danby *et al.*, "Observation of High-Energy Neutrino Reactions and the Existence of Two Kinds of Neutrinos", *Phys. Rev. Lett.* **9** (1962) 36.
7. Z. Maki, M. Nakagawa and S. Sakata, "Remarks on the Unified Model of Elementary Particles", *Progr. Theor. Phys.* **28** (1962) 870.
8. Y. Katayama, K. Matumoto, S. Tanaka and E. Yamada, "Possible Unified Models of Elementary Particles with Two Neutrinos", *Progr. Theor. Phys.* **28** (1962) 675.
9. S. L. Glashow, J. Iliopoulos and L. Maiani, "Weak Interactions with Lepton-Hadron Symmetry", *Phys. Rev.* **D 2** (1970) 1285.
10. K. Niu, E. Mikumo and Y. Maeda, "A Possible Decay in Flight of a New Type Particle", *Progr. Theor. Phys.* **46** (1971) 1644.
11. S. Ogawa, private communication.
12. T. Hayashi, E. Kawai, M. Matsuda, S. Ogawa and S. Shige-Eda, "A Possible Interpretation of the New Event in the Cosmic Ray Experiment", *Progr. Theor. Phys.* **47** (1972) 280; M. Kobayashi, H. Nitto and M. Nakagawa, "Quartet Models Based on Fundamental Particles with Fractional Charge", *Progr. Theor. Phys.* **47** (1972) 982; Z. Maki, T. Maskawa and I. Umemura, "Quartet Scheme of Hadrons in Chiral U(4) × U(4)", *Progr. Theor. Phys.* **47** (1972) 1682; H. Kondo, Z. Maki and T. Maskawa, "A Note on the Leptonic Decays of Charmed Mesons", Pr*ogr. Theor. Phys.* **47** (1972) 1060.
13. S. Weinberg, "A Model of Leptons", *Phys. Rev. Lett.* **19** (1967) 1264; A. Salam, "Weak and Electromagnetic Interactions", originally printed in "N. Svartholm: Elementary Particle Theory, Proceedings of the Nobel Symposium Held 1968 at Lerum, Sweden" (1968) 367–377; S. L. Glashow, "Partial Symmetries of Weak Interactions", *Nucl. Phys.* **22** (1961) 579.
14. J. H. Christenson, J. W. Cronin, V. L. Fitch and R. Turlay, "Evidence for the 2π Decay of the K_2^0 Meson", *Phys. Rev. Lett.* **13** (1964) 138.
15. J. J. Aubert *et al.* [E598 Collaboration], "Experimental Observation of a Heavy Particle J", *Phys. Rev. Lett.* **33** (1974) 1404; J. E. Augustin *et al.* [SLAC-SP-017 Collaboration], "Discovery of a Narrow Resonance in e^+e^- Annihilation", *Phys. Rev. Lett.* **33** (1974) 1406.
16. M. L. Perl *et al.*, "Evidence for Anomalous Lepton Production in e^+e^- Annihilation", *Phys. Rev. Lett.* **35** (1975) 1489.
17. S. Pakvasa and H. Sugawara, "CP Violation in Six Quark Model", *Phys. Rev.* **D14** (1976) 305.
18. J. R. Ellis, M. K. Gaillard and D. V. Nanopoulos, "Lefthanded Currents and CP Violation", *Nucl. Phys.* **B109** (1976) 213.
19. S. W. Herb *et al.*, "Observation of a Dimuon Resonance at 9.5-GeV in 400-GeV Proton-Nucleus Collisions", *Phys. Rev. Lett.* **39** (1977) 252.
20. F. Abe *et al.* [CDF Collaboration], "Observation of Top Quark Production in $\bar{p}p$ Collisions", *Phys. Rev. Lett.* **74** (1995) 2626 [arXiv:hep-ex/9503002]; S. Abachi *et al.* [D0 Collaboration], "Observation of the Top Quark," *Phys. Rev. Lett.* **74** (1995) 263 [arXiv:hep-ex/9503003].

21. A. B. Carter and A. I. Sanda, "CP Violation in B Meson Decays", *Phys. Rev.* **D23** (1981) 1567; I. I. Y. Bigi and A. I. Sanda, "Notes on the Observability of CP Violations in B Decays", *Nucl. Phys.* **B193** (1981) 85.
22. B. Aubert *et al.* [BABAR Collaboration], "Measurement of the CP-violating Asymmetry Amplitude sin 2β" , *Phys. Rev. Lett.* **89** (2002) 201802 [arXiv:hep-ex/0207042]; K. Abe *et al.* [Belle Collaboration], "An Improved Measurement of Mixing-induced CP violation in the Neutral B Meson System. ((B))", *Phys. Rev.* **D 66** (2002) 071102 [arXiv:hep-ex/0208025].
23. CKMfitter Group (J. Charles *et al.*), *Eur. Phys. J.* **C41** (2005), 1, updated results and plots available at: http://ckmfitter.in2p3.fr.
24. Y. Fukuda *et al.* [Super-Kamiokande Collaboration], "Evidence for Oscillation of Atmospheric Neutrinos", *Phys. Rev. Lett.* **81** (1998) 1562 [arXiv: hep-ex/9807003].
25. M. H. Ahn *et al.* [K2K Collaboration], "Indications of Neutrino Oscillation in a 250 km Long Baseline Experiment", *Phys. Rev. Lett.* **90** (2003) 041801 [arXiv:hep-ex/0212007]; ibid. "Measurement of Neutrino Oscillation by the K2K Experiment", *Phys. Rev.* **D74** (2006) 072003 [arXiv:hep-ex/0606032].
26. K. Eguchi *et al.* [KamLAND collaboration], "First Results from KamLAND: Evidence for Reactor Anti-neutrino Disappearance", *Phys. Rev. Lett.* **90** (2003) 021802 [arXiv: hep-ex/0212021]; T. Araki *et al.* [KamLAND collaboration], "Measurement of Neutrino Oscillation with KamLAND: Evidence of Spectral Distortion", *Phys. Rev. Lett.* **94** (2005) 081801 [hep-ex/0406035].

Portrait photo of Makoto Kobayashi by photographer Ulla Montan.

益川敏英

TOSHIHIDE MASKAWA

A BOY WISHES TO BECOME A SCIENTIST

I was born in 1940 as the second child in a family living in Nagoya, a city with a population of around a million inhabitants. My older sister died of tuberculosis before entering elementary school and so I was an only child until my second sister, who is seven years younger than me, was born after the War. I had a weak constitution and was a thin boy having poor digestion. My parents were worried about my health and repeatedly took me to be examined whenever they heard about doctors with a good reputation. In my infancy, therefore, I did not play with other children of the same age and was raised only among adults, so that I had a very precocious way of speaking. Thanks to this, later at elementary school I could get perfect scores in Japanese examinations designed to test pupils' ability to use words in practice, for example by constructing short sentences containing specified words. On the other hand I would get almost zero in reading and writing Chinese characters used in Japanese since I had not studied them.

After a new municipal library opened near my elementary school, I went there often and began to read books at random so that I gradually acquired the ability to read between the lines. Why did the author choose to write one way but not another even if they have almost the same meaning? In those days I got into the habit of thinking about the psychology of the book's author. This habit proved helpful to me later when I became a researcher. When discussing with my friends, I often find myself able to obtain more information from the same papers than they do. Sometimes however, I make an error by reading what is not written in the paper.

When I was in elementary and junior high schools I did not concentrate in class and could not be called a good pupil by any standard. For instance, at the end of my third year at junior high school there was a Japanese class in which the teacher handed out manuscript papers to the pupils to write an essay which was to be inserted in a graduation memorial collection of compositions. My classmates all wrote about their future ambitions; one wanted to become a carpenter following in his father's footsteps, another hoped to go to university to become a mechanical engineer and so on. But I wrote about the evolution of stars, about which I had been reading in a boys' magazine at that time. I had not heard the teacher's explanation that the essay was to be included in the graduation memorial volume.

At that time I had never thought about my future or had any definite goals. I went on to high school with no strong motivation, but simply because my friends did so. The ratio of students who went on to high school then was about 50%. There was a high-achieving girl in my class who had decided not to continue to high school. Although I was not particularly friendly with her, since our seats in the class were well separated, I remember feeling that it was one of life's absurdities when I heard her tell the teacher that she would not go on to high school.

There is an event engraved in my memory, which I experienced in the period before entering high school but after graduating from junior high school. I purchased the textbooks required for the high school and brought them back home. While I was browsing through the mathematics textbook, I noticed a strange character; the summation symbol Σ. As soon as I read the explanation I understood how to use it. I found that, by using the linearity of the Σ symbol, I could compute the sum of the n-th power of integers, $1^n + 2^n + \ldots + k^n$, for any value of n, in principle. I was so excited that I calculated the sum up to rather high powers n, although this took me a long time. Not surprisingly for a boy who had not yet entered high school, I did not have sufficient knowledge to devise a generating function for the sum.

After the war, my parents were engaged in a small business which required them to work together from early in the morning until late in the evening, so that they did not have the time to pay much attention to their children's study. Taking advantage of this, I played and played without studying. There was also another benefit from the fact that my parents were running their own business from home. They were mainly dealing with sugar as an ingredient for cakes. The ordinary sugar was packed in 30 kg bags of kraft paper like cement bags, and Cuban sugar in 100 kg bags of hemp. They also retailed sugar, subdividing it into 10 kg, 20 kg, and so on, so that the bags were stacked to be discarded. Our parents gave them to us children instead of an allowance. Since they could be sold if they were brought to a suitable place, I had some extra spending money – more than my friends did. I spent almost all my money on books. I still cannot get out of this habit. I buy books in bulk, not selecting them carefully, and read them later when I find them on my bookshelf.

When I was in high school, it had already been ten years since the war had ended and the world was becoming peaceful again. There still remained shortages of cultural materials, however. For example, the supply of new books was still insufficient, so that I had to walk around the area of second-hand bookstores every weekend afternoon with pocket money obtained by selling kraft and hemp bags for sugar. Initially the main genres of the books purchased with this money were detective and mystery stories and novels by Ryunosuke Akutagawa. Later I gradually began buying more mathematics books.

The first book I bought in mathematics was the "Theory of Functions" published as a volume in *New Mathematics Series* by Baifu-kan, Tokyo. I was very excited to see how a mathematics book was written, since I had seen no books other than school textbooks. I only knew the term "function", which had already appeared in the high school mathematics textbook. The book gave me a glimpse of the rich world of differentiable functions of complex variable and I felt dizzy as if wandering off into a foreign place, quite different from the world I had previously known.

During my last year of high school, the Soviet Union successfully launched the first artificial satellite, Sputnik I. After this event I began to calculate the orbits of satellites and rockets by using a slide rule and an abacus. From the relative position of the Moon and Earth, I predicted for my friends the day on which the Soviet Union would next launch a rocket. To confirm the validity of the prediction, I always listened to the short-wave broadcasts of Radio Moscow from 11:00 at night. Then I realised that there is an issue with the precision of the time shown by clocks. From my measurements I discovered that my watch had a systematic error of – 8 sec/day and a random error ± 2 sec/day. The problem was the seasonal variation of the systematic error. If the temperature rises, the balance wheel becomes larger and its moment of inertia also gets larger, so that the clock should become slower. The temperature dependence of this deviation was, however, the opposite of estimates. This annoyed me for a few years. I looked for and examined books about clocks whenever I visited big bookshops, but I could not find the answer. About five years later I found the answer at last. The thermal expansion of the balance wheel and change of the moment of inertia were actually compensated by an ingenious mechanical device. I learned a lesson from this experience. I knew that wall clocks have such a temperature correction device since it can be seen. But I didn't associate it with wristwatches. This taught me to think matters through carefully, taking as many relevant elements into account as possible.

I entered Nagoya University after one year of hard study, motivated by my desire to avoid succeeding my father and becoming a sugar merchant. The first class at the university was mathematical analysis by associate professor H. The first thing he said was: Suppose that there are two arbitrary positive numbers ε and a, then there always exists an integer N such that $N\varepsilon > a$ is realised. This is called Archimedes' axiom. Having declared "I will prove this", he began the lecture by explaining Dedekind's cut. What is this! Why isn't it all right that we calculate a/ε and take N to be the integer part of it plus 1? It was a Culture Shock.

Next was a biology class by professor T. When I took a seat in the front, a sheet of paper was sent to me from behind. It contained some challenges, saying, "Solve the following problems!" There were six mathematics problems written on the paper. I remember that they were problems, which required solving differential equations such as determining the catenary, the form of a suspended chain. In this way the circle of my friends increased and I was also able to meet some good teachers.

At that time at Nagoya University, the campus for freshmen and sophomores was separated from that for juniors and seniors. The campus for the general education course for the students in the first two years of study had previously been a high school in the old system of education and there were several teachers from that system remaining in the Physics Department.

When, during discussions with my friends, we encountered problems which we didn't understand, we used to go to the teachers' room to ask for help. When these visits became frequent, many teachers began to avoid us since they found us to be troublesome. There was however an exceptional teacher N who coolly answered our questions sitting back in an armchair. He was a young professor just past thirty. When we asked questions, he used to reply in a dignified manner, "I cannot immediately answer questions which are raised so suddenly. Study them yourselves. I will lend you this book if necessary."

This was the first time that I had met somebody referred to as a researcher. The teachers I had met until my high school days were just teachers, who taught what they knew. But I realised that teachers at the university are also required to do research, discovering novel things. The students who gathered around professor N began to form a relatively fixed group, doing many other things together. The members of this group came from the whole faculty of science, although most were students in the Physics Department. They called themselves the DEPHIO group taking the initials D(Dirac), E(Einstein), P(Pauling), H(Hilbert), I(Ingold), O(Oparin), who were great scientists representing each department. The activities of DEPHIO included, for instance, lodging together in a borrowed country house for ten days during the summer vacation and holding a seminar class at a 2,000 meter high hill of the North Japanese Alps, having carried Dirac's thick textbook up there. Although these activities were not directly relevant from the viewpoint of doing research, I think they were meaningful in encouraging solidarity among the members of the group and in helping them become researchers together. DEPHIO is no longer active, but the members still keep contact with each another.

At that time, I was mainly discussing mathematics with the group. I also continued to visit second-hand bookshops and found, for instance, books from a series of Mathematics lectures published by the publisher Iwanami before the War. Later I happened once to tell a mathematician that I had studied non-Archimedean valuation using one of those books. He was surprised and said, "I have heard that you are a fan of mathematics, Maskawa-san. But you are even studying such things!"

In our college days, the members of the DEPHIO group discussed together whenever they met. During the year 1960, Japan was politically in chaos since public opinion was split in two as to whether the Japan–U.S. Security Treaty should be concluded or not. The treaty could not be approved in the House of Councillors, but was soon automatically approved because of the dominance rule of the House of Representatives. The prime minister at the time then resigned.

Almost all classes were cancelled during this period because of a student strike. I attended all possible demonstrations, motivated by a young man's sense of justice. It was inadmissible to me that the clock in the House of

Councillors had been stopped to make time to enable the bill to be enacted and to prevent it from being scrapped.

There were no other periods later in my life in which I had as much free time as during those days. Although there were tasks to be performed, like going to demonstrations, distributing bills, collecting signatures and so on, there was also much free time in between. We used some of this time to hold outdoor seminars. Since everybody gathered in the university courtyard, it was easy to organise such seminars with little notice.

When I moved to the Higashiyama Campus as a junior student, I was still hesitating whether to study Mathematics or Physics. Mathematics in Japan at that time still remained under the strong influence of axiomatic Bourbakism. I did not know this definitely, but intuitively felt that this was the case. On the other hand, I felt that Physics was moving forward vigorously and so I submitted an application form for the graduate course in which I wrote that I wish to study theoretical physics. In fact, I had actually not yet decided my intended field of study quite so definitely even at that stage, as illustrated by the following event. When I was walking in the campus of the Faculty of Science at that time, a Professor in the Mathematics Department said to me, "You, Maskawa, will of course take an entrance examination for a Mathematics graduate course, won't you?" I answered, "No, I have just submitted the application form for a Physics graduate course." The professor looked upset hearing this totally unexpected answer. Probably, I had been saying until recently that I was planning to go on to the Mathematics graduate course.

In the graduate course of the Physics Department, there were about ten beginning students who intended to major in theory, of which about six wanted to study particle physics. But this over-concentration on particle physics was merely due to their ignorance of the various interesting fields in science which they might tackle. After one year of study in the graduate course they spontaneously scattered, finding the scientific fields which fit their personalities best, like astrophysics and nuclear physics.

The students majoring in theory were not assigned to any specific laboratory for their first year, but worked in turn for three month periods in a number of laboratories: particle physics, condensed matter physics, and so on. Thanks to this system, I had the chance to study condensed matter physics. During this period, in addition to my studies, I began a voluntary circle to study a *perceptron* with a few friends, since I thought that research of the brain and human consciousness was important but not yet fully understood theoretically. We read papers in that field but, unfortunately, none of us had any knowledge about the anatomy of the cerebrum. None of the members was sufficiently enthusiastic to transfer to the medical department in order to study it, and so the circle died out in course of time. I noticed then that the number of research papers about the perceptron was decreasing significantly. I understood the reason later. Nuclear submarines had become increasingly important strategically because of the Cold War. The perceptron became a subject of military research because of the possibility that it could be used to identify nuclear submarines by sound spectrograms.

In the Physics Department of Nagoya University, the first year students majoring in theoretical physics in the masters course had to attend the seminar on field theory as a compulsory subject for a year. In those days, all the particle physicists throughout the world believed for a variety of reasons that field theory would be replaced sooner or later by a future new theory. This was also the case in Sakata's laboratory. The curriculum of the field theory course was designed in the 1950s. It began with Dirac's quantization of fields using the variables of amplitude and phase. The course then continued to Heisenberg-Pauli's field theory and Fock space, and through Feynman's ingenious theory describing electromagnetic interaction as an action at a distance eliminating the photon picture, ending with Dyson's renormalisation theory supplemented with Salam's paper on b-divergence. This course made me pay attention to field theory, which no one took any notice of in the 1960s. Thanks to this, I got interested in theoretical problems related to weak interactions. When the importance of field theory was later appreciated, the situation had turned full circle and the curriculum of theoretical physics at the graduate school of Nagoya University was again at the forefront of world physics.

I entered the laboratory of Professor Sakata in 1964 and began my research in particle physics. However, my capriciousness did not change after this and I continued to collaborate with physicists in other fields like nuclear physics, condensed matter physics and so on.

Finally I would like to thank Professors Taichiro Kugo and Christopher Sachrajda who kindly translated my Japanese draft into this English form.

WHAT DOES CP VIOLATION TELL US?

Nobel Lecture, December 8, 2008

by

Toshihide Maskawa

Kyoto Sangyo University, Kyoto 603-8555, Japan.

I would first like to thank the Royal Swedish Academy of Sciences and the Nobel Foundation for awarding me this honour, of which I had never even dreamt.

I was born in 1940, the son of a furniture craftsman in the city of Nagoya, in Japan. My father wanted to change his job and was taking a correspondence course to become an electrical engineer, while he was a trainee furniture craftsman. However, he told me that he could not really understand sine and cosine, since he had not received a basic education. Eventually, though, he did manage a small furniture factory employing a few craftsmen and worked there himself. But this came to nothing because of the War, that reckless and miserable war which our country caused.

After the War, he displayed in front of his house the door hinges, wood screws and other pieces of furniture, which remained at hand. They sold quite well. Getting a taste for selling, he became a merchant dealing with sugar as an ingredient for cakes.

He still wanted to boast of his knowledge of electricity, but he could not find anyone suitable to explain it to. One day, though, he found a good target: his son.

In those poor days after the war, almost all the houses were without bathrooms and so people went to the public bath. On his way to and from the public bath, he boasted of his knowledge: Why do three-phase current motors rotate? Why don't the solar and lunar eclipses occur every month? He explained proudly that it was because of the revolution planes of the earth around the sun, and of the moon around the earth, which are tilted at an angle of 5 degrees.

This was the reason why I was a strange pupil at school. I had a poor record but could answer the teacher's questions when he digressed and spoke about subjects outside of the textbooks.

My parents neither observed their children carefully nor helped with their study. One day, my mother realised that she had never seen her son studying at home. So she told my teacher at a parent's association meeting. "Please give my son homework at least occasionally. Otherwise, he never studies at home." My teacher answered, "Your son has never done his homework despite the fact that I give him homework every day!"

Disastrous was that night. I got a two-hour lecture from my parents.

An event that gave me a strong wish to become a physicist happened after I went on to high school. I went to high school with no strong motivation. But, one day when I was in the first or second year, I found a newspaper article explaining that Professor Shoichi Sakata at Nagoya University had published a revolutionary theory [1] about the composite model for elementary particles in which he chose the proton, neutron and Lambda particles as the fundamental three constituents.

I was so childish at that time that I thought that all of science had already been completed in Europe during the 19th century. If Professor Sakata's theory had come from Tokyo, I would have thought it irrelevant to me. But scientific discoveries were being developed right in Nagoya, where I was living! I became eager to join in such research activity there. My father, however, wanted his son to succeed him in the family business. So I was given only one chance to take an entrance examination for the university, and could not fail it. I worked very hard to prepare for the entrance examination to Nagoya University.

Fortunately I was accepted. The lectures at the university were very different and much more stimulating than those in high school. The first class was about mathematical analysis. I learnt about an axiom from Archimedes explaining that, for any two positive numbers, ε and δ, there is an integer N such that $N \times \varepsilon > \delta$. And then the teacher continued the lecture to explain Dedekind's Cut. It was a Culture Shock. Everything that I experienced in the university, including those lectures, was fresh and stimulating. Each time I begin to study a new field, I was totally absorbed in it and told the people around me that I would research that subject in future.

When I was a senior, a Professor in the Mathematics Department told me, "You will, of course, take an entrance examination for a Mathematics graduate course, won't you?" When I answered, "No, I already sent the application form to a Physics graduate course," the professor looked upset hearing this totally unexpected answer. Probably, I had been saying that I was planning to go on to a Mathematics graduate course until just before.

This capriciousness of mine did not change even when I went to the physics graduate course in 1962. I thought that research on the brain was important and began a voluntary circle to study a PERCEPTRON with a few friends. However, when I had to write my Master's thesis in the second year, I was preparing a paper on particle physics in Prof. Sakata's laboratory.

In this period, research based on dispersion relations resulting from causality was popular in the world, and a "Bootstrap Model" advocated by G.F. Chew was in fashion. In Sakata's laboratory, the composite model [1] for 'elementary' particles was popular, which was originally proposed by Prof. Sakata in 1955.

At that time the symmetry between leptons and baryons concerning weak interactions which A. Gamba, R.E. Marshak and S. Okubo [2] had pointed out at the Kiev Conference in 1951 occupied the interest of the laboratory [3]; that is,

$$(\nu, e, \mu) \leftrightarrow (p, n, \Lambda).$$

Following this line, when the muon-neutrino ν_μ was found, the majority of the Sakata laboratory naturally began to research the quartet constituent models. Indeed, in 1964, a quartet constituent model was proposed by Ziro Maki [4], the associate professor at the laboratory.

Although I did not write papers during this period, I was very fascinated by spontaneously broken chiral symmetry after reading Professor Nambu's papers [5, 6]. I was also interested in developments in current algebra and PCAC (partially conserved axial-vector current). Extending my interest in this direction, I encountered important papers like: "The Axial Vector Current in Beta Decay" by M. Gell-Mann and M. Levy (1960) [7], and "Question of Parity Conservation in Weak Interactions" by T.D. Lee and C. N. Yang (1956) [8].

As the keystones of my research in particle physics, those papers strengthened my interest in observing how particles emerge through their weak interactions.

Although I did not publish it, I calculated several physical quantities in the framework of the Nambu-Jona-Lasinio (NJL) model [6]. For instance, I computed the pion decay constant f_π, but felt it rather small given the fact that it has a typical quantity characterising the strong interaction. I persistently examined how the decay constant could be decreased by adjusting the free parameters in the NJL model: the coupling constant and the UV cutoff. I could get no definite answer, since the contribution from the momentum region near the UV cutoff was most dominant. But, I recognised the importance of renormalisability from this experience.

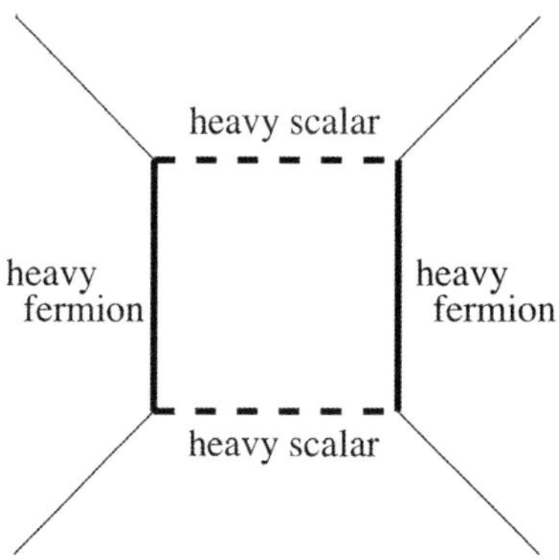

Figure 1. Box diagram.

Although this research wasn't being paid much attention in Japan, there were physicists around the world who were examining higher order effects of weak interactions. If weak interaction is described by the current-current type 4-fermion interaction, then the higher order effects diverge. So these authors replaced the 4-fermion interaction by a box diagram as shown in Fig. 1, in which heavy scalar bosons and heavy fermions propagate in the intermediate state. This reduces to the usual 4-fermion interaction in the heavy limit of scalar and fermion masses. It did not yet provide a satisfactory model, but it did allow the possibility of making the weak interaction renormalisable by scalar boson mediation. I was also paying attention to papers which persistently examined the higher order effects of the current-current interaction.

The laboratory subscribed to only one physics journal, and so many people wanted to read the latest issue. Partly to resolve this problem, Sakata's laboratory also had a journal club in which members took turns reading and reporting on the new issue. When my turn came for the first time, the journal contained the paper by Fitch and Cronin [9] reporting the observation of CP violation. This made a strong impression on me and raised the question: What is this? However, I had no clue at all how to attack the problem and thus laid it aside for eight years.

In the latter half of the 60s, my interest was mainly focused on chiral symmetry and chiral dynamics. In this period I found, for instance, in the Nambu-Jona-Lasinio model that the pion decay constant is rather small for a typical constant characterising the strong interaction, and that pion squared mass is the quantity of the first order in chiral symmetry breaking. I still didn't write any papers about this. I regard them as etudes for a physicist, just as painters draw variations of changing poses or compositions.

The idea of Spontaneous Symmetry Breaking in Nambu and Jona-Lasinio's paper was very attractive, but there are no massless particles interacting strongly in reality. So many physicists begin to focus their efforts on evading the Nambu-Goldstone (NG) Theorem [11, 12]. Higgs was among them. The reason why the NG massless boson appears is that there are only two form factors proportional to $\gamma_5\, p_\mu$ and $\gamma_\mu \gamma_5$ in the axial current, and the current conservation condition leads to the existence of a massless pole in one of the two form factors. This was what the NG theorem describes. Then Higgs [10] appeared and developed his arguments on the Coulomb gauge where manifest Lorentz invariance is lost, so that one more form factor can appear and one can evade the NG theorem.

It was the paper on the Higgs mechanism with which not only the massless NG boson disappears but also the gauge boson becomes massive. I could not follow his logic. Soon, fortunately, Kibble [13] appeared and made clear the mechanism in a Lorentz covariant manner, which I could understand. Soon, the electro-weak unified theory by S. Weinberg [14] and A. Salam [15], as well as the Faddeev-Popov's paper [16] clarifying the Feynman rule for Yang-Mills theory, appeared.

In connection with the four-quark model and from my interest in the higher order effects of weak interaction, I noticed a paper by GIM (Glashow, Iliopoulos and Maiani) [17] in 1969 and called the attention of some of my colleagues to it. This reminded me of the four-quark model once again, although all the graduates of Sakata laboratory already knew about it.

There is a paper from this period, which was cited as "Z. Maki and T. Maskawa, RITP-146" [18] in my paper on CP violation with M. Kobayashi [19]. It was written just before the CP violation paper and the title was

"Hadron Symmetries and Gauge Theory of Weak and Electromagnetic Interactions". This paper is little known since I never announced that it was published in *Progress of Theoretical Physics*, 49 (1973), 1007. If the strong interaction chooses a specific direction in the four-quark model, then its interference with the weak interaction would have easily caused CP violation. People came to know after 1974 that the strong interaction is described by QCD (Quantum ChromoDynamics) and that it has no specific direction other than the quark mass terms specified by the Higgs field. This was, however, unknown at that time in 1972, so we needed the analysis of the strong interaction given in the paper above.

When the electro-weak unified gauge theory appeared, I felt intuitively that the time had come to take up the CP violation problem. After finishing the graduate course, I became an assistant professor for three years in Nagoya University and then moved to Kyoto University in 1970. Kobayashi also came to Kyoto as an assistant professor two years later, in April 1972.

We have a series of holidays called Golden Week in the beginning of May in Japan. After the Golden Week, Kobayashi and I started a new collaboration. It was a good opportunity for us to work again in the same laboratory. I felt that the time was ripe for attacking the CP violation problem as stated above and proposed it as the theme of our collaboration. Kobayashi had also been approaching this problem in Nagoya.

We started to study CP violation based on the quartet quark model in the electro-weak unified gauge theory. First we had to determine how the left- and right-handed components of the four quarks transform under the electro-weak symmetry $SU(2) \times U(1)$:

A) $4 = 2 + 2$
B) $4 = 2 + 1 + 1$
C) $4 = 1 + 1 + 1 + 1$

There are these three cases for each of left- and right-handed components, so that we have to consider 9 cases in all. In all the cases except for case (A, B) where the left-handed component is A) and the right-handed B), we see that CP violation cannot occur if we take into account the freedom to arbitrarily choose the phases of the quark fields. The last case (A, B) was an exception.

I thought that it seemed to work since all the other cases failed and asked Kobayashi whether he agreed. Kobayashi put his head a little to one side and answered "I will examine this in detail overnight." He probably knew the result already, but answered like that for safety. The next day he told me, "That case does not work since it predicts the wrong sign for g_V / g_A contradicting the experiment."

During this time, I was chief secretary of the labour union at Kyoto University. There was a growing problem related to the dismissal of secretaries in our Faculty of Science. I could not concentrate on my research while pretending not to see these dismissals, since the secretaries really supported our young researchers. I discussed physics with Kobayashi in the morning and worked at the union in the afternoon. Coming back home in the evening, I enjoyed dinner with my family and talked with them about the events of the day. After taking a bath I had time to think about my research. This was how I was living back in those days.

After Kobayashi eliminated the possibility of the case (A, B), a search for the way out began, attempting to explain how CP violation could be understood as a four quark model. We fought a tough struggle for a month or so. And one day, while taking a bath, I thought: No good model exists which can explain CP violation in the four-quark model if we stick to the renormalisable theory of weak interaction. So I made up my mind to write a negative paper, which proved that CP violation cannot be explained in the four-quark model in a framework of renormalisable electro-weak theory.

Deciding to finish this work by writing a poor paper and thinking up an excuse, I got out of the bath tub. Probably this may have freed me from sticking to the four-quark model. Suddenly, I recognised that it must work well in the six-quark model. It was already clear from our many trials and calculations up until then that a complex phase could remain there.

In the framework of unified electro-weak gauge theory, there is a scalar field called the Higgs field. When it develops a non-vanishing vacuum expectation value, the quark fields acquire their masses. If we rewrite the quark fields into those, which have definite masses, then the charged current operators in the weak interaction develop a unitary matrix U. If a complex phase appears in the matrix elements of U, CP violation occurs.

This explanation may give the wrong impression that the complex phase always leads to CP violation. Actually, since the quark fields q_i are complex, we have the freedom to change their phases arbitrarily as $q_i \to e^{i\alpha_i} q_i$. Clearly this leaves physics intact, but the phases of the matrix elements of U are changed. Recall that the charged weak current has the structure, complex conjugate of up-type quark fields, $\bar{u}_i$ times down-type quark fields d_j times the i–j matrix elements U_{ij} of U: $\bar{u}_i U_{ij} d_j$. Here i, j run from 1 to N if there exist N generations of quarks.

Therefore the phase change of the $2N$ quark fields,

$$u_i \to e^{ia_i} u_i \, , \; d_j \to e^{ib_j} d_j \; , \quad (1)$$

induces the phase change of the matrix element

$$U_{ij} \to e^{-i(a_i - b_j)} U_{ij} . \qquad (2)$$

Therefore, among the $2N$ phase changes for $2N$ quark fields, $2N - 1$ degrees of freedom can be used to change the phase of U, since a common phase transformation for all quark fields does not change U.

The $N \times N$ unitary matrix U, which appears in the charged weak current when the $SU(2) \times U(1)$ doublet quark fields are rewritten in terms of the mass eigenstate quark fields, generally has $N(N + 1)/2$ degrees of freedom for the complex phase. If we subtract the above freedom of quark field phase change, then we are left with

$$\frac{N(N+1)}{2}-(2N-1)=\frac{(N-1)(N-2)}{2}$$

as the net number of physical complex phases in the matrix U.

This number vanishes for $N = 1$ and $N = 2$, which correspond to the two and four-quark cases, respectively, explaining that we have no complex phases in the matrix U and thus no CP violation for the two and four-quark cases. However, when we go to the $N = 3$ generation case, i.e., the six-quark model, then this number is 1 and CP violation can occur for the first time!

We have thus completed our work. The origin of CP violation was revealed partially at least.

It, however, took a long time: more than 30 years and vast efforts of many experimenters to really verify this theory. I would like to thank here all the people in the world who supported this grand project of humanity.

REFERENCES

[1] S. Sakata, "On a composite model for new particles", *Prog. Theor. Phys.* **16** (1956), 686.
[2] A. Gamba, R. E. Marshak and S. Okubo, "Proceedings of the Ninth Conference on High Energy Nuclear Physics", *Proc. Nat. Acad. Sci.* **45** (1959), 881.
[3] Z. Maki, M. Nakagawa, Y. Ohnuki and S. Sakata, "A unified model for elementary particles", *Prog. Theor. Phys.* **23** (1960), 1174.
[4] Z. Maki, "The forth baryon, Sakata model and modified B-L symmetry.1", *Prog. Theor. Phys.* **31** (1964), 331.
[5] Y. Nambu, "Quasi-particles and gauge invariance in the theory of superconductivity", *Phys. Rev.* **117** (1960), 648.
[6] Y. Nambu and G. Jona-Lasinio, "Dynamical model of elementary particles based on an analogy with superconductivity. I", *Phys. Rev.* **122** (1961), 345; **124** (1961), 246.
[7] M. Gell-Mann and M. Levy, The axial vector current in beta decay", *Nuovo Cim.* **16** (1960), 705.
[8] T. D. Lee and C. N. Yang, "Question of parity conservation in weak interactions", *Phys. Rev.* **104** (1956), 254.
[9] J. H. Christenson, J. W. Cronin, V. L. Fitch and R. Turlay, "Evidence for the 2π decay of the K_2^0 Meson", *Phys. Rev. Lett.* **13** (1964), 138.
[10] P. W. Higgs, "Broken symmetries, massless particles and gauge fields", *Phys. Lett.* **12** (1964), 132; "Spontaneous symmetry breakdown without massless bosons", *Phys. Rev.* **145** (1966), 1156.
[11] J. Goldstone, "Field theories with superconductor solutions", *Nuovo Cim.* **19** (1961), 154.
[12] J. Goldstone, A. Salam and S. Weinberg, "Broken symmetries", *Phys. Rev.* **127** (1962), 965.
[13] T. W. B. Kibble, "Symmetry breaking in non-Abelian gauge theories", *Phys. Rev.* **155** (1967), 1554.
[14] S. Weinberg, "A model of leptons", *Phys. Rev. Lett.* **19** (1967), 1264.
[15] A. Salam, "Weak and Electromagnetic Interactions" in *Elementary Particle Theory* ed. by N. Svartholm (Almqvist & Wiksell, 1968), 367–377.
[16] L. D. Faddeev and V. N. Popov, "Feynman diagrams for the Yang-Mills fields", *Phys. Lett.* **B 25** (1967), 29.
[17] S. L. Glashow, J. Iliopoulos and L. Maiani, "Weak interactions with lepton-hadron symmetry", *Phys. Rev.* **D 2** (1970) 1285.
[18] Z. Maki and T. Maskawa, "Hadron symmetries and gauge theory of weak and electromagnetic interactions", *Prog. Theor. Phys.* **49** (1973), 1007.
[19] M. Kobayashi and T. Maskawa, "CP violation in the renormalizable theory of weak interaction", *Prog. Theor. Phys.* **49** (1973), 652.

Portrait photo of Toshihide Maskawa by photographer Ulla Montan.

Physics 2009

Charles Kuen Kao

"for groundbreaking achievements concerning the transmission of light in fibers for optical communication"

and the other half jointly to

Willard S. Boyle and George E. Smith

"for the invention of an imaging semiconductor circuit – the CCD sensor"

THE NOBEL PRIZE IN PHYSICS

Speech by Professor Joseph Nordgren of the Royal Swedish Academy of Sciences. Translation of the Swedish text.

Your Majesties, Your Royal Highnesses, Ladies and Gentlemen,

This year's Nobel Prize in Physics rewards work that has had a decisive impact both on our daily lives and on scientific progress. Today we take it for granted that at any time, we can phone someone on the other side of the globe and have a conversation while enjoying the same sound quality as if we were talking to someone in the same room. Likewise, we take it for granted that we can view photos and video reports of events on other continents at the same instant as they unfold. Our computers are linked up to the world-wide Internet, which has become an almost indispensable element of modern society. It provides efficient communication channels between people and quick access to the information we need at the moment.

Various forms of optical communication were used in the past to send information across long distances. Chains of beacons signalled approaching misfortune, and the optical telegraph could be used for more complex messages. But these forms of optical communication quickly became obsolete as electrical communication in the form of telegraphy, telephony and radio were developed.

Optical communication has made a comeback as a superior communication channel. Light has many thousands of times greater capacity to convey information than copper cables and radio waves. This is related to the fact that the frequency of light waves is much higher than that of electrical waves, enabling the transmission of much faster variations in the signal. Early in the second half of the 20th century, researchers tried to utilise this advantage by transmitting light in optical fibers across long distances, but they encountered great difficulties because attenuation in the fiber was so great that practically all the light had faded away after a few hundred metres. It was not until after Charles Kao showed the way in 1966 towards successful transmission using ultra-pure glass fibers that this new communication technology made its breakthrough. Only four years later, the optical fibers that Kao predicted had become possible, and today his vision of a future communication-based society has become a reality.

Video technology and digital photography play a very prominent role in our daily lives, thanks to television and digital cameras. In science and technology, such tools are irreplaceable today. A large proportion of the information flowing through the global fiber-based Internet consists of images and video material in digital form. One technology that has been of crucial importance in making this possible originated from an invention made in 1969 by Willard Boyle and George Smith. They designed an electronic circuit that can store and read out information about how electrical charges are distributed across its surface – electrical charges that together form an image. The charge-coupled device, or CCD, was born.

Through its simple structure and ingenious function, the CCD enables very efficient and sensitive recording, storage and read-out of high-resolution images in electronic form. As an image sensor, the CCD usually serves as the electronic eye of advanced digital cameras and TV cameras. It is also a very central component of many scientific and medical devices, for example astronomical telescopes and instruments used for "key-hole" surgery.

Dr Kao, Dr Boyle, Dr Smith,

You have been awarded the 2009 Nobel Prize in Physics for your groundbreaking contributions to optical communication and image recording. On behalf of the Royal Swedish Academy of Sciences, it is my privilege and pleasure to convey to you the warmest congratulations for your outstanding work. I now ask you to step forward to receive your Nobel Prizes from the hands of His Majesty the King.

Charles Kao.

CHARLES KUEN KAO

FAMILY BACKGROUND

The Kao family comes from a township called Zhangyan in the Jinshan district near Shanghai, China[1]. As landowners, the family would have been considered wealthy. The sons of each generation would be well educated in the style of the times. My knowledge of the family's genealogy goes back only to Grandfather.

Grandfather Kao Hsieh[2] also went by the courtesy name Kao Ch'ui Wan[3]. He was a literary man, famous for his beautiful poems, which he would render in Chinese calligraphy; the combination of poetry with calligraphy is an art form of the East. As a Confucian scholar, he was a collector of books and also a prominent member of the *Nan She* (Southern Society). Other family members were also active in the Society; its aim during the 1911 Chinese Revolution was to help bring down the ruling Qing dynasty. A museum now erected in the town exhibits his work and also gives a history of the political activities the man participated in. He was of a liberal bent.

Grandfather had four sons and two daughters. My father, Kao Chun Hsin, was not the eldest of the sons. The eldest would have stayed in the town to look after the family properties and affairs; that is where those duties always fell. He was the third son. These sons were growing up as modern times were coming to China. After a good education in Shanghai, my father went to study for a year at the Michigan Law School. Before he left for the U.S.A., a marriage was arranged to a petite lady from one of the families in the social circle. She was left behind to wait for the return of the adventuring husband. Coming from an equally modern and intellectually accomplished family, the new wife was well educated and a poet herself.

1 Much of the family information is reconstructed by Gwen Kao, based on her own knowledge and memories of Charles' family background.

2 In modern pinyin Gao Xie.

3 In modern pinyin Gao Chuiwan.

Figure 1. Charles Kao in 1942, with his younger brother and cousins in Shanghai, China. Second from left: young Charles Kao.

Those were the years when scions of the few wealthier middle-class families in China ventured out to Paris, to London, to New York, to further their experiences and studies. When they returned to China, they were welcomed with great enthusiasm.

On his return from the U.S.A., Kao Chun Hsin, a young man then only in his mid-twenties, was appointed to sit as the Chinese judge in the Court for International Law. He shared the bench with established judges from Western countries. With this prestigious appointment, he and his young wife moved to Shanghai, where they joined the social life of the city.

Figure 2. In 1942, young Charles Kao (second from left) with his family.

Figure 3. Charles Kao in 1938, with his younger brother.

The first child born to the couple was a daughter, followed two years later by a son. Misfortune struck. In a measles epidemic, both children were stricken, and both died from complications, the elder child ten years old, the younger eight. The mother was small boned and of a delicate, fragile appearance. Childbirth would not have been easy. In the years after the tragedy, she had miscarriages one after another. Finally in 1933, I was born, for my parents happily a healthy child. My younger brother Timothy was added to the family four years later.

Because of the earlier loss of the two elder siblings, my brother and I lived a very pampered and protected life. Nursemaids kept constant watch. With my parents busy at dinner parties and social events, we only met them as if for a daily royal audience. Later, home tutors came to give us lessons. The basic lesson plans were readings from the well-known classics, the *Four Books*[4], which we learned to recite by memory. A second tutor taught English.

Finally, when I was ten, I was sent off to school. The driver dropped me off inside the school playground, telling me to wait and someone would tell me where to go. I had never seen so many kids running about so wildly in a crowd and stood wide-eyed.

A bell rang and soon the playground was empty, but for one solitary child. A kindly looking lady appeared and took me to my class. Maybe it was the home tutoring, or the late start to formal schooling, or an overly cautious and protective upbringing, but in any case I never became a talkative person. As an adult I am not always comfortable in social gatherings with small talk. I must have inherited my father's gentle nature.

4 The *Great Learning*, the *Doctrine of the Mean*, the *Analects* of Confucius, and the *Mencius* were the core texts of Confucian learning.

The primary school I attended in Shanghai was a very liberal one, established by scholars who had return from an education in France. The children of leading families were enrolled there, including the son of a well-known man, believed to be a top gangster of the underworld!

Figure 4. Kao's old home in Shanghai, China.

By now the family had a home that was situated inside the French Concession. The International Settlements were in an area within the city where various Western powers held authority. These areas were generally kept free from the rough and tumble of the metropolis, providing a haven from the human poverty of China. The locals nicknamed this area *shili yangchang* (a ten-mile strip of foreign spectacles). So when the Japanese invaded China in the 1930s, these areas were off limits for the invaders.

The family was shielded from the horrors that occurred. With the chaos of war, the law courts were suspended, and social gatherings dried up. My parents stayed home more during this period, and we built a closer relationship with them. Together we played bridge and card games to pass the hours.

The cessation of war with Japan did not bring peace. Soon the Red Army, who had been fighting the Nationalist Government, was at the city gates. My father made the decision to leave. The topic had been discussed often over the family games of bridge and the various options mulled over. To move to Chungking[5], or to join relatives in Taiwan, or perhaps there were relatives who lived in Hong Kong too?

5 Now known as Chongqing.

In 1948, gathering a few belongings, the family boarded a ship that sailed out of Shanghai on a grey dismal wet day. The famous waterfront at the Bund slowly receded out of sight and that was the last we saw of our home city for many years.

A short sojourn in Taipei, the capital of Taiwan, persuaded my father that Hong Kong would be a better haven. With help from maternal relatives in Hong Kong, the family found a small apartment and my brother and I were enrolled into St. Joseph's College where our cousins also studied. As English was the preferred medium of instruction, the local dialect of Cantonese was not a requirement and I never felt the need to learn the tongue.

At 48, my father felt he was too old to study for the law exams that would qualify him to practice in Hong Kong. Instead he took on a job as legal adviser to a company and also taught a few Chinese law courses at local colleges. He became known in Hong Kong as the expert interpreter of the old Chinese laws. There were still some cases of family inheritance and such matters that would need expert advice of this type.

I did well academically, but did not apply myself much to athletics or sports during my five years at St. Joseph's College. My ex-classmates, in later life, reported that they remember me as a quiet person who did not join in the rough and tumble of boy play. I managed almost straight A's in the school matriculation examinations, which qualified me to apply for entry to the University of Hong Kong. However the University was still in some disarray after the war and not all faculties were functioning. I wished to study electrical engineering. Britain beckoned and the British Council in Hong Kong was helpful. In 1953 I boarded a P & O liner bound for England.

I enrolled at Woolwich Polytechnic in London, to sit for the A-level exams which I passed easily. I was so much at home at Woolwich Polytechnic that instead of applying for entry to the other more prestigious Colleges of London University, I continued on to study for my bachelor's degree there. I graduated in 1957 with a B.Sc. in Electrical Engineering. In those days the degrees were awarded as a First, Second, Pass or Fail. As I spent more time on the tennis court than with my books, my degree was a Second.

It was necessary to immediately find a job. Financing my studies had been a heavy burden for my father. I joined Standard Telephones & Cables (STC), a British subsidiary of International Telephone & Telegraph Co (ITT) in North Woolwich, a factory located on the opposite bank of the Thames. As a trainee, I was rotated through different sections for a year before I decided that I liked the work in the microwave division. After two more years, I felt it was time to move on, and applied for a lectureship at Loughborough Polytechnic.

During my three years at STC, I met and married Gwen, a fellow engineer who worked in the lab one floor above my work bench. So we drove to Loughborough for the interview, I got the job, and we immediately set about to look for a home. Finding newly built homes, we put down a deposit on one and drove back to hand in our notices to STC.

HOWEVER, IT WAS NOT TO BE

Figure 5. In 1965, the then young scientist Charles Kao doing an early experiment on optical fibers at the Standard Telecommunication Laboratories in Harlow, U.K.

Someone in senior management had noticed how I had worked on the microwave projects and felt the promise I showed should not be lost to the company. The offer came immediately to allow me to transfer to the research lab, STL, in Harlow. To add incentive for me to stay, a job would be found for Gwen too at the new location. Loughborough had to be pacified and the lawyers got to work on that, and to get the house deposit refunded. The offer was too good to pass up, and my future was destined to take on this new course. I stayed with ITT Corp, working for the next thirty years at various locations, in the U.K., the U.S.A. and Europe.

During these years, in 1967 my parents emigrated to join us in the U.K.. They were then both in their late 60s and elderly, but still able to live independently. They were happy to see a grandson, born in 1961, and a granddaughter, born in 1963.

In 1970, The Chinese University of Hong Kong, CUHK, came calling. Would I come to the institution to set up an electronics department? STL agreed to give a two-year leave of absence, which then became four years. I was able to see the first batch of students graduate, and to establish a graduate program as well. During this time, 1970–1974, annual summer leaves were taken to return to STL to keep abreast of developments in the research field of optical fibers. It was also an opportunity to see my parents, who had remained in Harlow.

By 1974, the project I had seeded, with the now famous 1966 paper published in the *IEE Proceedings*, had progressed to the pre-production development stage. An industry had grown up around it and it was full speed ahead to revolutionize the telecommunications systems around the world. ITT wanted me back to be part of the team for this endeavor.

So the young family was uprooted again and the move this time was to the ITT plant in Roanoke, Virginia, in the U.S.A.. I was promoted to Chief Scientist, then Vice-President and Director of Engineering in charge of the electro-optical products division at ITT. During my years in the U.S.A., I continued to travel to other research labs around the world, to discuss progress, to encourage work, to keep abreast of the leading edge of the developments. It was a busy time. I always dropped by to visit my parents in the U.K..

In 1976, my mother died at the age of 76 years. My widowed father, who visited Roanoke once or twice, preferred his life in the U.K. and this situation continued until finally, in his eighth decade, he moved in to live with us, when we were back in Hong Kong again.

Figure 6. Charles Kao showed an early fascination with the properties of sand.

By the 1980s optical fibers were being laid across the world in vast quantities, and the industry had evolved into a giant. The capacity for communications had grown exponentially. In 1982 I was appointed Executive Scientist at ITT in charge of all research and development activities and I moved to the Advanced Technology Center in Connecticut, U.S.A.. This post was specially created and allowed me the freedom to do anything I considered important for ITT. In order to chart the waters on the ultimate limits of optical communication technology, I pioneered the Terabit Optoelectronics Technology Project to explore the technologies that could reach terabits per second of transmission capacity. The project involved a consortium of ten universities and institutions, and aimed at a goal that was three orders of magnitude higher than the then state-of-the-art. In 1985 I was appointed the Director of Corporate Research at ITT. During this time, many innovations by creative minds were evolving to make use of the increased capacity for sending information over the system. The internet was born.

In 1986, CUHK again came calling. This time the request was for me to accept the position as President of the University, the contract to begin in 1987. In British terminology, this position is known as the Vice-Chancellor[6] of the University.

The second move to Hong Kong was made as ITT Corp began the process of selling off all its technical divisions in the U.S.A. to Alcatel, a French business. After thirty years it was a sad moment to part company with colleagues of long standing.

I spent nine years as the third Vice-Chancellor of CUHK. It was an interesting and challenging time, both for Hong Kong as it prepared for the resumption of sovereignty by China, and for the higher education sector, which saw a massive expansion and improvements in quality. During those years I sought to establish research more prominently as part of the normal work of the faculty, set up contacts with many leading institutions in the U.S.A. and the U.K., and set the University on a path to become a more well-known institution, competitive with the best institutions. I was especially pleased with the establishment of the Faculty of Engineering, in which information technology played a major role. A Faculty of Education was also established during those years. A number of research institutes also came into being. The University nearly doubled in size during those few years, and a fourth undergraduate college was established.

At the University, my role was to create space for people to grow. What I had done essentially was create situations where people would like to take on responsibilities. This had enabled the University to grow as a whole: everybody would be contributing what they should because they feel it is their responsibility and the environment allows them to do so. I created the space at the right time for talent within the University to perform as it should, thus taking it to a new level of development.

The rapid expansion of the tertiary education sector and the increased government funding that came with it have allowed us to do a lot of things to become a top university. The most satisfying change was a scholarly atmosphere on campus – people are pursuing important things because they believe such things are important.

6 The Chancellor being the Governor (after the resumption of sovereignty by China, the Chief Executive) serving in a largely ceremonial role.

It was perhaps natural that my advice was sought and I became involved in technology issues in the community, especially in relation to information technology. I was able to play a part in setting up an Internet exchange in Hong Kong (then a bit of a novelty and even now providing useful service), and in promoting the establishment of the Hong Kong Science Park.

My father died in 1996, just prior to my retirement from the University – a year before the return of colonial Hong Kong back to China. My father's ashes were brought back to England to be laid to rest with my mother – in a cemetery in Harlow. After a year of lecture tours around S. E. Asia, I stayed in Hong Kong, setting up my own company for consultancy. I was appointed to a number of companies as non-executive director.7 In 2009, I moved to California to be closer to my two children.

VIGNETTES FROM CHILDHOOD AND WORKING LIFE MEMORIES

I had a fight with Lo, in my classroom. With our calligraphy brushes, we attacked and daubed each other's faces and hands with black ink. The teacher was horrified. Lo, my mother warned me, was the son of a bad man. I was to be careful not to fight with the boy again. Otherwise school was fine. We sang French songs. The calligraphy teacher was very encouraging. This stroke is beautiful and will be marked with a red dot. But this one is leaning and unbalanced, it will get two red dots!

I made a good friend in primary school who liked to play with me at home. We got our driver to buy all sorts of chemicals. Reading from scientific magazines, together we experimented and made mud balls from phosphorous to throw at stray cats and dogs. The mud balls would explode with a bang and frighten the animals out of their wits! From glass bottles and such like I would fashion the funnels and beakers needed. I had bottles of cyanide and concentrated acids. One day as I was boiling up some nitric acid, the bottle exploded and the concentrated acid splashed onto my young brother's trousers. It burnt the cloth entirely away but fortunately did not land on any of his skin! My parents were furious and confiscated all my chemicals, including the cyanide. I wonder where they disposed of the stuff.

One day my father brought home a big round yellow disc of food that he told us was called 'cheese' that foreigners ate. Food was in short supply so we ate it, but it tasted very funny. When we went outside for some walks we sometimes passed a big tall building with Japanese soldiers standing at the door. My parents told us to walk past quickly as people inside were killed there, and to bow to the soldiers.

It was sad to leave Shanghai, but I was fourteen years old and ready for new adventures. The war with Japan was also exciting. As we cowered under the desks, we could hear the aerial dogfight in the skies above. Peering from under, I could even catch a glimpse of the American planes chasing and diving after the Japanese planes!

7 Note added by Gwen Kao: In 2005, Alzheimer's disease was confirmed, and at the moment there is no cure for it.

So the waterfront of Shanghai faded away and we went to Hong Kong. The schooling there was easy and the book work was not hard at all. The class went on a field trip and some of my classmates got lost in the hills until it was pitch dark and they could see "tigers"! I was not in that group, but now, whenever we have a class re-union, the boys will relate their adventures with gusto and lots of laughs. So my years in Hong Kong passed by swiftly and I was to embark on my next adventure all by myself.

On the boat to England to study electrical engineering, I shared a cabin with three others. Two worked for the Hong Kong Government and were going to England to attend some courses relevant to their work, meteorology and water treatment respectively. The third was a professor of mathematics. During the sea trip, which lasted six weeks, the professor taught me quantum mechanics. He also took some other young students and myself under his wings. He escorted us to his friend's home when we stopped at Singapore and taught us all to eat hot curry and to wash the fiery food down with beer.

I was nineteen years old and had never before left my family to be on my own. Everything was a new experience. To put my foot down on the soil at Port Said so I could tell myself I had been to Africa, to sail pass the Straits of Gibraltar and say I had seen Europe and the Sahara, to find England a cold and grey land! Little did I think eleven long years would pass by before I would see my parents again.

The British Council staff met me and arranged lodgings for me with a landlady in a house in Plumstead Commons, near the Woolwich Poly. Her house was old and large, with rooms for several lodgers. We all ate breakfast and dinner together under the stern eyes of the landlady. Food was still scarce after World War II and the slices of meat served for dinner were so thin, they were transparent when held up to the light! After such a sparse meal, we all trooped out for 'fresh air', but the real reason was to buy fish and chips, which we scoffed down hungrily on the walk back up the hill. The habit remains; I still love fish and chips.

After graduation in 1957 from the Polytechnic, I joined STC and had to walk through the tunnel under the River Thames every day to work. My first project was to build an amplifier and I got out my books to study the theories. My boss came by and said to put away the books, just do it. School work is to exercise the brain so it can think intelligently. There was no need for more revamping of theories!

I met my wife-to-be at work. She was an engineer in the coil section. We were married in 1959 and our children were born, a son in 1961, and a daughter in 1963. By then we were living in Harlow and working at STL.

The research was enthralling work and in 1966 I published the now famous ground-breaking paper, "Dielectric-fibre Surface Waveguides for Optical Frequencies". This research was to spawn a whole new industry over the next twenty years.

People asked me if the idea came as a sudden flash, eureka! I had been working since graduation on microwave transmissions. The theories and limitations were ground into my brain. I knew we needed much more bandwidth and thoughts of how it could be done were constantly in my mind.

Transmission of light through glass is an old, old idea. It had been used in past years to shine light for entertainment, for decoration, for short distances for surgery, through a glass rod, but it had not been possible to use it over the long distances required for telephony. Light passing through a rod of glass just fades out to nothing after a very short distance of a few feet. Efforts by many research laboratories to find a way to transmit light over long distances were desperate as the public, prompted by media reports of hopeful technical advances, were expecting more and more exotic services.

I played around with what was causing the failure of light to penetrate glass. With the invention of the laser in the 1950s and subsequent developments, there was an ideal source of light that could send out pulse of light in a digital stream of noughts and ones, represented by off and on states of the pulse.

Ideas do not always come in a flash, but by diligent trial-and-error experiments that take time and thought.

AWARDS AND DISTINCTIONS RECEIVED BY CHARLES KUEN KAO

1. Morey Award of the American Ceramic Society, in 1976. Citation: for outstanding contribution to glass science and technology.
2. Stuart Ballantine Medal of the Franklin Institute, U.S.A., in 1977. Citation: for conceptual work on optical fiber communication systems.
3. Rank Prize of Rank Trust Fund, U.K., in 1978. Citation: for pioneering work in optical fibre communication.
4. Morris H. Liebmann Memorial Award of IEEE, U.S.A., in 1978. Citation: for making communication at optical frequencies practical by discovering, inventing, and developing the material, techniques and configurations for glass fiber waveguides and, in particular, for recognizing and proving by careful measurements in bulk glasses that silicon glass could provide the requisite low optical loss needed for a practical communication system.
5. L.M. Ericsson International Prize of Ericsson Foundation, Sweden, in 1979. Citation: for fundamental contributions to the long distance transmission of information through optical fibers.
6. Gold Medal of the Armed Forces Communications and Electronics Association, U.S.A., in 1980. Citation: for contribution to the application of optical fiber technology to military communications.
7. Alexander Graham Bell Medal of IEEE, U.S.A., in 1985. Citation: for pioneering contributions to optical fiber communications.
8. Marconi International Fellowship of the Marconi Foundation, U.S.A., in 1985. Citation: for contributions toward a revolution in communication techniques in the form of optical fiber technology.
9. Columbus Medal of the city of Genoa, Italy, in 1985. Citation: for significant scientific discovery.
10. C & C Prize of the Foundation for Communication and Computer Promotion, Japan, in 1987. Citation: for pioneering work in optical fiber communication.
11. Doctor of Science, *Honoris Causa*, of The Chinese University of Hong Kong, in 1985. Citation: for outstanding contributions to optical fibre communications.
12. Faraday Medal of IEE, U.K., in 1989. Citation: for recognition of continuing outstanding work on optical communication including early work on establishing the

feasibility of optical fibre communication systems and defining the parameters involved.

13. International Prize for New Materials of the American Physical Society, U.S.A., in 1989. Citation: for contribution to the materials research and development that resulted in practical low loss optical fibers, one of the cornerstones of optical communications technology.
14. Doctor of Science, *Honoris Causa*, the University of Sussex, U.K., in 1990. Citation: for making important contributions to the understanding and application of opto-electronic techniques for communication.
15. Degree of Honorary Doctor, Soka University, Japan, in 1991. Citation: for promoting science and culture.
16. Doctor of Engineering, *Honoris Causa*, University of Glasgow, U.K., in 1992. Citation: for recognition of his achievements in the field of engineering.
17. Gold Medal of The International Society for Optical Engineering (SPIE), U.S.A., in 1992. Citation: for recognition of his pioneering work and many contributions to the field of fiber optic communications.
18. Commander of the Most Excellent Order of the British Empire (CBE), in 1993. Citation: for his distinguished academic career and his pioneering work on optical fibres.
19. Doctor of Science, *Honoris Causa*, University of Durham, U.K., in 1994. Citation: for recognition of his pioneering work in the field of telecommunications.
20. Gold Medal of Engineering Excellence, of the World Federation of Engineering Organizations (WFEO), in 1995. Citation: for recognition of his pioneering work and development in fiber optics which have deep significance for all mankind.
21. Doctor of the University of Griffith University, Australia, in 1995. Citation: for recognition of his distinguished service to the world of international learning and university education.
22. Japan Prize in the field of Information, Computer and Communication Systems, of The Science and Technology Foundation of Japan, in 1996. Citation: for recognition of his pioneering research on wide-band, low-loss optical fibre communications.
23. Prince Philip Medal of the Royal Academy of Engineering, U.K., in 1996. Citation: in recognition of his pioneering contributions towards the invention of the optical fibre for telecommunications and for his leadership in its engineering and commercial realization; and for his distinguished contribution to higher education in Hong Kong.
24. Doctor of Telecommunications Engineering, *Honoris Causa*, The University of Padova, Italy, in 1996. Citation: for recognition of his pioneering work in the field of telecommunications engineering.
25. Doctor of Science, *Honoris Causa*, University of Hull, U.K., in 1998.
26. Doctor of Science, *Honoris Causa*, Yale University, U.S.A., in 1999. Citation: for recognition of his achievements as an inventor and an educator.
27. Charles Stark Draper Prize of the National Academy of Engineering, U.S.A., in 1999. Citation: for recognition of the conception and invention of optical fiber for communications and for the development of manufacturing processes that made the telecommunications revolution possible.
28. Doctor of Science, *Honoris Causa*, The University of Greenwich, U.K., in 2002.
29. Doctor of Science, *Honoris Causa*, Princeton University, U.S.A., in 2004.
30. The Nobel Prize in Physics, the Royal Swedish Academy of Sciences, Sweden, in 2009. Citation: for groundbreaking achievements concerning the transmission of light in fibers for optical communication.

SAND FROM CENTURIES PAST: SEND FUTURE VOICES FAST*

Nobel Lecture, December 8, 2009

by

CHARLES K. KAO

The Chinese University of Hong Kong, Hong Kong, China.

INTRODUCTION

First of all I would like to express my gratitude to the Royal Swedish Academy of Sciences and its Nobel Committee for Physics for awarding me the 2009 Nobel Prize in Physics. It is not often that the award is given for work in applied science. The practical nature of that work, and the time that has since elapsed, means that much has been done over the years, by many people, and the collective result has had a significant impact on society. Therefore this lecture will be about both the research leading up to the key discoveries, and some of the succeeding developments since.

In the 43 years since the publication of my paper with Hockham on glass fiber cables in 1966 [1], the world of telephony has changed vastly. The skepticism of the early years has gradually been subdued by the steady advance of the optical fiber revolution. In the 1970s, while I was working on the pre-production stage research on optical fibers at the ITT Corporation in Roanoke, Virginia, U.S.A., I received two letters: one contained a threatening message accusing me of releasing an evil genie from its bottle; the other, from a farmer in China, asked for a means to allow him to pass a message to his distant wife to bring his lunch. Both letter writers saw a future that has since become history.

THE EARLY DAYS[1]

In 1960, I joined Standard Telecommunications Laboratories Ltd. (STL), a subsidiary of ITT Corporation in the U.K., after having worked as a graduate engineer at Standard Telephones and Cables in Woolwich for some time. Much of the work at STL was devoted to improving the capabilities of the existing communication infrastructure, with a focus on the use of millimeter wave transmission systems.

---

* This lecture was prepared for Charles Kao by Gwen Kao, with the help of Lian Kuan Chen, Kwok Wai Cheung, Melody Lee, Wing Shing Wong and Kenneth Young. The Lecture was delivered on December 8, 2009 by Gwen Kao on behalf of Charles Kao.

1 The following paragraphs are taken essentially verbatim from Ref. [2, 3]

Millimeter waves at 35 to 70 GHz could have a much higher transmission capacity. But the waters were uncharted and the challenges enormous, since radio waves at such frequencies could not be beamed over long distances due to beam divergence and atmospheric absorption. The waves had to be guided by a waveguide. And in the 1950s, research and development work on low-loss circular waveguides in the HE_{11} mode was started. A trial system was deployed in the 1960s. Huge sums were invested, and more were planned, to move this system into the pre-production stage. Public expectations for new telecommunication services such as the video phone had heightened.

I joined the long-haul waveguide group led by Dr. Karbowiak at STL. I was excited to see an actual circular waveguide [4] (Figure 1). I was assigned to look for new methods for microwave and optical transmission. Ray optics and wave theory were used together to gain a better understanding of waveguide problems – then a novel idea. Later, Dr. Karbowiak encouraged me to pursue a doctorate while working at STL. So I registered at University College London and completed the dissertation "Quasi-Optical Waveguides" in two years.

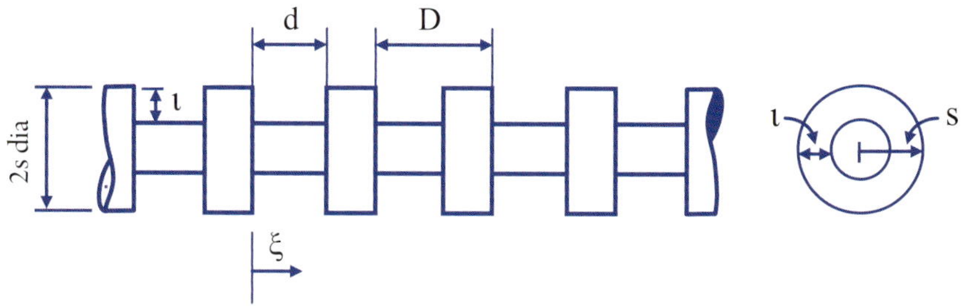

Figure 1. A circular waveguide (after Ref [4]). © [1954] IEEE

The development of the laser in the 1960s gave the telecom community a great dose of optimism that optical communication could be just around the corner. Coherent light was to be the new information carrier with capacity 10^5 times higher than point-to-point microwaves – based on the simple comparison of frequencies: 3×10^{15} Hz (3000 THz) for light versus 3×10^9 Hz (3 GHz) for microwaves.

The race between circular microwave waveguides and optical communication was on, with the odds heavily in favor of the former. In 1960, optical lasers were in their infancy, demonstrated at only a few research laboratories, and performing much below the needed specifications. Optical systems seemed a non-starter. But I still thought the laser had potential. I said to myself: "How can we dismiss the laser so readily? Optical communication is too good to be left on the theoretical shelf." I asked myself two obvious questions:

1. Is the ruby laser a suitable source for optical communication?
2. What material has sufficiently high transparency at such wavelengths?

At that time only two groups in the world were starting to look at the transmission aspect of optical communication, while several other groups were working on solid-state and semiconductor lasers. Lasers emit coherent radiation at optical frequencies, but using such radiation for communication appeared to be very difficult, if not impossible. For optical communication to fulfill its promises, many serious problems remained to be solved.

THE KEY DISCOVERY[2]

In 1963 I was already involved in free-space propagation experiments: The rapid progress of semiconductor and laser technology had opened up a broader scope to explore optical communication realistically. With a HeNe laser beam directed to a spot some distance away, the STL team quickly discovered that the distant laser light flickered. The beam danced around several beam diameters because of atmospheric fluctuations. The team also tried to repeat experiments done by other research laboratories around the world. For example, we set up con-focal lens experiments similar to those at Bell Laboratories: a series of convex lenses were lined up at intervals equal to the focal length. But even in the dead of night when the air was still and even with refocusing every 100 meters, the beam refused to stay within the lens aperture.

Bell Laboratories' experiments using gas lenses were abandoned due to the difficulty of providing satisfactory insulation while maintaining the profiles of the gas lenses. These experiments were struggles in desperation, to control light traveling over long distances.

At STL the thinking shifted towards dielectric waveguides. Dielectric means a non-conductor of electricity; a dielectric waveguide is a waveguide consisting of a dielectric cylinder surrounded by air. Dr. Karbowiak suggested that I and three others should work on his idea of a thin film waveguide [5–6] (Figure 2). But thin film waveguides failed: the confinement was not strong enough and light would escape as it negotiates a bend.

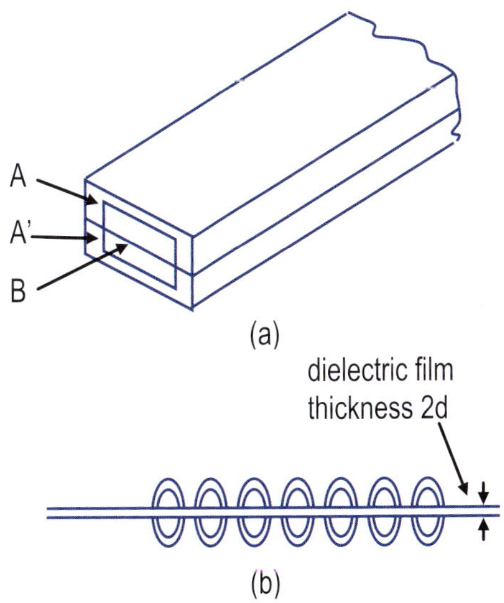

Figure 2. Thin film waveguide (after Ref [6]). © [1965] IEEE

2 The following paragraphs are also taken essentially verbatim from Ref. [2, 3]

When Dr. Karbowiak decided to emigrate to Australia, I took over as the project leader and recommended that the team should investigate the loss mechanism of dielectric materials for optical fibers. A small group worked on methods for measuring material loss of low-loss transparent materials. George Hockham joined me to work on the characteristics of dielectric waveguides. With his interest in waveguide theory, he focused on the tolerance requirements for an optical fiber waveguide, in particular, the dimensional tolerance and joint losses. We proceeded to systematically study the physical and waveguide requirements in glass fibers.

In addition, I was also pushing my colleagues in the laser group to work towards a semiconductor laser in the near infrared, with emission characteristics matching the diameter of a single-mode fiber. A single-mode fiber is an optical fiber that is designed for the transmission of a single ray or mode of light as a carrier. The laser had to be made durable and to work at room temperatures without liquid nitrogen cooling. So there were many obstacles. But in the early 1960s, esoteric research was tolerated so long as it was not too costly.

Over the next two years, the team worked towards its goals. We were all novices in the physics and chemistry of materials and in tackling new electromagnetic wave problems. But we made very credible progress in considered steps. We searched the literature, talked to experts, and collected material samples from various glass and polymer companies. We also worked on the theories and developed measurement techniques to carry out a host of experiments. We developed an instrument to measure the spectral loss of very low-loss material, as well as one for scaled simulation experiments to measure fiber loss due to mechanical imperfections.

I zeroed in on glass as a possible transparent material. Glass is made from silica – sand from centuries past – which is plentiful and cheap. The optical loss in transparent materials is due to three mechanisms: (a) intrinsic absorption, (b) extrinsic absorption, and (c) Rayleigh scattering. The intrinsic loss is caused by the infrared absorption of the material structure itself, which determines the wavelength of the transparency regions. The extrinsic loss is due to impurity ions left in the material and the Rayleigh loss is due to the scattering of photons by the structural non-uniformity of the material. For most practical applications such as windows, the transparency of glass was entirely adequate, and no one had studied absorption down to such levels. After talking with many people, I eventually formed the following conclusions.

1. Impurities, particularly transition elements such as Fe, Cu, and Mn, have to be reduced to parts per million (ppm) or even parts per billion (ppb). However, can impurity concentrations be reduced to such low levels?
2. High-temperature glasses are frozen rapidly and should have smaller micro-structures and more even distribution of non-homogeneities than low-temperature glasses such as polymers, and therefore a lower scattering loss.

Our ongoing microwave simulation experiments were also completed. The characteristics of the dielectric waveguide were fully defined in terms of its modes, its dimensional tolerance both for end-to-end mismatch and for its diameter fluctuation along the fiber lengths. Both theory and simulated experiments supported our approach.

Hockham and I wrote a paper entitled "Dielectric-Fibre Surface Waveguides for Optical Frequencies" and submitted it to the *Proceedings of the Institute of Electrical Engineers* (IEE). After the usual review and revision, it appeared in July 1966 – the date now regarded as the birth of optical fiber communication.

Dielectric-fibre surface waveguides for optical frequencies

Figure 3. The 1966 paper.

THE PAPER

The paper started with a brief discussion of the mode properties in a fiber of circular cross section. It then quickly zeroed in on the material aspects, which were recognized to be the major stumbling block. At the time, the most transparent glass had a loss of 200 dB/km, which would limit transmission to about a few meters – this is very obvious to anyone who has ever peered through a thick piece of glass: nothing can be seen. But our paper pointed out that the intrinsic loss due to scattering could be as low as 1 dB/km, which would have allowed propagation over practical distances. The culprit is the impurities: mainly ferrous and ferric ions at these wavelengths. Quoting from the paper:

> "It is foreseeable that glasses with a bulk loss of about 20 dB/km at around 0.6 micron will be obtained, as the iron-impurity concentration may be reduced to 1 part per million."

In layman terms, if one has a sufficiently "clean" type of glass, one should be able to see through a slab as thick as several hundred meters. That key insight opened up the field of optical communications.

The paper considered many other issues:

- The loss can be reduced if the mode is chosen so that most of the energy is actually outside the fiber.
- The fiber should be surrounded by a cladding of lower index (which later became the standard technology).
- The loss of energy due to bends in the fiber is negligible for bends larger than 1 mm.
- The losses due to non-uniform cross sections were estimated.
- The properties of a single-mode fibre (now the mainstream especially for long-distance and high-speed transmission) were analyzed. It was explained how dispersion limits bandwidth; an example was worked out for a 10 km route – a very bold scenario in 1966.

The paper concluded with the following [1]:

> "The realization of a successful fiber waveguide depends, at present, on the availability of suitable low-loss dielectric material. The crucial material problem appears to be one which is difficult but not impossible to solve. Certainly, the required loss figure of around 20 dB/km is much higher than the lower limit of loss figure imposed by fundamental mechanisms."

Basically all of the predictions pointed accurately to the path of development, and we now have 10^{-2} of the loss and 10^{5} times the bandwidth then forecast – the revolutionary proposal in the 1966 paper was in hindsight too conservative.

CONVINCING THE WORLD

The substance of the paper was presented at an IEE meeting on January 27, 1966 [7]. Most of the world did not take notice – except for the British Post Office (BPO) and the U.K. Ministry of Defence, who immediately launched major research programs [8, 9]. By the end of 1966, three groups in the U.K. were studying the various issues involved: I myself at STL; Roberts at BPO; Gambling at Southampton in collaboration with Williams at the Ministry of Defence Laboratory [10].

In the next few years, I traveled the globe to push my idea: to Japan, where enduring friendships were made dating from those early days; to research labs in Germany, in the Netherlands and elsewhere to spread the news. It was obvious to me that until more and more jumped on the bandwagon, the use of glass fibers would not take off. There was widespread skepticism, but I remained steadfast. The global telephony industry is huge, too large to be changed by a single person or even a single country, but I was stubborn and I remained optimistic; slowly others were converted into believers.

The experts at first proclaimed that the materials were the most severe of the intrinsic insurmountable problems. Gambling [9] wrote that British Telecom had been "somewhat scathing" about the proposal earlier, and Bell Laboratories, which could have led the field, took notice only much later. Approaches were made to many glass manufacturers to persuade them to produce the clear glass required. There was eventually a response from Corning, where Bob Maurer led the first group that later produced the glass rods and developed the techniques to make the glass fibers to the required specifications.

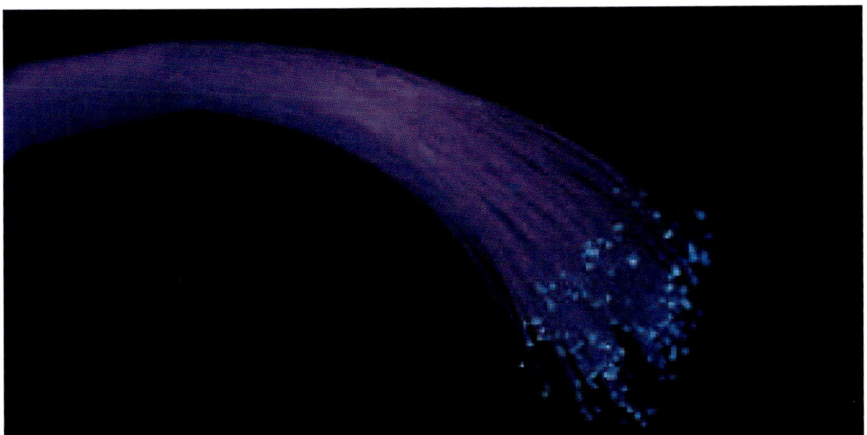

Figure 4. Glass fiber.

Meanwhile, I continued to work on proving the feasibility of glass fibers as the medium for long-haul optical transmission. There were a number of formidable challenges. The first was the measurement techniques for low-loss samples that were obtainable only in lengths of around 20 cm. The problem of assuring surface perfection was also formidable. Another problem is end-surface reflection loss, caused by the polishing process. There was a measurement impasse that demanded the detection of a loss difference between two samples of less than 0.1%, when the total loss of the entire 20 cm sample is only 0.1%. An inexact measurement would have been meaningless.

In 1968 and 1969, I and my colleagues at STL, T. W. Davies, M. W. Jones and C. R. Wright, published a series of papers [11–13] on the attenuation measurements of glass that addressed the above problems. At that time, the measuring instruments called spectrophotometers had a rather limited sensitivity – in the range of 43 dB/km. The measurement was very difficult: even a minute contamination could have caused a loss comparable to the attenuation itself, while surface effects could easily be ten times worse. We assembled a homemade single-beam spectrophotometer that achieved a sensitivity of 21.7 dB/km. Later improvements with a double-beam spectrophotometer yielded a sensitivity down to 4.3 dB/km.

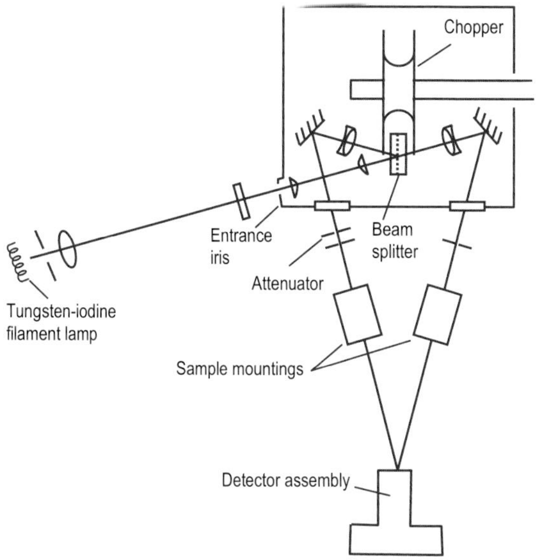

Figure 5. Double-beam spectrophotometer (after Ref [12]).

The reflection effect was measured with a homemade ellipsometer. Measurements were made with fused quartz samples made by plasma deposition, in which the high temperature evaporated the impurity ions. With the sensitive instrument, the attenuation of a number of glass samples was measured and, eureka, the Infrasil sample from Schott Glass showed an attenuation as low as 5 dB/km at a window around 0.85 μm (Figure 6) – at last proving that the removal of impurity would lower the absorption loss to useful levels.

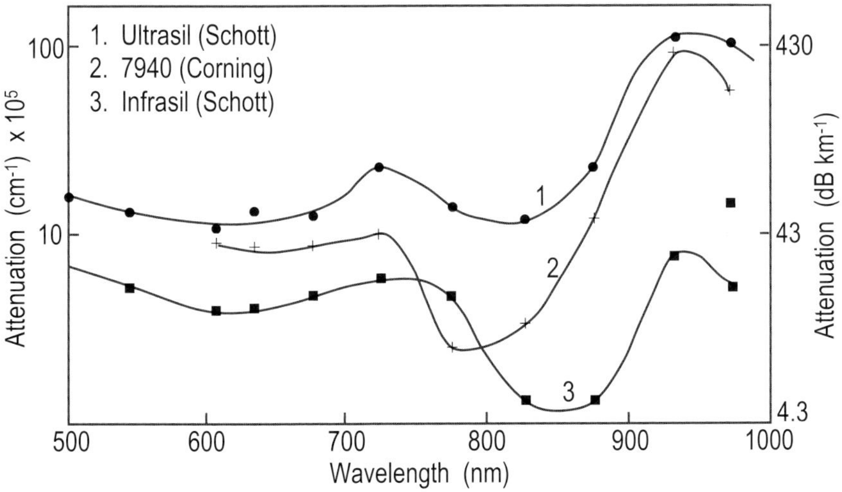

Figure 6. Glass attenuation vs. wavelength (after Ref [12]).

This was really exciting because the low-loss region is right at the GaAs laser emission band. The measurements clearly pointed the way to optical communication – compact GaAs semiconductor lasers as the source, low-cost cladded glass fibers as the transmission medium, and Si or Ge semiconductors for detection [12]. The dream no longer seemed remote. These measurements apparently turned the sentiments of the research community around. The race to develop the first low-loss glass fiber waveguide was on.

In 1967, at Corning, Maurer's chemist colleague P. Schultz helped purify the glass. In 1968, his colleagues D. Keck and F. Zimar helped draw the fibers. By 1970, Corning had produced a fiber waveguide with a loss of 17 dB/km at 0.633 μm using a Ti-diffused core with silica cladding, using the Outside Vapor Deposition (OVD) method [15]. Two years later, they reduced the loss to 4 dB/km for a multimode fiber by replacing the Ti-doped core with a Ge-doped core.

Bell Laboratories finally created a program in optical fiber research in 1969. Their work on hollow light pipes was finally stopped in 1972. Their millimeter wave research program was wound down and eventually abandoned in 1975.

IMPACT ON THE WORLD

Since the deployment of the first-generation, 45 Mb/s fiber optic communication system in 1976, the transmission capacity in a single fiber has rapidly increased: we now talk about terabits per second. In order to understand the fundamental limits of fiber-optic communication, the Terabit per Second Optoelectronic Project was launched [16] during 1982–85, involving ten research organizations. The target technology, three orders of magnitude higher than the then state-of-the-art, was considered impossible at the time.

But the transmission capacity has gradually increased a millionfold from the tens of Mb/s in the early days to many tens of Tb/s in recent years. Data can be carried over millions of km of fibers without going through repeaters, thanks to the invention of the optical fiber amplifier and wavelength division multiplexing. So that is how the industry grew and grew.

The world has been totally transformed because of optical fiber communication. The telephone system has been overhauled and international long distance calls have become easily affordable. Brand new mega-industries in fiber optics including cable manufacturing and equipment, optical devices, network systems and equipment have been created. Hundreds of millions of kilometers of glass fiber cables have been laid, in the ground and in the ocean, creating an intricate web of connectivity that is the foundation of the World Wide Web. The Internet is now more pervasive than the telephone used to be. We browse, we search, we hold net conferences, we blog, we watch videos, we shop, we socialize online. The information revolution that started in the 1990s could not have happened without optical fibers. Over the last few years, fibers have been laid all the way to our homes. All-optical networks that are environmentally green are being contemplated. The revolution in optical fiber communication has not ended – it might still just be at the beginning.

CONCLUSION

The world-wide communication network based on optical fibers has truly shrunk the world and brought human beings closer together. I hardly need to cite technical figures to drive this point home. The news of the Nobel Prize reached me in the middle of the night at 3 am in California, through a telephone call from Stockholm (then in their morning) no doubt carried on optical fibers; congratulations came literally minutes later from friends in Asia (for whom it was evening), again through messages carried on optical fibers. Too much information is not always a good thing: we had to take the phone off the hook that night in order to get some sleep.

In the last twenty years, creative minds have added many ingenious ways to make use of the added bandwidth that has resulted from the use of optical fibers. Optical communication is by now not just a technical advance, but has also caused major changes in society. It will continue to change the way people learn, the way they live and relate to each other, as well as the way they work. For example, manufacturing of all the bits and pieces of a single product can now take place over a dozen locations around the world, providing huge opportunities for people worldwide. The ready accessibility of information has obviously led to more equality and wider participation in public affairs. The evolution continues. Who knows what other forward thinkers will dream of in the future. The unused amount of bandwidth beckons and tempts. Many words, indeed many books have been written about the information society, and I do not wish to add to them here – except to say that it is beyond the dreams of the first serious concept of optical communication in 1966, when even 1 GHz was only a hope.

I would like to thank ITT Corporation, where I developed my career for 30 years, and all those who climbed on to the bandwagon with me in the early days, since without the legions of believers the industry would not have evolved as it did. I planted the seed; Bob Maurer watered it and John MacChesney grew its roots.

REFERENCES

[1] K. C. Kao and G. A. Hockham, "Dielectric-fibre surface waveguides for optical frequencies," *Proc. IEE*, Vol. 113, pp. 1151–1158, 1966. (Republished in *Proc. IEE*, Vol. 133, Pt. J, pp. 191–198, 1986.)

[2] K. C. Kao, *Autobiography of Charles K. Kao* 《潮平岸闊——高錕自述》(translated by 許迪鏘) Joint Publishing, 2005.

[3] K. C. Kao, "Optical Communication Systems," pp. 3–100, in K. Young (ed), Opto-electronics, Lectures at the 1982 International Summer School at The Chinese University of Hong Kong, 26 July – 6 August 1982, The Chinese University of Hong Kong Press, 1982.

[4] H. E. Barlow and A. E. Karbowiak, "An experimental investigation of the properties of corrugated cylindrical surface waveguides," *Proc. IEE*, Vol. 101, Part III, pp. 182–188, 1954.

[5] K. C. Kao, "Electromagnetic wave propagation on a double-layer dielectric film," *Elect. Lett.*, Vol. 1, No. 2, pp. 35–37, 1965.

[6] A. E. Karbowiak, "New type of waveguide for light and infrared waves," *Elect. Lett.*, Vol. 1, No. 2, pp. 47–48, 1965.

[7] R. Dyott, "Some memories of the early years with optical fibres at the British Post Office: a personal account," *Proc. IEE*, Vol.133, pp. 199–201, 1986.

[8] W. A. Gambling, "Optical fibres: the Southampton scene," *Proc. IEE.*, Vol. 133, pp. 205–210, 1986.

[9] STC Press Office, "STL develops techniques aimed at communication by guided light. British team's work described at IEE Meeting in London", January 1966.

[10] D. Williams and K. C. Kao, "Pulse communication along glass fibers," *Proc. of IEEE.*, Vol. 56, No. 2, pp. 197–198, 1968.

[11] K. C. Kao and T. W. Davies, "Spectrophotometric studies of ultra low loss optical glasses I," *J. Sci. Instrum.* (J. Phys. E), Vol. 1, pp. 1063–8, 1968.

[12] M. W. Jones and K. C. Kao, "Spectrophotometric studies of ultra low loss optical glasses II," *J. Sci. Instrum.* (J. Phys. E), Vol. 2, pp. 331–5, 1969.

[13] C. R. Wright and K. C. Kao, "Spectrophotometric studies of ultra low loss optical glasses III," *J. Sci. Instrum.* (J. Phys. E), Vol. 2, pp. 331–5, 1969.

[14] J. Lytollis, C. K. Kao and G. I. Turner, "Infra-red optical communication systems," *Infrared Physics*, Vol. 8, No. 1, pp. 123–130, 1968.

[15] F. P. Kapron, D. B. Keck and R. D. Maurer, "Radiation losses in glass optical waveguides," *Appl. Phys. Lett.*, Vol. 17, pp. 423–5, 1970.

[16] C. K. Kao, "1012 bit/s optoelectronics technology," *Proc. IEE.*, Vol. 133, pp. 230–236, 1986.

Portrait photo of Professor Kao by photographer Ulla Montan.

WILLARD S. BOYLE

I was born on August 19, 1924 in Amherst, Nova Scotia and raised in the village of Wallace until I was about two years old. At this time, my family moved to Chaudiere, a small logging community in northern Quebec. My father was the local physician in the community and my mother took on the task of home schooling me. She believed in the Socratic method of teaching, asking me questions about my work for the day that required very detailed explanations as answers. She was a curious woman, and through her teaching I developed a strong curiosity as well. When I was fourteen, I began my formal education at Lower Canada College in Montreal. My peers were a stark contrast to the logging men I'd known growing up and I quickly earned the nickname Butch.

Following Lower Canada College, I continued to pursue my scientific interests at McGill University. Then, in 1943, I had some time away from my studies, when I joined the Royal Canadian Navy. From there I applied to the Fleet Air Arm and was trained to land Spitfire fighter planes on aircraft carriers. When the war was coming to an end, I was anxious to return home to school. The transition back to class was not easy, but I found it very helpful that my best friend from the Navy was also at McGill and we were able to work together to adapt back to our old lives. During this same fall I met my wife Betty, who made me believe in love at first sight. We married in 1946 and had our first son, and in 1947 we had our first daughter and I completed my Bachelor of Science. With two young children at home only twenty-one months apart, I continued my studies at McGill, completing my Masters in Science in 1948 and Doctorate in Physics in 1950.

After completing my Doctorate, I remained at McGill in Canada's Radiation Laboratory for one year. I then spent two years teaching physics at the Royal Military College of Canada in Kingston, Ontario. Here Betty and I purchased our first small house for our growing family while welcoming our next son in 1952. About this time, I was met with several new opportunities for employment. In 1953, after careful consideration, I decided the academic community available through Bell Labs would become my new home. While work life became very busy, in 1955 we welcomed our youngest daughter into the world.

My time with Bell Labs was always challenging, and this led to several important discoveries. In 1962, I worked with Don Nelson to create the first continuously operating ruby laser. Also in 1962, I became the Director of Space Science and Exploratory Studies at Bellcomm, a division of Bell Labs in Washington, D.C. While at Bellcomm I supported the Apollo space program and was able to aid in the selection of lunar landing sites. I returned to Bell Labs in 1964. Then in 1969, during a brainstorming session with George Smith, we created the charged-coupled device. During my time at Bell Labs I worked on about 18 patents.

I retired from Bell Labs in 1979 as the Executive Director of the Communication Science division. Retired at 55, I sailed my 33-foot boat for six leisurely weeks up the inland waterway from New Jersey, the New York Harbor and up to Quebec and down the St. Lawrence to the house we had built in Wallace. I was accompanied by five or six different crews made up of former assistants, friends and family, and a cat and a dog. Following this I have continued to live in Wallace, Nova Scotia and Lac Tremblant, Quebec. I have been graced with ten wonderful grandchildren and one great-grandchild.

I have been active in several advisory capacities and have won several awards (listed below) for my work.

- The Ballantine Medal of the Franklin Institute, 1973.
- Morris Lieberman Award of the IEEE, 1974.
- Progress Medal of the Photographic Society of America.
- Breakthrough Award by the Device Research Conference of the IEEE.
- Co-winner, C&C prize of the NEC Foundation, Tokyo, 1999.
- Edwin H. Land Medal, Optical Society of America, 2001.
- Canadian Science & Engineering Hall of Fame, 2005.
- Charles Stark Draper Prize, National Academy of Engineering, 2006.

CCD – AN EXTENSION OF MAN'S VISION

Nobel Lecture, December 8, 2009

by

WILLARD S. BOYLE

Bell Laboratories, Murray Hill, NJ, U.S.A. (retired)

I'm sure that you've all heard that this event starts with a phone call at five o'clock in the morning. There's a friendly voice at the end that says, "You have just won the Nobel Prize." I thought I'd just tell you the things that they don't tell you at that time. That is, that you're being offered a job and there's no salary involved. The working hours are typically twelve hours a day. However, there are certain offsets. That is, that overnight you'll become world famous and that you can always hope that perhaps there's no extension of this service.

It is a great honor to be awarded a Nobel Prize. This is a wonderful experience for my wife Betty and me. We received congratulations by email, phone and post, many from old friends we had not seen for some time.

I found that my career at Bell Telephone Labs thrived because of the environment, which encouraged cooperative research, offered opportunities for access to sophisticated equipment, and fellowship. The fellowship was very important when it coupled the skills of a theoretical physicist with an experimentalist. It is not surprising that George Smith's and my Nobel Prizes represented the seventh Nobel awards for Bell Labs.

Lester Germer was my first supervisor at Bell Labs. He was the Germer of the *Davisson and Germer Experiment* that is sometimes referred to in introductory texts on physics. During some early work on thermionic cathodes, they were able to show data that the electron could behave as either a wave or a particle. This was a key to understanding many atomic scale phenomena in the world of quantum mechanics.

Germer's main interest outside of the laboratories was rock climbing and sometimes mountain climbing when he was on extended vacations. The new boys in his laboratory were expected to participate in his sport at least once. My test came one day; we drove to his favorite spot, about ninety miles north of our Murray Hill, New Jersey lab. There it was – a fifty-foot sheer cliff! The face of the cliff made a ninety-degree angle with the road. He started up the cliff, hand over hand. He was laughing at me as I stood at the bottom of the cliff with both feet still on the walkway.

During the nineteen-forties, a major problem began to appear in the telephone-switching network. As the number of subscribers increased, the complexity of the network increased even faster. Some of the switching was done manually. It had been jokingly said, that soon everyone in the country would be working for the telephone company as telephone operators. The real problem was the number of electromagnetic relays that were needed. They were big, expensive, and unreliable. I once saw the machine that made them. It was the size of a small steam locomotive and sounded like one. Plastic, wire, and metal plate went in one end and relays came out the other.

Faced with the problem, the president of Bell Labs, Mervin Kelly, hired three new scientists: a physicist, a theoretical physicist, and a Ph.D. in electrical engineering. I do not know what he said to them, but I don't believe he told them to design a better mechanical relay. We are talking now about Brattain, Bardeen, and Schockley, who went on to explore the *Transistor Effect* and, in the process, established the whole field of digital electronics.

George Smith and I spent a great deal of time thinking and dreaming about possible new electronic devices. We considered both new materials and new structures. For example, the semimetal bismuth was considered because of the small effective mass of its electrons. For various other reasons, however, it was not promising. Our boss, Jack Morton, called me frequently on our picture phones. He was keen about utilizing the then-new magnetic bubble devices. Could we do something similar in silicon technology? It was not easy to ignore Jack Morton. He was a man of strong personalities. The picture phone was a large piece of equipment that stood on our desks, well before the invention of the CCD. I tried to get out of his view, but as I sank down in my chair, Morton would shout, "Sit up straight so I can see you! If your lab is not more productive, I may have to cut your budget."

The invention of the CCD took place one afternoon over one of our frequent brainstorming sessions at the blackboard. We began drawing a diagram, and before it was finished, we knew we had something special. After a few weeks of work, George asked the "shop" to make a model of our device. Somewhat to our surprise, the very first model worked as we had hoped. The first 3-bit device was born. George and I described this discovery one afternoon at a meeting of the Institute of Electrical Engineers. It had been rumored that Bell Labs would be presenting something special at this meeting and it was attended by many interested people from the West Coast.

Some people have a succinct way of summarizing events in science and technology. Jim Early, late Director at Fairchild Semiconductor said, "The transistor worked with the sense of sound while the CCD worked with the sense of sight."

Over the years, George and I have received congratulatory letters. One was from a group of twenty celestial telescope directors. They said the CCD had made revolutionary improvements in the performance of their telescopes. Then, one night while I was watching television, I saw a live photo of the planet Mars. There were boulders and banks of sand. It could have been a desert on Earth. I felt a vicarious achievement with mankind's new extension of his vision into planetary space.

Finally, I've given you a little outline as to how things got started, but I may have left out the most important phase of this starting altogether, and that's in the beginning of my education. We lived in Northern Quebec and the nearest school was thirty miles away, so my mother took on the task of home schooling me. She spoke to some friends, received some instructions from the provincial school board, and found some interesting books that perhaps I might find useful. The two key books were both authored by Lancelot Hogben, *Science for the Citizen* and *Mathematics for the Million.* My mother was not a teacher, but she believed in the Socratic method and from time to time she would ask questions she found in these books and others until she was satisfied that I was making progress. Indeed, it was a good education, with a good introduction to Calculus, Egyptian history, and Radio Engineering, forming a strong background for future studies. This continued until I began public school in the ninth grade. Before that, she began to worry. I had no other children to interact with and had only lumberjacks as friends. We would go out sledding and play, but her concern led her to taking me on trips into Montreal, the nearest town. We would go to university lectures at McGill University and as the lecture proceeded she'd be sitting alongside me and would discuss what we had learned about the topics in our work. At the end of the lecture, she would nudge me and whisper, "Listen to what they are saying, you know that. So why don't you stand up and tell them who you are." If she were alive today she might say, "I guess you don't have to tell them so often, now that you have a Nobel Prize."

Portrait photo of Dr. Boyle by photographer Ulla Montan.

George E. Smith

GEORGE E. SMITH

I was born in White Plains, New York in 1930. I grew up in several different states and attended a variety of primary and secondary schools. Upon graduation from high school in 1948, I joined the U.S. Navy where I spent four years as an aerographer's mate (weatherman), part of it during the Korean War. While stationed in Miami, Florida, I managed to take enough courses at the University of Miami to qualify as a starting sophomore at the University of Pennsylvania in 1952. Since I served during the Korean War, I qualified for the GI Bill, which helped pay tuition. I majored in Physics and graduated with honors in 1955. I married right after graduation. I was accepted into the graduate program at the University of Chicago with a job as teaching assistant and graduated there in 1959 with a Ph.D. in Physics. During that time I also received grants from the National Science Foundation and Bell Telephone Laboratories. Upon graduation, I was offered a job with the Research area of Bell Labs and accepted it. I never bothered to interview at any other institution.

I was assigned to a new department, headed by Bill Boyle, where I started out continuing along the lines of my thesis topic, studying the electronic properties of semimetals. I branched out into other fields as well, including thermoelectric cooling materials and low temperature electronic devices. During the five years in that position I generated many papers and patents. Bill was then promoted to Director of the exploratory semiconductor device development laboratory and he offered me the position as head of a new department entitled Device Concepts. Several bright, imaginative people were assigned to my group and I was given the mandate to hire more of the same. Many fields were pursued including junction lasers, semiconducting ferroelectrics, electroluminescence, transition metal oxides, and the silicon diode array camera tube. In 1969, Bill and I invented the Charge Coupled Device and much of my time was then spent in that field. In addition, my department was renamed the VLSI Device Department where responsibilities covered the physics of devices made with submicron lithography and their use in high performance digital and analog circuits.

In the realm of academic recognition, I have been a member of Pi Mu Epsilon, Phi Beta Kappa, and Sigma Xi. I was made a fellow of the Institute of Electrical and Electronic Engineers, Fellow of the American Physical Society and a member of the National Academy of Engineering. I hold 31 US patents and am the author of over 40 papers. I was founding editor of the IEEE publication "Electron Device Letters".

My major technical accomplishment, of course, was the inception of the Charge Coupled Device with Willard S. Boyle. We hold the basic patent

(US 3,858,232) and published the first paper disclosing the device concept accompanied by a paper on its experimental verification. A following invention of the Buried Channel Charge Coupled Device (US patent 3,792,322) significantly improved the performance of the original CCD. These accomplishments have been recognized in the following awards:

- 1973 *Ballantine Medal of the Franklin Institute* – "For the invention of the Charge-Coupled Device structure, a conceptually simple semiconductor technology with significant application to image sensing, serial memory and signal processing."
- 1974 *Morris N. Liebman Award of the Institute of Electrical and Electronic Engineering*– "For the invention of the Charge-Coupled Device and leadership in the field of MOS device physics."
- 1986 *Progress Medal of the Photographic Society of America* – "For the Charge-Coupled Device's applications to electronic imaging devices."
- 1999 *IEEE Device Research Conference Breakthrough* – "For pioneering work in the field of Buried-Channel Charge Coupled Devices".
- 1999 *Computer and Communications Prize* – "For the Invention of the Charge Coupled Device (CCD)".
- 2001 *Edwin Land Medal* – "For the invention and development of the Charge-Coupled Device, a contribution that has had extraordinary impact on image creation and utilization".
- 2006 *Charles Stark Draper Prize* – "For the invention of the Charge-Coupled Device (CCD), a light-sensitive component at the heart of digital cameras and other widely used imaging technologies."

And, of course,

- 2009 *Nobel Prize in Physics* – "For the invention of an imaging semiconductor circuit – the CCD sensor."

Additional honors include:

- 2006 Honored for the invention of the CCD by US Senate Resolution 478
- 2006 Induction into the National Inventors Hall of Fame
- 2006 Induction into the Imaging Hall of Fame
- 2008 Induction into the New Jersey Inventors Hall of Fame

Also, I received the: IEEE Electron Devices Society Distinguished Service Award (1997).

On the personal side, I had always wanted to sail and purchased my first boat right after joining Bell Labs and sailed it and subsequent boats on Barnegat Bay, about halfway down the coast of New Jersey. My wife also enjoyed sailing but she passed away in 1975. I then commenced to raise children aged 10, 11 and 14 as a single parent. No comment necessary. Two years later, I became partners with Janet Murphy who also loved sailing and we had many adventures sailing a small (22 foot) cabin boat from Northeast Harbor, Maine to Beaufort, North Carolina. We both decided to retire (she was a teacher) early and sail around the world. To do this, we had a seagoing 31-foot Southern Cross, named Apogee, semi custom built for the task in Bristol, Rhode Island in 1983. After two shakedown trips to Bermuda, we retired in 1986 and started around the world. We did not get back until 2003, although we occasionally flew home for a short visit. We now live in our home on a lagoon leading to Barnegat Bay. Apogee is moored to our dock in the back yard and we still enjoy sailing on the bay.

REFERENCES

1. W. S. Boyle and G. E. Smith, "Charge Coupled Semiconductor Devices," Bell Sys. Tech. J., 49, 587, (1970).
2. G. F. Amelio, M. F. Tompsett, and G. E. Smith, "Experimental Verification of the Charge Coupled Device Concept," Bell Sys. Tech. J., 49, 593, (1970).

THE INVENTION AND EARLY HISTORY OF THE CCD

Nobel Lecture, December 8, 2009

by

GEORGE E. SMITH

Bell Laboratories, Murray Hill, NJ, U.S.A. (retired)

To try to indicate how the invention of Charge Coupled Devices in 1969 by W. S. Boyle and myself came about, it is first necessary to describe three important technologies being pursued at the time. Magnetic bubbles, the silicon diode array vidicon for the Picturephone, and MOS semiconductor technology were all contributing factors to the invention of the device.

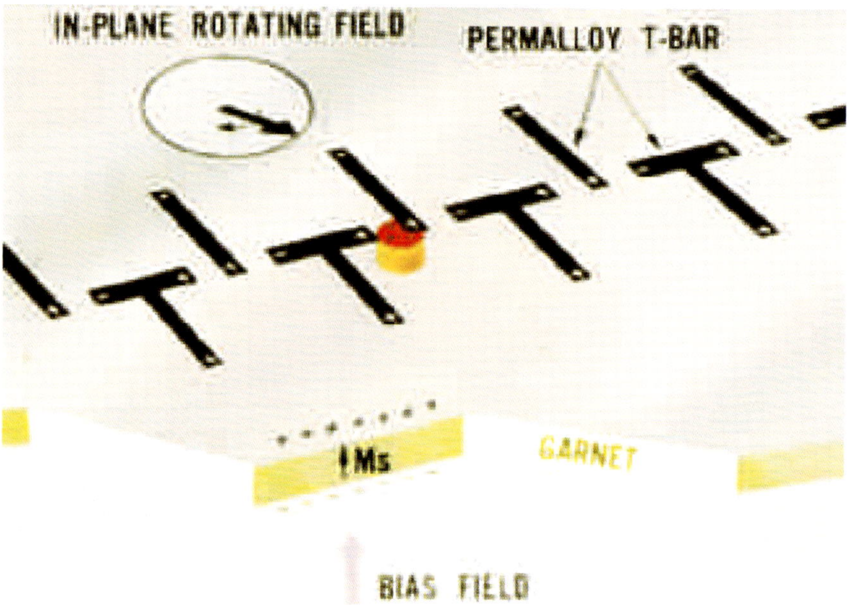

Figure 1. Magnetic Bubble serial shift register.

Magnetic bubbles were formed by placing a special magnetic material, typically garnet, in a magnetic field so that circular magnetic domains could be formed as shown in Figure 1. A permalloy pattern was deposited on the surface which could capture the domain, and when a rotating magnetic field was applied in the plane of the slice, the bubble would be shifted along from one site to the next. With an input device to inject the presence or absence of bubbles representing 0s or 1s and a detection device at the end, one had a serial memory device.

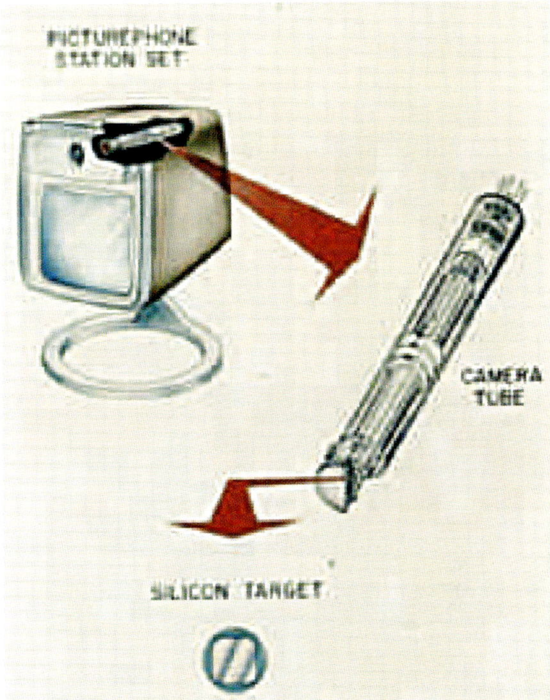

Figure 2. The Picturephone set and silicon diode array camera tube.

The second driving technology being worked on, which just happened to be in my department, was the silicon diode array camera tube, shown in Figure 2. This was being developed for the Picturephone, which AT&T – Bell Labs' parent and the major telephone provider at the time – thought would be a huge market once people discovered they could see as well as hear one another over the existing twisted pair phone lines. The initial imaging device considered was the standard commercial vidicon with an antimony trisulfide coating on glass accessed by an electron beam as the photosensitive element. This device was notorious for its lack of robustness and longevity. Instead, we were working on a tube with a silicon slice with millions of diodes placed on the surface, which were accessed by an electron beam in a similar manner. The silicon was an improvement; however, we strongly hoped to find an all solid state device to eliminate the electron beam.

The third field being vigorously pursued was MOS transistor technology. This was the technology which finally became the basis of all modern integrated circuits. The Metal Oxide Semiconductor capacitor which forms the gate of the MOS transistor is shown in Figure 3. Basic studies of this structure were also being carried out in my department.

To understand the atmosphere at the time, one must look at the structure of the Electronics Division at Bell Labs, which was under Vice President Jack Morton. This was split into two divisions, one for semiconductor devices and one for all others. Bill Boyle was Executive Director of the semiconductor part and I was a Department Head under him. Jack Morton was anxious to speed up the development of magnetic bubbles as a major memory technology and there was talk of transferring resources from Bill's division to the other where the bubble work was being done. For this not to happen, Morton demanded that Bill's division come up with a semiconductor device to compete with bubbles. To address this demand, on October 17, 1969, Bill and I got together in his office. In a discussion lasting not much more than an hour, the basic structure of the CCD was sketched out on the blackboard, the principles of operation defined, and some preliminary ideas concerning applications were developed. The train of thought evolved as follows.

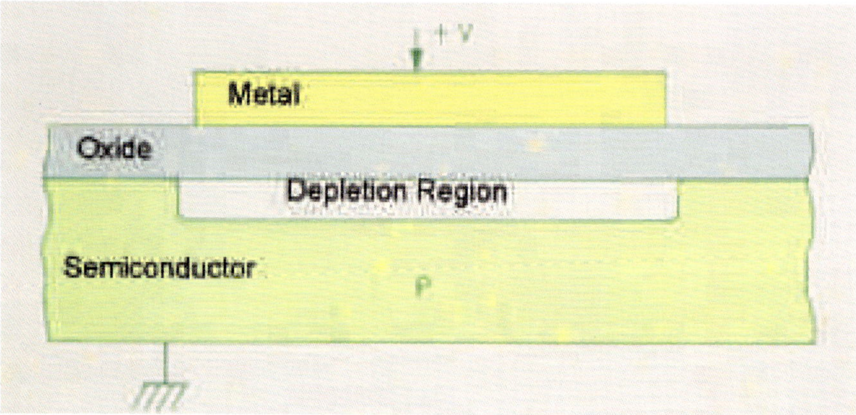

Figure 3. The basic MOS structure.

First, the semiconductor analogy of the magnetic bubble is needed. The electric dual is a packet of charge. The next problem is how to store this charge in a confined region. The structure which came to mind, of course, was the simple MOS capacitor shown in Figure 3. Charge can be introduced into this depletion region with the amount of charge stored being the magnitude of the signal. To understand this better, a plot of electron energy vs. distance into the structure is shown in Figure 4. As charge is introduced into the depletion region, the potential at the surface rises until the maximum allowable charge is reached. Any further charge added will flow into the substrate.

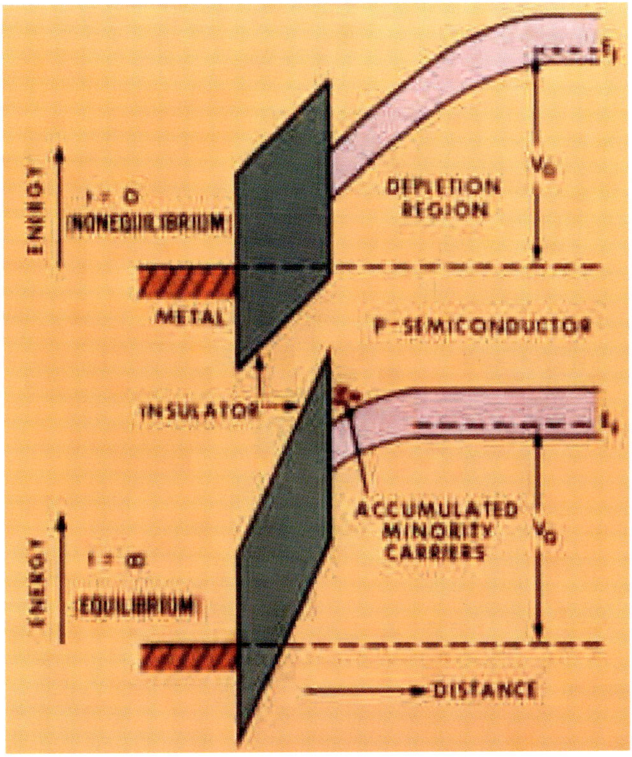

Figure 4. A plot of electron energy vs. distance through the MOS capacitor.

The last problem was to shift the charge from one site to another, thereby allowing manipulation of the information. This is solved by simply placing the MOS capacitors very close together as shown in Figure 5, one with charge and the second empty. In order to pass the charge from one to the next, one simply applies a more attractive voltage to the second, causing its depletion region to overlap the first and the charge to flow along the surface to the silicon-silicon dioxide interface of the second capacitor.

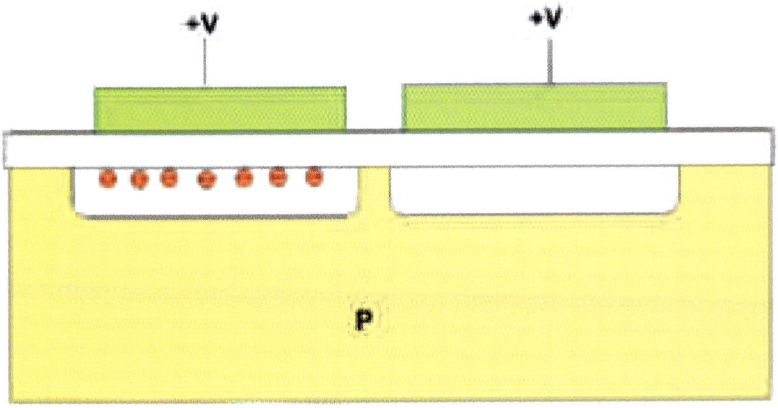

Figure 5. Two closely spaced MOS capacitors.

The original structure using this mechanism to make a shift register is shown in Figure 6. Many MOS capacitors are placed closely together in a row and connected to a three phase voltage source. The top figure shows the storage phase with Va applied to one set of electrodes and a smaller rest voltage Vb applied to the other two. One site has charge and the other has none. The second figure shows the transfer phase where a larger voltage, Vp, is applied to the adjacent plates to transfer charge from one to the next. The last two show resetting the voltages to the initial state with the charge information shifted by one site. This is continued to an output device at the end of the row in order to read the stored information. Many other storage and transfer schemes are possible. The charge can be injected into the device electrically at the beginning of the row making a shift register or supplied by light incident on a structure with empty cells. Then the amount of charge which accumulates by the absorption of photons is determined by the intensity of the light, and the resulting charge pattern can then be read out in shift register fashion after a suitable integration time.

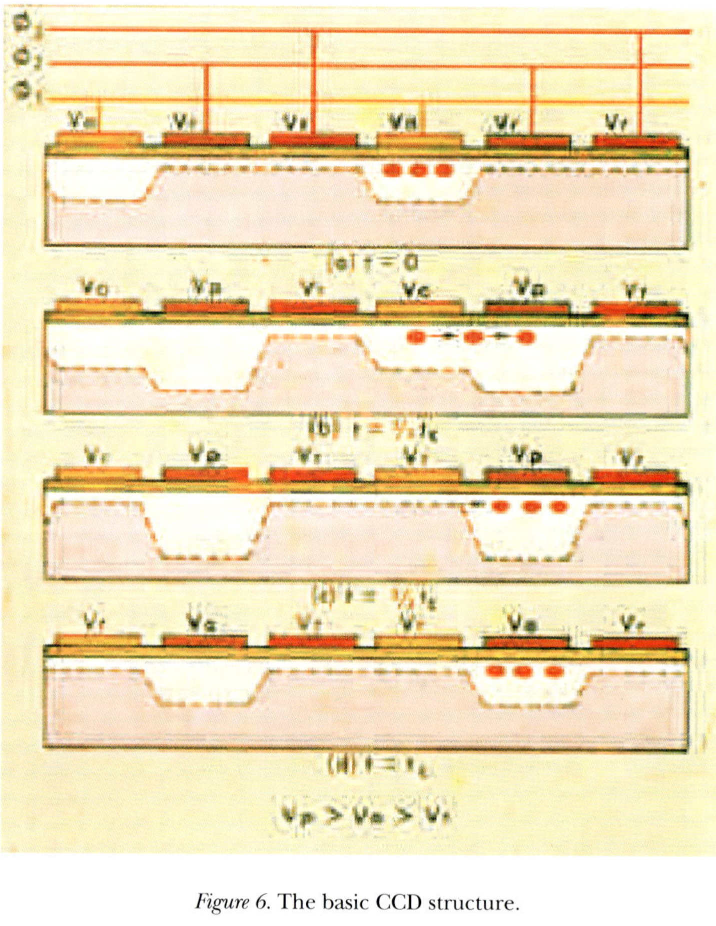

Figure 6. The basic CCD structure.

This completed the basic invention. It should be stressed that the basic unit of information in the device was a discrete packet of charge and not the voltages and currents of circuit based devices. The CCD is indeed a functional device and not a collection of individual devices connected by wires.

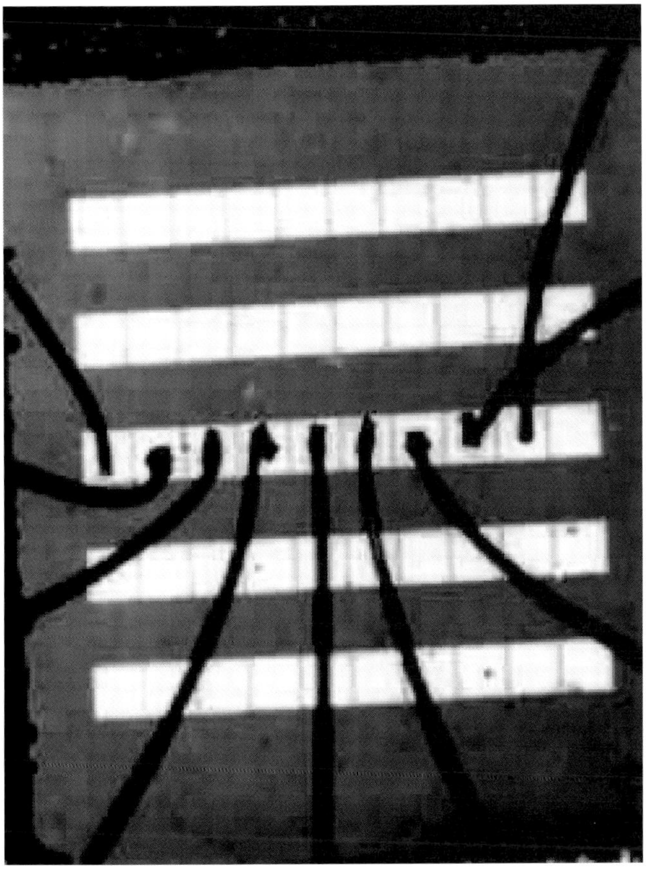

Figure 7. The first CCD device.

Finally, it was decided to go ahead and fabricate a device to show experimental feasibility. In less than a week, masks were made and devices were fabricated and tested. This first simple structure is shown in Figure 7. Charge was introduced in the first MOS capacitor by thermal generation and then transferred to the output by applying voltages to the plates where it was detected by pushing the charge into the substrate and measuring the substrate current.

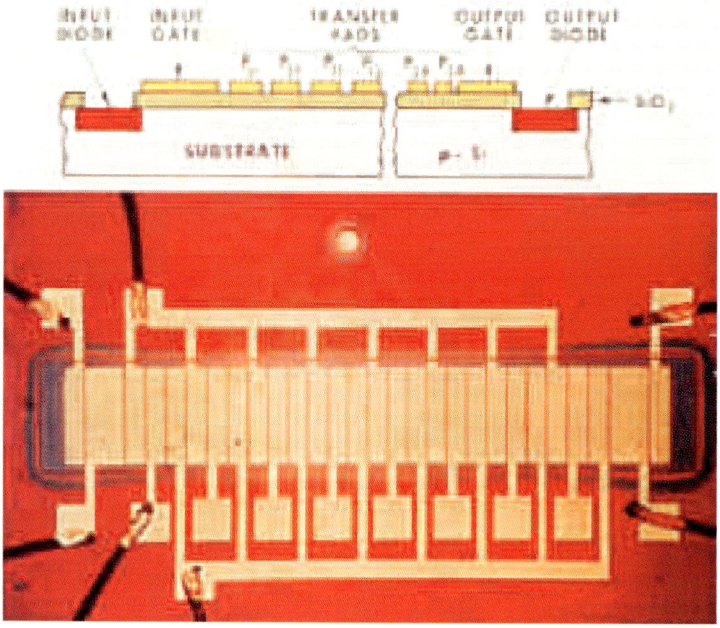

Figure 8. The first integrated device.

The first device was very crude but charge transfer was successfully demonstrated and this was followed by the first integrated structure, which is shown in Figure 8. This device had a three-phase metallization and diffused input and output successfully demonstrating that it could be operated as a serial memory, the first driving force of the invention. It is no surprise that we tried to use the eight bit CCD as a linear scanning imaging device and the first rather crude image is shown in Figure 9.

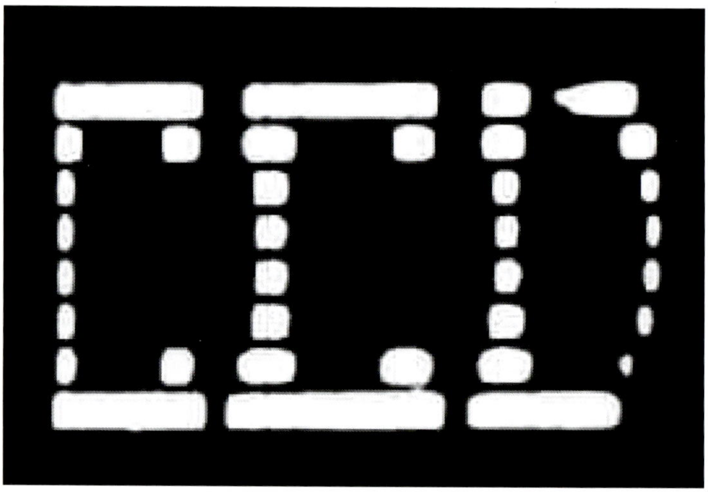

Figure 9. Oscilloscope display of the output of the 8-bit device used as an imager.

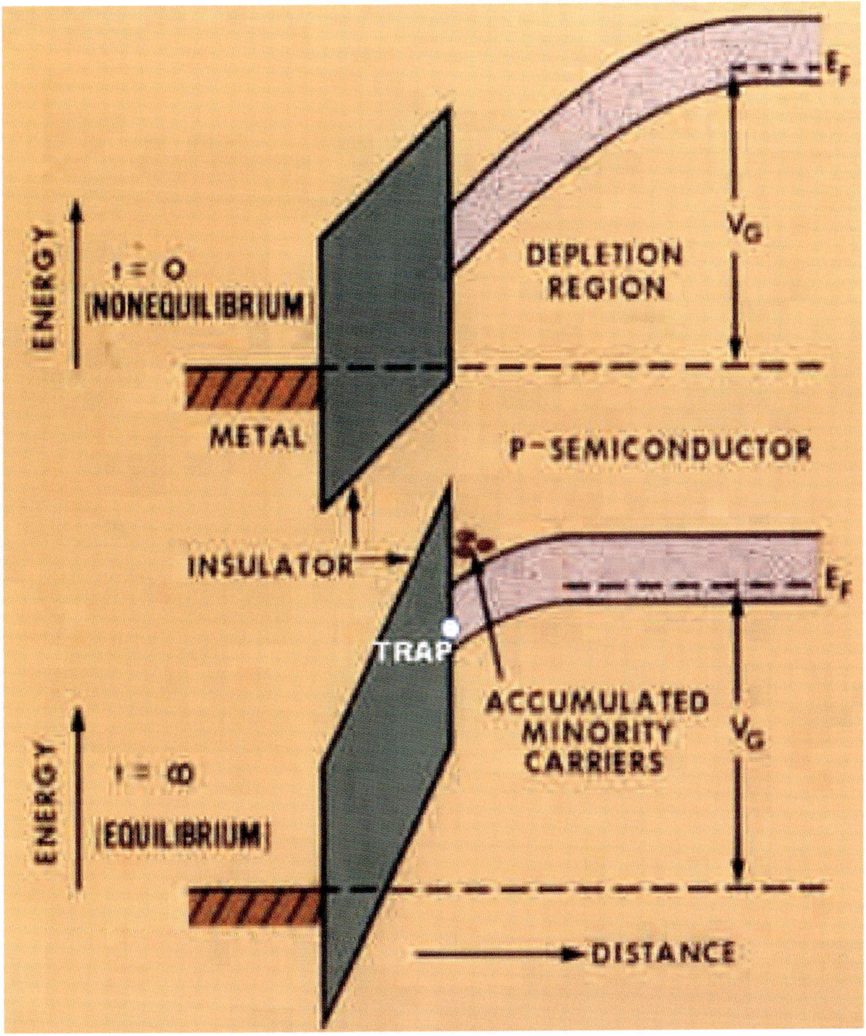

Figure 10. Traps at the Si-SiO$_2$ interface.

Following the initial experiments, it was evident that the main problem with the device was charge transfer inefficiency, the inability to transfer all of the charge from one element to the next. The main reason for this was the trapping of charge in traps at the silicon-silicon dioxide interface, see Figure 10. The trapped electrons would emit at a later time causing smearing of the image. So Bill and I got together again and invented the buried channel CCD, which placed the stored charge in the interior of the semiconductor where there was relatively little trapping. The structure is shown in Figure 11 where a lightly doped n layer has been added to the original structure. Once the layer has been depleted by transferring the electrons to the output diode, the resulting potential is shown. Electrons in the channel region will now accumulate in the valley created in the interior of the silicon and kept away from the surface traps.

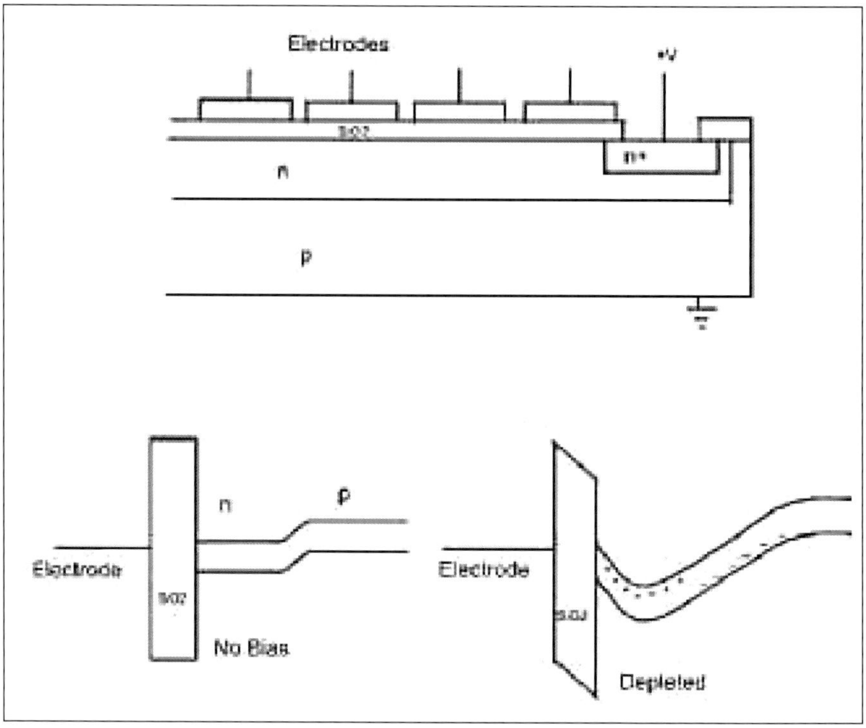

Figure 11. The buried channel CCD.

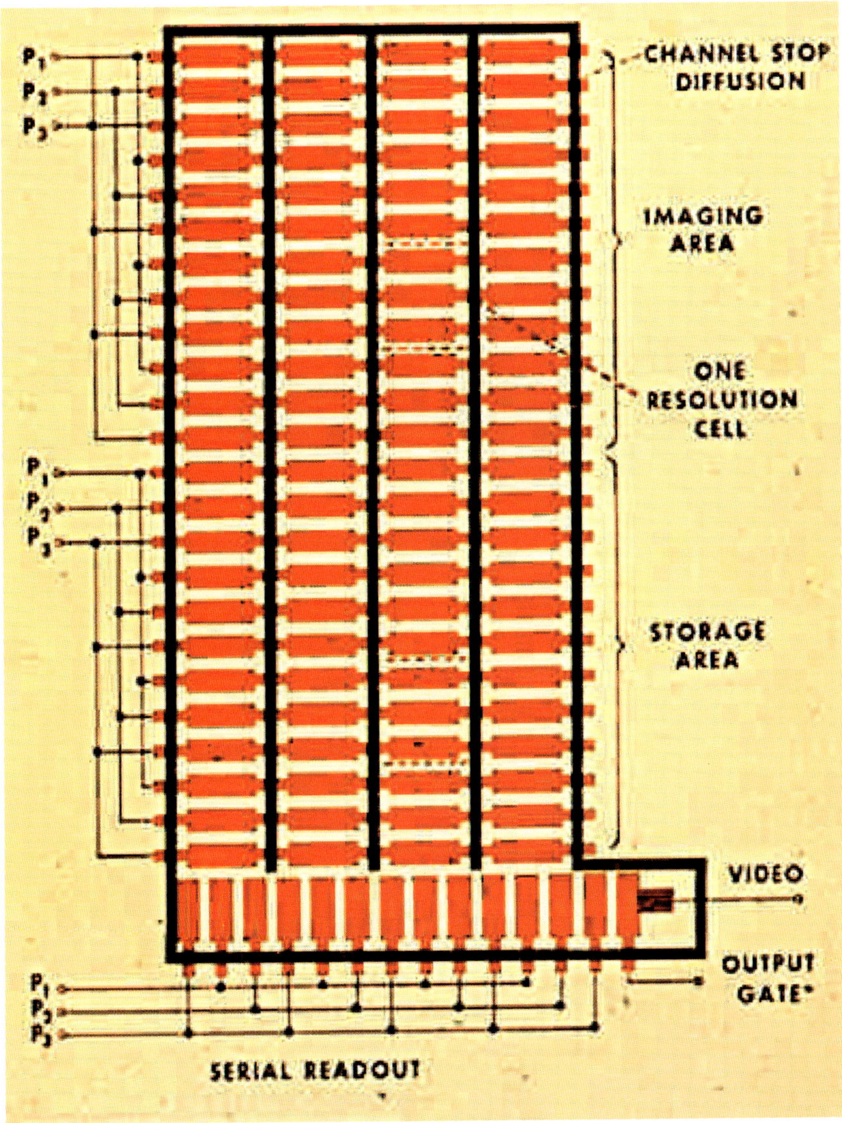

Figure 12. An area imaging CCD geometry.

A period of rapid development followed both at Bell Labs and other companies. One major activity was to make an area imaging device for video applications. Many different schemes were devised. The one we chose for the Picturephone application is shown schematically in Figure 12. Linear CCDs are formed side by side and split into an imaging and storage area. During a frame time, the image is taken in the upper region and then transferred rapidly to the lower where it is read out a line at a time by the serial readout while the next frame is being taken. The chip that we made for the Picturephone is shown in Figure 13 along with a self-contained experimental camera. Successful testing of the device is shown in Figure 14.

Figure 13. The CCD Picturephone imaging chip and a self-contained camera.

Figure 14. W. S. Boyle and G. E. Smith, 1970.

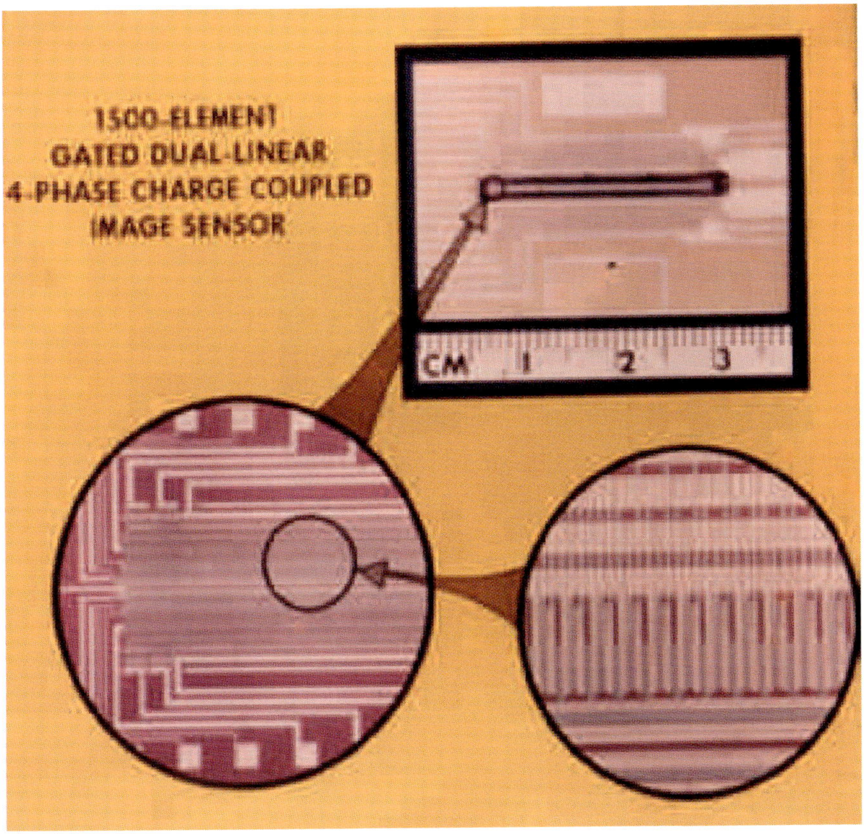

Figure 15. An early linear imaging CCD.

Another device that we developed in the early days was a linear scanning device shown in Figure 15 for possible use in facsimile type applications. Many other companies initiated strong development efforts after the initial announcement of the invention was made in early 1970 and the field rapidly became very active. An aerial photo taken with a linear scanning device made by Fairchild is shown in Figure16.

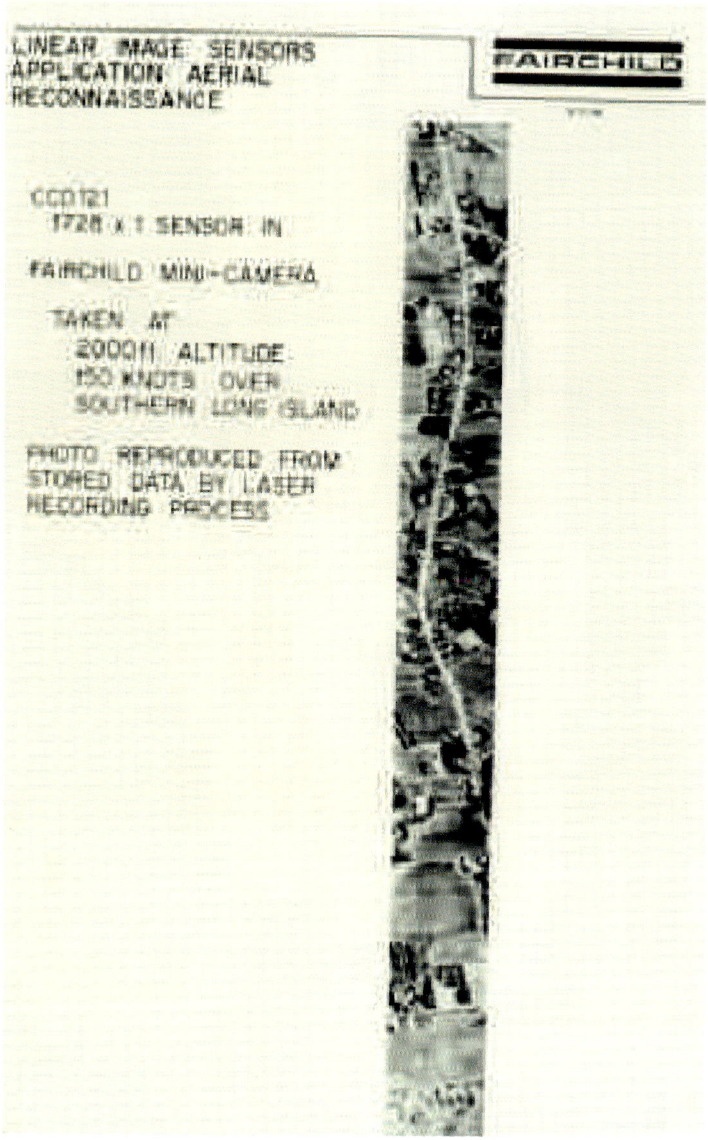

Figure 16. An aerial photograph taken with a Fairchild linear CCD.

Table 1. Advantages over photographic film

90% Quantum Efficiency (Film is 5%)
Long integration times (at 160°K) 20 electrons/hr/pixel)
Large dynamic range105 electrons/pixel
Repeatability eliminates systematic errors

Perhaps the most impressive scientific applications were in astronomy, where the impressive advantages over using photographic film listed in Table 1 led to many new discoveries. An early four mega pixel CCD for use in an astronomical telescope is shown in Figure 17. A comparison between a state of the art photographic image, Figure 18, compared to an image taken by a CCD, Figure 19 of the same region, exemplifies this. Such improvements led to such things as experimental verification of dark matter in the universe.

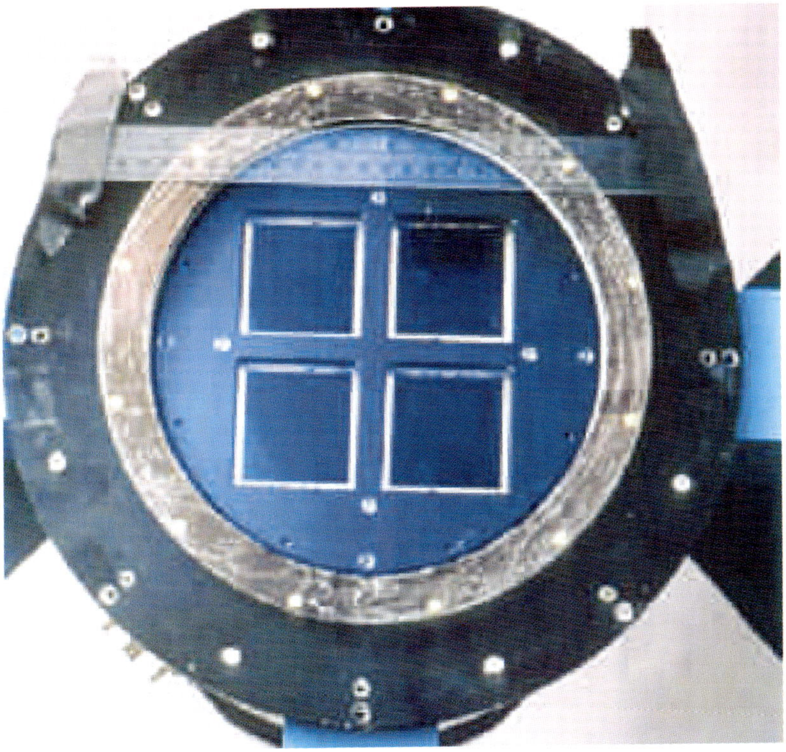

Figure 17. An early four 1 Megabit CCD camera for an astronomical telescope.

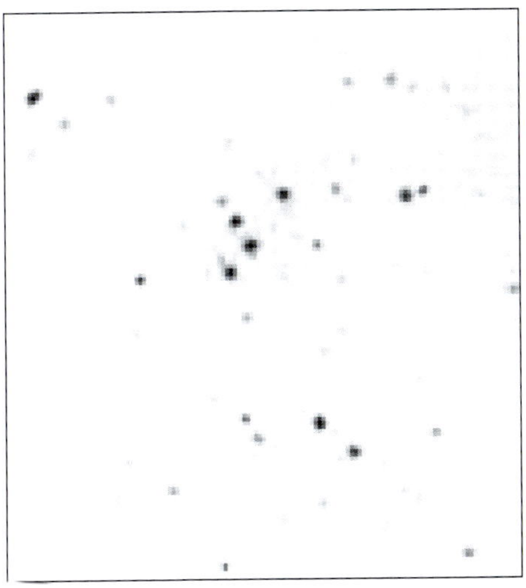

Figure 18. A film photograph of the sky.

Figure 19. A CCD image of the same portion of the sky.

In summary, CCDs were born in the Si-SiO_2 revolution and, because of their unique properties, created their own revolution in widespread imaging device applications listed in Table 2.

Table 2. Imaging device applications of CCD.

Imaging device applications
Digital cameras
TV cameras
Scanners
Medical devices
Fax machines
Bar code readers
Satellite surveillance
Earthly surveillance
Astronomy

REFERENCES

1. W. S. Boyle and G. E. Smith, "Charge Coupled Semiconductor Devices," *Bell Sys. Tech. J.*, 49, 587, (1970).
2. G. F. Amelio, M. F. Tompsett, and G. E. Smith, "Experimental Verification of the Charge Coupled Device Concept," *Bell Sys. Tech. J.*, 49, 593, (1970).
3. W. S. Boyle & G. E. Smith, *US Patent 3,792,322 – Buried Channel Charge Coupled Devices,* issued February 12, 1974.

Portrait photo of Dr. Smith by photographer Ulla Montan.

Physics 2010

Andre Geim and Konstantin Novoselov

"for groundbreaking experiments regarding the two-dimensional material graphene"

THE NOBEL PRIZE IN PHYSICS

Speech by Professor Per Delsing of the Royal Swedish Academy of Sciences.
Translation of the Swedish text.

Your Majesties, Your Royal Highnesses, Ladies and Gentlemen,

The 2010 Nobel Prize in Physics is awarded for research concerning a new ultra-thin material with exceptional properties. This material is called graphene and is the first of a new class of materials. Surprisingly, it can be produced in small quantities with the help of ordinary office materials such as pencils and adhesive tape.

Carbon is probably the most important element in nature. It is the basis of all life that we know of. The most common form of pure carbon is graphite, which we find in pencils, for example. If carbon is subjected to high pressure, it can also assume the form of a diamond, which is considerably more expensive and is undoubtedly found in some of the jewellery in this hall.

Graphite consists of many one-atom-thin layers of carbon stacked on top of each other. One single such layer is what we today call graphene. A graphene layer looks something like chicken wire. We say that it has a hexagonal structure, but graphene is much thinner – actually as thin as one single atom. Each graphene layer is very strong, but in graphite these layers are only weakly connected to each other. Because of this, it is easy to split graphite, which is exactly what happens when we write with a pencil. When we drag a pencil across paper, flakes containing many layers of carbon break loose. Naturally some of the flakes are thinner than others, and it is conceivable that a tiny fraction of the flakes actually consist of single carbon layers. Most of you in this hall have thus produced graphene-like layers when you have written with a pencil.

Graphene is representative of an entire class of two-dimensional crystalline materials. By two-dimensional, we mean that atoms are added to the length and width of such a material but not to its height. This means that electrons in the material can only travel in two dimensions, but not in the third. The electrons in graphene behave in an unusual way. It actually looks as if they have no mass. In physics, this gives rise to interesting phenomena such as an unusual form of the so-called quantum Hall effect, a phenomenon which resulted in a Nobel Prize in 1985. Another example is so-called Klein tunnelling. This effect was predicted by the Swedish physicist Oskar Klein in 1929. It had not been observed earlier in other systems, but in graphene it was observed last year.

Graphene also has other exceptional properties. For example, it is 100 times stronger than steel. If we imagine making a hammock out of graphene that is one square metre in size, even though it is only one atom thin it will be able to hold a newborn infant or a cat without breaking. Such a hammock would weigh about one milligram, about the same as one of the cat's whiskers. Graphene is also a good conductor of electricity, and it conducts heat 10 times better than silver. In addition graphene is transparent, flexible and very stretchable.

It had been known for a long time that graphite consists of hexagonal carbon layers, and the behaviour of electrons in graphene was calculated as early as in 1947 by Philip Wallace, but few scientists believed it would be possible to isolate graphene in a way that would enable electrical measurements of single layers. It was therefore surprising when this year's Laureates, Andre Geim and Konstantin Novoselov, together with their collaborators published their first article about graphene in October 2004. With the help of inventive methods, including the use of ordinary adhesive tape, they succeeded in isolating thin carbon layers and transferring them to a suitable surface. With the help of different microscopes, they were able to show that some of the layers were only one atom thin. They were also able to etch the samples into a suitable shape and connect electrodes. They made electrical measurements and were able to show that the material actually had the expected properties. In subsequent research, many teams – and not least the Laureates themselves – have performed new experiments and studied a number of exciting properties of this new carbon material.

The graphene research field is still young, and it is too early to single out what applications will be the most important, but there are great hopes that the exceptional properties of graphene can be utilised in many areas. Examples of possible applications are touch screens, solar cells, fast transistors, gas sensors and lightweight super strong materials.

Professor Geim, Professor Novoselov,

You have been awarded the 2010 Nobel Prize in Physics for your groundbreaking experiments regarding the two-dimensional material graphene. On behalf of the Royal Swedish Academy of Sciences, it is my honour and pleasure to convey to you the warmest congratulations for your outstanding work. I now ask you to step forward to receive your Nobel Prizes from the hands of His Majesty the King.

Andre Geim

ANDRE GEIM

"I don't think anyone should write their autobiography until after they're dead."
Samuel Goldwyn

Several years ago I was on a trekking trip in the Jordanian desert with a large group of Brits. We were camping and, as usual, there was not much to do in the evenings, so we filled the hours by sitting around a campfire, playing the popular British game 'Call My Bluff'. In it a player makes several statements only one of which is true, and the rest of the group have to guess which one it is. All other statements are called 'bluffs'. I teased my fellow hikers with statements like 'I was born in a Mediterranean climate', 'I was a lieutenant in the Red Army', 'I have won an Ig Nobel prize', 'I climbed several five kilometre high mountains', 'I fell down a 100 m deep crevasse without a rope', 'I was called 'Russian' for the first time at the age of 32', 'At my university I studied intercontinental ballistic missiles', 'I was a bricklayer north of the Arctic Circle', 'I knew Mikhail Gorbachev personally' and so on. What surprised me was that all but the last statement were dismissed by most of the group as 'bluffs', while people found it easy to believe that it is typical for any Russian to know their political leaders personally. I won every single game because the truth was a complete opposite: Apart from knowing Gorbachev (whom I only ever saw on TV) all the other statements were true. This made me think for the first time that, perhaps, my life had not been as trivial as I thought.

Still, with reference to the epigraph, I am not dead yet. I think it is too early for me to write an autobiography, as doing so somehow implies that one's life story is finished. I am only fifty-two and plan to actively continue my research work. However, I am a law-abiding citizen (of course!) and, according to the rules of the Nobel Foundation, I must provide an autobiography. So, below I have conceded a sort of one, a literary exercise. Although I do not expand on any of the non-bluff statements above, the reader is still likely to find my life path atypical. I do not know whether this somehow influenced my way of doing things or it is just a separate story, having little in common with my research career.

The timeline of this autobiography ends in 1987 when I received a PhD. After that point, my scientific biography is given in the Nobel lecture 'Random walk to graphene'.

SOVIET TAXONOMY

I was born on October 21, 1958 in the small Black Sea resort of Sochi, the second son of Nina Bayer and Konstantin Geim. The first seven years of my life I spent there with my grandmother Maria Ziegler and grandfather Nikolai Bayer. I remember little of my grandfather because he died when I was only six, but my grandmother was my best friend and an important part of my life until the university years, when I left home. At the age of seven, it was time to go to school and, reluctantly, I had to leave Sochi and go to live with my parents and my elder brother Vladislav in the city of Nalchik where they worked. Nalchik is the capital of a small Republic of Kabardino-Balkaria in the foothills of the Caucasus Mountains and can be found on the world map as a host to the Europe's highest peak Elbrus and in proximity to the infamous Chechnya. For the next ten years I spent my school time there but returned to Sochi every year to stay with my grandmother during the summer months.

At this point, it is probably right to mention my ethnic origins, because for certain groups of people in the Soviet Union ethnicity was a very important factor and often defined their life choices and eventually their life path. I belonged to one such group. Despite the great ethnic diversity of the Soviet population (the official census of 1989 listed over 100 ethnicities), the authorities managed to keep track of each and every one of them by having a special line in the Soviet passport ('line 5: nationality'). In my passport this line stated 'German'. This is because my father came from the so-called Volga Germans, descendants of colonists from Germany who settled on the Volga River banks in the 18th century. My mother's bloodline was also mostly German. I have long believed that my maternal grandmother Maria was Jewish, but according to my brother's recent research into family history, her father was also German. Therefore, to the best of my knowledge, the only Jew in the family was my great-grandmother, with the rest on both sides being German.

A note is needed here to explain why I devote so much space to explaining my ethnicity. Firstly, of course, the word 'German' in my Soviet passport had a very real effect on my life, as the reader will find out below. Secondly, the issue of my ethnicity unexpectedly surfaced again after the announcement of the Nobel Prize – suddenly there have been a lot of discussions whether this prize is British, Dutch, Russian, German or Jewish. To me these discussions seem silly. Having lived and worked in several European countries, I consider myself European and do not believe that any further taxonomy is necessary, especially in such a fluid world as the world of science.

SKELETONS IN THE OLD CHEST

My knowledge of our family history is rather sketchy and, for a Western person, it is perhaps difficult to understand why. The reason goes back to well before I was born. In Stalin's time, family history was a dangerous subject to discuss, and stories were not passed from generation to generation because parents deliberately concealed their history from the children in order to protect them. A telling example can be found in the many documents that I had to fill when applying to university, for a job and so on. Among such documents there was always a questionnaire asking whether you had relatives abroad, whether any of your relatives were prisoners in forced labour camps (the infamous Gulag) or were prisoners of war. I always answered 'No' to all those questions, in good faith, believing this answer to be true. It was only in the late 1980s that I learned that nearly everyone in my family, including my father and grandfather, spent many years in the Gulag, that some of the family were prisoners in German concentration camps, and that I had an uncle living in Bavaria. This was deliberately and successfully concealed from me during my first 30 years of life.

Below is what I have learned since then from my few living relatives. My grandfather Nikolai Bayer was a professor at Kharkov University who specialised in aerial cartography. In 1946, documents were found by the Soviet Army in post-war Poland which revealed that after the First World War, he was a junior minister in Petliura's short-lived Ukrainian nationalist government. This anti-Bolshevik past, together with his German ethnicity and the fact that at the time he was compiling maps of Eastern Siberia, was apparently enough reason to accuse him of passing state secrets to the Japanese and send him to a northern Gulag camp near Vorkuta. He was released only in 1953, after Stalin's death.

When I was born, my father was forty-eight years old and already had quite a long and difficult history behind him as well, which I managed to learn from him bit by bit over many years. Until his last years, he avoided discussing it, even when I asked, and those bits came out mostly accidentally. Before the Second World War he was a young professor at Saratov State University, lecturing in physics and maths. However, when the war broke out in Europe, being an ethnic German became a political crime and he was sent to a Gulag camp in Siberia, where he spent many years building a hydroelectric power station and a railway. In 1949 he was allowed to join his family, who in the meantime had been deported to Novosibirsk.

An episode I vividly remember from my early years is finding a box of old medals at the bottom of an old chest hidden in my grandparents' garden shed in Sochi. One of them was the Cross of St George, an award of high military distinction in the Russian Empire (before the revolution). I showed my findings to my grandmother. Being confronted, she explained that the Cross belonged to her father, who served as an army surgeon in World War I, whereas other decorations were related to the nobility status of her grandfather, a descendant of German aristocrats. In the 19th century her family lived in Poland (then a part of the Russian Empire), where they took part in the 1863 uprising and consequently were deported to Siberia, which was to become such a familiar place to my forebears a century later. The next time I tried to find those medals, they were long gone. It was only many years later that I found that my grandma Maria threw them all away immediately after the episode. Incomprehensible as it sounds to us today, this kind of behaviour became imprinted in the DNA of people who lived through the Stalinist terror. She was afraid I would talk about the medals to my friends and, if the story got around, the whole family would be in trouble. This happened in Khrushchev's times, when the terror receded, but 'bourgeois' reminders were still deemed unacceptable by 'the proletariat' until the 1990s.

By the time I went to school, the mentality of Stalin's time was largely gone from the Soviet system. Except for some remnants, such as the 'nationality' line and all those family questionnaires, young people like me were largely unaware of the recent terror. The only time I really suffered because of my ethnicity was when trying to get into a top university, as described below. Otherwise, it was just being occasionally called 'fascist' in the playground, or a 'bloody Jew' ('жид' or 'Zhid'), because a foreign name was often associated with being a Jew (in Russian, the word жид sounds very offensive). Maybe because of the latter I am particularly keen to emphasise that some small portion of my blood is likely Jewish.

SCHOOLING AS USUAL

Despite this sombre family history, I myself was lucky enough to be born late and had a happy childhood. My best childhood memories are associated with my birthplace, Sochi. My grandma Maria was a meteorologist and I spent my first years of life on the beach, around the weather station where she worked. My mother was a head of quality control and my father chief engineer at a very large vacuum-electronics factory (chief engineer would be equivalent to a CEO in the West). After two decades, many people in Nalchik still remember him as a hard-working and influential person. Perseverance and hard work are the qualities I probably inherited from him. My parents' occupations placed our family in the top layer of technocrats in the Soviet Union. They were not within the communist party elite who enjoyed all the perks of the Soviet system and, as ethnic Germans, they could not possibly be. Nevertheless their status allowed the family a relatively comfortable existence.

My school in Nalchik was called a specialist English language school and considered to be the best in town. Despite its name, the teaching of English was not its strongest point. Looking back and comparing how we were taught English then and how I was taught Dutch 30 years later, the notion of English specialisation in my old school seems nothing but laughable. On the other hand, mathematics was taught at an extremely high level, especially in senior forms, thanks mostly to our maths teacher, Valenida Sedneva. I may not have realised this at the time but, when I looked at my old exam papers several years later and was already a student at an elite university, I was amazed at how tough and challenging those papers were. Some of them required not only powers of recall but also imaginative and non-standard thinking. Physics and chemistry were taught at a good level, too. I once won a regional chemistry Olympiad, which however was not so much due to my love of the subject as to the fact that in a couple of days I managed to memorise a whole chemistry dictionary some 1,000 pages long (happily forgotten in the following few days).

I also fondly remember Olga Peshkova, our teacher of Russian and Literature. Despite getting excellent marks in these two subjects, I did not excel in either of them. Still, I like to think that her lessons were helpful in learning – eventually – how to write research papers in a clear and concise way. There is nothing else particularly remarkable to mention about my schooling, except for the brain-washing Soviet propaganda that penetrated every aspect of our lives at that time. As a counterbalance, schoolchildren often listened to the Voice of America and similar radio stations, and this small rebellion helped us to develop a healthy scepticism about many things (albeit not all) that the propaganda told us. Of course, like everyone around me, I played my due role of a disciplined Soviet pupil.

FAILING THE FIRST HURDLE

At the age of 16, I graduated from school with a gold medal, a distinction given to those who achieved the perfect score in all subjects (typically, the top 5%). My parents encouraged me to go to the best possible university and my sights were set on a couple of elite universities in Moscow. At school I was doing well in all exact sciences, including physics and chemistry, but my strongest subject was maths. However, my parents persuaded me that pure maths would not offer good career prospects. Hence, my decision was to study physics. The very top university for physics in Russia was (and still is) the Moscow Institute of Physics and Technology (Phystech). However, the entrance examinations to Phystech were famously competitive and extremely tough and, as I grew up in a provincial town, I believed they were beyond my ability. So, I chose to go to another leading university, Moscow Engineering and Physics Institute (MIFI). In the way of preparation I solved problems from sample MIFI and Phystech exam papers and felt ready, even if still not very confident. Little did I know that the main obstacle for me would turn out to be my ethnicity.

The first exam in MIFI was written maths, and I was pretty confident that I solved all the problems correctly and would get an 'excellent' (the marking system in Russian schools and universities consists of four grades: 'excellent', 'good', 'satisfactory' and 'fail'). However, I then found it was only a 'satisfactory' and, even worse, my mark for the oral maths was a 'fail'. I attributed this failure to poor preparation and my inexperience in sitting real tests: problems at my oral exam seemed a lot harder than those from the sample MIFI papers that I did at home. So I decided to go home, continue to study and take my chances a year later.

That gap year turned out to be very important for me. My parents were supportive and found a job for me at the factory where they worked, as a technician responsible for calibration of measurement equipment, and also paid for tutoring in maths, physics and Russian literature (these were standard entrance exams at my chosen universities). After a couple of weeks I found that I knew maths better than my tutor (who was considered to be the best in the town), so these tutorials stopped. On the other hand, my physics tutorials were the best I could wish for. My tutor was a physics professor from Nalchik's University, Valery Petrosian. I thoroughly enjoyed every lesson. We solved many problems from old exam papers either from Phystech or, even harder, from international Olympiads. But even more helpful was the way he taught me to deal with physics problems: it is much easier to solve a problem if you first guess possible answers. Most problems at Phystech level require understanding of more than one area of physics and usually involve several logical steps. For example, in the case of a five-step solution, the possibilities for dealing with the problem quickly diverge and it may take many attempts before you get to the final answer. If, however, you try to solve the same problem from both ends, guessing two or three plausible answers, the space of possibilities and logical steps is much reduced. This is the way I learned to think then and I am still using it in my research every day, trying to build all the logical steps between what I have and what I think may be the end result of a particular project. After a couple of months, my tutor no longer asked me to write up a solution. Instead, I just explained verbally the way I would solve a particular problem – all the logical steps required to get to its end without describing routine details. This allowed us to go through the problems at lightning speed.

I also learned an important lesson from my tutorials in Russian literature. My tutor said that what I was writing was good but it was clear from my essays that I tried to recall and repeat the thoughts of famous writers and literature critics, not trusting my own judgement, afraid that my own thoughts were not interesting, important or correct enough. Her advice was to try and explain my own opinions and ideas and to use those authoritative phrases only occasionally, to support and strengthen my writing. This simple advice was crucial for me – it changed the way I wrote. Years later I noticed that I was better at explaining my thoughts in writing than my fellow students.

ENEMY OF THE STATE

After this year of intensive preparations I felt I knew enough and was much more confident than the previous year and ready for MIFI. I easily solved all the problems in the written maths exam (which again was first), polished the presentation and expected an 'excellent' mark. However, at the next exam (oral maths) I was told that the mark was only 'good', and the examiner refused to explain what was wrong or to show me the script, even though it was right there in front of him. He gave me three further maths problems, the hardest I had ever seen. I managed to solve one, partially solved the second one, with a minor mistake, and provided the correct answer to the third one. However, I could not explain how I came up with this answer. It just appeared in my head and I still remember it now: the answer was 998. The mark I got for these efforts was 'satisfactory', which was clearly not enough to be admitted to the university. In addition to the rather harsh treatment from the examiner, I noticed more odd things about the exam – apart from me, not one single person in the same room (about 20 candidates) managed to get even a 'satisfactory' mark; they all failed. Even more curiously, the names of all the candidates were either Jewish- or foreign-sounding. I went to look at the lists of people in other examination rooms and most of the names sounded Russian, with a very few exceptions.

Even for someone as naïve as I was at 17, it was clear that there was a policy in place to fail certain ethnic minorities. In hindsight this can be easily explained because this particular university specialised in nuclear physics and, at that time, if you were a Jew or a German, you were assumed to be a potential emigrant who would learn 'state secrets' and then go abroad. That was always considered a threat in the Soviet Union. So in a sense it was clearly a policy, and even an understandable policy, but not much advertised. Several years later I found that there were a few Jewish people who attended and successfully graduated from MIFI. To achieve this, their parents had to go to KGB representatives at MIFI (they were present in every Soviet organisation at the so-called First Departments) and persuade them that their children were reliable Soviet citizens and had no intention of leaving the country. Apparently, these tactics did work but neither I nor probably my parents even suspected that it was needed. Or, maybe my parents were too aware of the true lies in my family questionnaires.

ACCIDENTAL PHYSICIST

This was the first time I experienced discrimination at an official level and it was quite a shock. Fortunately, there was still a week left to try my luck at another university. I said to myself 'what the hell' and applied to Phystech. The way I was treated there was a shocking experience in itself, as it was so different from MIFI. The examiners were friendly and even helpful, the exam problems interesting and the whole environment welcoming. I felt as if by mistake someone put me in a wrong room, away from a firing squad of examiners. Perhaps, this was the case.

My examination marks were comfortably above the threshold required for admission, even though I got only one 'excellent' mark out of four exams, with the rest 'good'. I felt that I could have done better but my MIFI experience was still fresh, and the memories of those failed exams kept coming back, affecting my concentration and sometimes my judgement of the difficulty of the problems. This was especially apparent in my oral physics exam, which I still remember well. The first problem given to me seemed easy and I quickly solved it, but the examiner said 'It's a wrong answer'. I tried to protest, and it took us a few minutes to understand that I had solved a much harder problem than the one he gave me; even though the answer to the problem I actually solved was correct, it was still wrong. Incredibly, the same story happened with the second problem. So, when giving me the third one, the examiner repeatedly asked whether I was sure that I understood what was being asked.

The last hurdle at Phystech was an admissions interview and I was scared that the question of my ethnicity would arise again and they might not accept me despite my good marks. It was well known that, on the basis of the interview, sometimes candidates with marks just below the threshold were accepted and those with marks above rejected. The ethnic question did arise in the form of 'How is your German?' I answered 'Barely' and started thinking what else to add. One of the panel members (Seva Gantmakher, as in the Gantmakher effect) quickly interjected saying 'Then he is not a real German'. As it turned out, this remark, as well as his following interventions, influenced all of my further life by putting me on the path of solid state physics.

Like many would-be students of that age, I dreamed of doing astrophysics or particle physics and aspired to solve 'the greatest mysteries of the universe'. But there was a rumour among Phystech candidates that saying so was considered to be very naïve by interviewers. I remembered that but did not want to cheat. So, when asked about my aspirations, I said that I wanted to study neutron stars (true) because I wanted to understand how matter behaved at extremely high densities (an excuse, not to sound so naïve). A prompt reply from Seva was 'Good, you can then study high-pressure physics at our Institute [of Solid State Physics].'

Another memory of that interview is being asked to estimate the weight of the earth's atmosphere (it was customary to give candidates some tricky mental problems to solve). I spent most of my three minutes multiplying the numbers in my head (atmospheric pressure multiplied by the surface area of the earth divided by gravity, all in SI units) and when I gave an answer in trillions of trillions of kg, everyone was surprised because I was only expected to give a general answer, not a specific number.

This is how I entered Phystech. In the end, my rejection from MIFI turned out to be a blessing in disguise because Phystech was a two-notch higher level university. The only reason I did not go there first was because I did not believe I was up to it. Basically, circumstances forced my first choice on me rather than my second one!

MOTHER OF ALL GRILLING

Phystech is quite an exceptional university, not only by Russian standards, where it is considered crème de la crème, but also with respect to any other university I know. The only reason that it is not found in any world league tables is that it is a purely teaching university. (Teaching and research are traditionally separated in Russia – research is done mainly at the Academy of Sciences and teaching at universities). In addition to the very rigorous student selection, a well-known reason for Phystech being so good was that, unlike other Soviet universities, all specialist and some general courses were taught by practising scientists from the Academy institutes from all over the Moscow region. Of course, in the West it is a standard to have active researchers giving undergraduate courses, but in Russia it is an exception.

Even more importantly, as Phystech students we were forced to think and find logic in everything we studied, as opposed to just memorising facts and formulas. This was largely due to Phystech's examination style: when it came to specialised subjects, many of the exams we took every year were open-book. This meant that there was no need to remember formulas, as long as one knew where to find them. Instead, the problems were challenging, requiring combinations of different subject areas and thus teaching us to really understand science rather than merely to memorise it.

From the moment of its establishment, Phystech was led by prominent Soviet scientists such as Kapitsa, Landau and many others. Among my own lecturers and examiners were many eminent scientists such as Emmanuel Rashba, Vladimir Pokrovski, Viktor Lidskii, Spartak Belyaev, Lev Pitaevskii, Isaak Khalatnikov and Lev Gorkov, to name but a few. I have to admit that their names did not tell me much at the time, which was helped by the fact that I was not very good at attending lectures. I rediscovered some of the names only recently, when I saw their signatures in my old exam certificates, which Phystech put on the web after the Nobel Prize announcement.

The workload at Phystech was heavy and the courses extremely challenging. It is probably enough to say that our standard textbooks for quantum mechanics, statistical physics, electrodynamics and classical mechanics were from the Landau-Lifshitz Theoretical Physics Course. Perhaps they are not the best textbooks for undergraduate students, but they are a good indication of the expected level of achievement. Not all students managed to cope with the psychological pressure imposed by this teaching style and some dropped out not only because of bad marks but, more often, because of nervous breakdowns. I personally knew several students who developed suicidal tendencies and psychiatric problems. My own sanity was perhaps saved by the amount of alcohol that I and some of my friends consumed after each exam to release the accumulated stress.

The first two and a half years of foundation courses were particularly tough. After that the pressure subsided as we moved on to specialist courses. From year three we started attending lectures at the so-called base institutes of the Academy of Sciences. In my case it was the Institute of Solid State Physics in Chernogolovka, chosen at the above-mentioned interview due to my love for high-density neutron stars. From year five, we also started working in research labs – not on some specially designed undergraduate projects but on real ongoing projects, where we worked as part of an academic research team. Year six was a Master's year and 100 % research based. After that, the normal route (if you wanted to stay in academia) was two years of research probation and, if you were successful, you were eligible for a PhD studentship, which lasted another 3 years. It was an 11-year long process to get a PhD – 6 years at Phystech plus 5 years leading to a 'viva', or oral defence of one's thesis.

For me personally, only the first half year at Phystech was a struggle. I came from a provincial town, while some of my classmates were graduates of elite Moscow schools specialising in physics and maths. Quite a few were winners of international Olympiads in physics or mathematics. The first few months were essentially designed to bring everyone to the level of those guys; they were nearly a year ahead of the rest of us in formal topics, especially maths. Only after I got all the highest marks in the first set of mid-year exams did I start feeling confident enough in this wunderkind environment and was able to relax somewhat. Despite all the pressure and grilling, every single one of us who managed to graduate from Phystech have great memories of those hard years and are most proud of our alma mater.

GO WITH THE TOP FLOW

I graduated from Phystech with a so-called 'red diploma', which meant within the top 5 to 10% of my class. Out of 50 or so final exam marks, I got only two 'good'. One of them was for a course on "political economy of socialism", which I attributed without much shame to my inability to find any logic in the subject. By contrast, I got 'excellent' for the political economy of capitalism and to this day have fond memories of reading *Das Kapital* by Karl Marx, whom I occasionally quote to tease, or perhaps shock, my Western colleagues. My second 'good' was for the course on superconductivity taught by Lev Gorkov himself, who also was my examiner. Oddly for Phystech, he did not allow us to use textbooks during the exam (shame on him), and I made a mistake in one of the derivations. This is funny because in the 1990s, when I was already a professor in the Netherlands, superconductivity became my research subject.

Despite my exam success, I do not believe I particularly stood out among the students in my class. In my year there were one or two students with only 'excellent' marks, and some were digging deeper and understood the courses better than I did. At that time I did not really try my best; I worked just hard enough to guarantee myself maximum marks and stay at the top of the class. I was successful at that but it did not take all of my time or effort. In fact, in my university years I was not at all an exemplary student. With excellent marks, I normally was entitled to a scholarship awarded every half a year, but it was quite regularly (four or five times) withdrawn as a punishment for missing some mandatory lectures, being late from holiday breaks, organising those after-exam parties that sometimes saw some people end up in a hospital and similar misbehaviour. Missing lectures was generally allowed (unless it was a political subject) and I managed to miss most of them. I learned from textbooks and attended group tutorials, unless I disliked particular tutors. I would not recommend this style of learning to aspiring students as a recipe for success, but it may well suit some people as it suited me and a few other students in my class.

My attitude of doing alright to reach a goal but not doing my utmost persisted through all the university and PhD years. I only started to really enjoy physics and do my absolute best, for its own sake, much later when I became an independent researcher.

FROM THE SUBLIME TO THE RIDICULOUS

The topic of my Master's project was electronic properties of metals, which I studied by exciting electromagnetic waves (so-called helicons) in spherical samples of ultrapure indium. From the helicon resonances I could extract information about the resistivity of those samples. The competitive edge of this research was the extreme purity of the indium I was working with, such that at low temperatures electrons could shoot over distances comparable with the sample diameter (~1 cm). After graduating I started working towards my PhD in the same laboratory, as was customary for many Phystech graduates. Looking back, those five years of doing PhD seem remarkably uneventful in terms of the science I was doing.

My first year as a PhD student was signified by an event that was to become a rather regular perturbation in my life: moving from one institute to another. This was when my PhD supervisor, Victor Petrashov, moved from the Institute of Solid State Physics to the newly established Institute of Microelectronics Technology. Although the two were only 200 m apart, it meant a serious disruption of work, losing some equipment and setting everything up again. Initially, I did the metal physics research with some enthusiasm but it gradually faded away as I realised that no one, except perhaps my supervisor, was interested in what I was doing. Nevertheless, educationally, those years were very important for developing experimental skills and making my fingers 'green'. This experience played a crucial role in my further research career, including the graphene story. In this respect I owe a lot to Victor, whom I count as one of the most skilful experimentalists I ever met. With the help of a shoestring and sealing wax he could do amazing things, and a shoestring and sealing wax was what, in those days, we typically had in research labs in Chernogolovka.

I meet quite a few people who feel nostalgia for the 'golden era' of Soviet science, but I myself never saw those times even in Chernogolovka, which was a rather elitist academic place. My recollection is that the arrival of almost any material important for research, be it copper wire or GE varnish, was a cause for celebration, almost on a par with the arrival of a multimillion piece of equipment in the West. Once Victor was lucky to borrow a US-made lock-in amplifier to do some measurements, which we usually had to do using a Soviet equivalent (the word 'equivalent' does not describe the entirety of the difference). In just a couple of weeks I was able to get results that I could not dream of with the 'equivalent'. The availability of resources (or the lack of them) essentially dictated what I could possibly do. I believe experimentalists who claim to have witnessed 'the greatness of Soviet science' either belonged to the select few who had benefactors among the top academicians or more likely fool themselves, choosing to believe that the skies were bluer in the old days.

Having said this, it is true that in the Soviet Union there was a huge difference between being an experimentalist and being a theorist. The theory school was extremely strong, especially what people referred to as 'Landau theory school'. Those guys did things at the highest possible level. The roots of this strength were partly in education but also in the way Soviet theorists worked. I witnessed it by attending many research seminars. A lot of time was spent in discussions and heated debates, where there were no questions that could not be asked and no authority that could not be questioned. In the West, this style is still remembered well by those who 'experienced' Soviet scientists in the 1980s and 1990s. It could be a dreadful experience for the participants, but sometimes I really miss this style. The nostalgia usually appears after coming across certain papers in today's scientific literature: If they were to be first presented at such seminars, even the authors would not dare to put them in print. Those debates were very influential and allowed people to learn more quickly and to develop a broad and informed view of many areas of physics. I myself benefited greatly from such seminars and consider them the second most important part of my education in Chernogolovka. Many of the seminars I attended were organised by Seva Gantmakher. His care for detail and breadth of experimental knowledge were a great example for me and my fellow students.

Despite the great atmosphere in theory departments, even theorists suffered from the state of Soviet science and in the late 1980s many of the best of them moved to the West. I do not think that better living conditions were the only reason for this brain drain: Theoretical ideas do not come out of vacuum; they are often born in interaction with experimentalists, as experimental results serve as a trigger for new ideas. This was completely lacking in Chernogolovka, because new results were hard – if at all possible – to get with the existing equipment. By the time of my PhD, Soviet experimental science had decayed to the point where it was considered that the most appropriate route to reach the top of fame and glory for an experimentalist was to confirm a theory produced by an eminent Soviet theoretician. Indeed, many experimentalists in Chernogolovka were doing just that.

This was my scientific life. Parallel to that, there was another life, busy with events. Chernogolovka is a nice Moscow suburb, quiet and peaceful, surrounded by forest. Life was generally pleasant, even though my living conditions were austere to the extreme – for most of my years there I lived in a residence hall, sharing a room with two other young researchers. One of my roommates was Sergey Dubonos who over the years became my regular co-author and also played an important role in the graphene paper recognised by the Nobel award. In addition to research, my other hobbies were mountaineering and white-water canoeing. Every year I spent more than a month in the mountains and on the rivers in different corners of the Soviet Union, from the Caucasus to Central Asia, sometimes managing to fit in as many as four trips in a year. Those travel experiences were often shared with Max Maximenko and Phystech friend Stas Ionov. It was at this time that I met my wife, Irina Grigorieva, who was also working towards a PhD at the neighbouring Institute of Solid State Physics. She later became my collaborator and significantly contributed to the graphene work.

In a way Chernogolovka offered ideal conditions for scientists – there were hardly any distractions, which allowed us to concentrate on research. Except for queuing for hours for sausages and cheese (which had become a regular scene in the 1980s), most of our time was spent in the labs. Even without much enthusiasm, my research advanced at a steady pace, with a few papers published and due progress made. But it was only when I became an independent researcher, and especially after moving to the West in 1990, that I started to do my real best and the pace of my life changed dramatically, as described in my Nobel lecture 'Random walk to graphene'.

RANDOM WALK TO GRAPHENE

Nobel Lecture, December 8, 2010

by

ANDRE K. GEIM

School of Physics and Astronomy, The University of Manchester, Oxford Road, Manchester M13 9PL, United Kingdom.

If one wants to understand the beautiful physics of graphene, they will be spoiled for choice with so many reviews and popular science articles now available. I hope that the reader will excuse me if on this occasion I recommend my own writings [1–3]. Instead of repeating myself here, I have chosen to describe my twisty scientific road that eventually led to the Nobel Prize. Most parts of this story are not described anywhere else, and its timeline covers the period from my PhD in 1987 to the moment when our 2004 paper, recognised by the Nobel Committee, was accepted for publication. The story naturally gets denser in events and explanations towards the end. Also, it provides a detailed review of pre-2004 literature and, with the benefit of hindsight, attempts to analyse why graphene has attracted so much interest. I have tried my best to make this article not only informative but also easy to read, even for non-physicists.

ZOMBIE MANAGEMENT

My PhD thesis was called "Investigation of mechanisms of transport relaxation in metals by a helicon resonance method". All I can say is that the stuff was as interesting at that time as it sounds to the reader today. I published five journal papers and finished the thesis in five years, the official duration for a PhD at my institution, the Institute of Solid State Physics. *Web of Science* soberly reveals that the papers were cited twice, by co-authors only. The subject was dead a decade before I even started my PhD. However, every cloud has its silver lining, and what I uniquely learned from that experience was that I should never torture research students by offering them "zombie" projects.

After my PhD, I worked as a staff scientist at the Institute of Microelectronics Technology, Chernogolovka, which belongs to the Russian Academy of Sciences. The Soviet system allowed and even encouraged junior staff to choose their own line of research. After a year of poking in different directions, I separated research-wise from my former PhD supervisor, Victor Petrashov, and started developing my own niche. It was an experimental system that was both new and doable, which was nearly an oxymoron, taking into account the scarce resources available at the time at Soviet research institutes. I fabricated a sandwich consisting of a thin metal film and a superconductor separated by a thin insulator. The superconductor served only to condense an external magnetic field into an array of vortices, and this highly inhomogeneous magnetic field was projected onto the film under investigation. Electron transport in such a microscopically inhomogeneous field (varying on a submicron scale) was new research territory, and I published the first experimental report on the subject [4], which was closely followed by an independent paper from Simon Bending [5]. It was an interesting and reasonably important niche, and I continued studying the subject for the next few years, including a spell at the University of Bath in 1991 as a postdoctoral researcher working with Simon.

This experience taught me an important lesson: that introducing a new experimental system is generally more rewarding than trying to find new phenomena within crowded areas. The chances of success are much higher where the field is new. Of course, the fantastic results one originally hopes for are unlikely to materialise, but, in the process of studying any new system, something original inevitably shows up.

ONE MAN'S JUNK, ANOTHER MAN'S GOLD

In 1990, thanks to Vitaly Aristov, director of my Institute in Chernogolovka at the time, I received a six month visiting fellowship from the British Royal Society. Laurence Eaves and Peter Main from Nottingham University kindly agreed to accept me as a visitor. Six months is a very short period for experimental work, and circumstances dictated that I could only study devices readily available in the host laboratory. Available were submicron GaAs wires left over from previous experiments, all done and dusted a few years earlier. Under the circumstances, my experience of working in a poverty-stricken Soviet academy was helpful. The samples that my hosts considered practically exhausted looked like a gold vein to me, and I started working 100 hours per week to exploit it. This short visit led to two *Phys. Rev. Letters* of decent quality [6,7], and I often use this experience to tease my younger colleagues. When things do not go as planned and people start complaining, I provoke them by proclaiming 'there is no such thing as bad samples; there are only bad postdocs/students'. Search carefully and you will always find something new. Of course, it is better to avoid such experiences and explore new territories, but even if one is fortunate enough to find an experimental system as new and exciting as graphene, meticulousness and perseverance allow one to progress much further.

The pace of research at Nottingham was so relentless and, at the same time so inspiring, that a return to Russia was not an option. Swimming through Soviet treacle seemed no less than wasting the rest of my life. So at the age of thirty-three and with an *h*-index of 1 (latest papers not yet published), I entered the Western job market for postdocs. During the next four years I moved between different universities, from Nottingham to Copenhagen to Bath and back to Nottingham. Each move allowed me to get acquainted with yet another topic or two, significantly broadening my research horizons. The physics I studied in those years could be broadly described as mesoscopic and involved such systems and phenomena as two-dimensional electron gases (2DEGs), quantum point contacts, resonant tunnelling and the quantum Hall effect (QHE), to name but a few. In addition, I became familiar with GaAlAs heterostructures grown by molecular beam epitaxy (MBE) and improved my expertise in microfabrication and electron-beam lithography, technologies I had started learning in Russia. All these elements came together to form the foundation for the successful work on graphene a decade later.

DUTCH COMFORT

By 1994 I had published enough quality papers and attended enough conferences to hope for a permanent academic position. When I was offered an associate professorship at the University of Nijmegen, I instantly seized upon the chance of having some security in my new post-Soviet life. The first task in Nijmegen was of course to establish myself. To this end, there was no start-up and no microfabrication to continue any of my previous lines of research. As resources, I was offered access to magnets, cryostats and electronic equipment available at Nijmegen's High Field Magnet Laboratory, led by Jan Kees Maan. He was also my formal boss and in charge of all the money. Even when I was awarded grants as the principal investigator (the Dutch funding agency FOM was generous during my stay in Nijmegen), I could not spend the money as I wished. All funds were distributed through so-called 'working groups' led by full professors. In addition, PhD students in the Netherlands could formally be supervised only by full professors. Although this probably sounds strange to many, this was the Dutch academic system of the 1990s. It was tough for me then. For a couple of years, I really struggled to adjust to the system, which was such a contrast to my joyful and productive years at Nottingham. In addition, the situation was a bit surreal because outside the university walls I received a warm-hearted welcome from everyone around, including Jan Kees and other academics.

Still, the research opportunities in Nijmegen were much better than in Russia and, eventually, I managed to survive scientifically, thanks to help from abroad. Nottingham colleagues (in particular Mohamed Henini) provided me with 2DEGs that were sent to Chernogolovka, where Sergey Dubonos, a close colleague and friend from the 1980s, microfabricated requested devices. The research topic I eventually found and later focused on can be referred to as mesoscopic superconductivity. Sergey and I used micron-sized Hall bars made from a 2DEG as local probes of the magnetic field around small superconducting samples. This allowed measurements of their magnetisation with accuracy sufficient to detect not only the entry and exit of individual vortices but also much more subtle changes. This was a new experimental niche, made possible by the development of an original technique of ballistic Hall micromagnetometry [8]. During the next few years, we exploited this niche area and published several papers in *Nature* and *Phys. Rev. Letters* which reported a paramagnetic Meissner effect, vortices carrying fractional flux, vortex configurations in confined geometries and so on. My wife Irina Grigorieva, an expert in vortex physics [9], could not find a job in the Netherlands and therefore had plenty of time to help me with conquering the subject and writing papers. Also, Sergey not only made the devices but also visited Nijmegen to help with measurements. We established a very productive modus operandi where he collected data and I analysed them within an hour on my computer next door to decide what should be done next.

A SPELL OF LEVITY

The first results on mesoscopic superconductivity started emerging in 1996, which made me feel safer within the Dutch system and also more inquisitive. I started looking around for new areas to explore. The major facility at Nijmegen's High Field Lab was powerful electromagnets. They were a major headache, too. These magnets could provide fields up to 20 T, which was somewhat higher than 16 to 18 T available with the superconducting magnets that many of our competitors had. On the other hand, the electromagnets were so expensive to run that we could use them only for a few hours at night, when electricity was cheaper. My work on mesoscopic superconductivity required only tiny fields (< 0.01T), and I did not use the electromagnets. This made me feel guilty as well as responsible for coming up with experiments that would justify the facility's existence. The only competitive edge I could see in the electromagnets was their room temperature (T) bore. This was often considered as an extra disadvantage because research in condensed matter physics typically requires low, liquid-helium T. The contradiction prompted me, as well as other researchers working in the lab, to ponder on high-field phenomena at room T. Unfortunately, there were few to choose from.

Eventually, I stumbled across the mystery of so-called magnetic water. It is claimed that putting a small magnet around a hot water pipe prevents formation of scale inside the pipe. Or install such a magnet on a water tap, and your kettle will never suffer from chalky deposits. These magnets are available in a great variety in many shops and on the internet. There are also hundreds of articles written on this phenomenon, but the physics behind it remains unclear, and many researchers are sceptical about the very existence of the effect [10]. Over the last fifteen years I have made several attempts to investigate "magnetic water" but they were inconclusive, and I still have nothing to add to the argument. However, the availability of ultra-high fields in a room T environment invited lateral thinking about water. Basically, if magnetic water existed, I thought, then the effect should be clearer in 20 T rather than in typical fields of <0.1 T created by standard magnets.

With this idea in mind and, allegedly, on a Friday night, I poured water inside the lab's electromagnet when it was at its maximum power. Pouring water in one's equipment is certainly not a standard scientific approach, and I cannot recall why I behaved so 'unprofessionally'. Apparently, no one had tried such a silly thing before, although similar facilities existed in several places around the world for decades. To my surprise, water did not end up on the floor but got stuck in the vertical bore of the magnet. Humberto Carmona, a visiting student from Nottingham, and I played for an hour with the water by breaking the blockage with a wooden stick and changing the field strength. As a result, we saw balls of levitating water (Figure 1). This was awesome. It took little time to realise that the physics behind was good old diamagnetism. It took much longer to adjust my intuition to the fact that the feeble magnetic response of water ($\sim 10^{-5}$), billions of times weaker than that of iron, was sufficient to compensate the earth's gravity. Many colleagues, including those who worked with high magnetic fields all their lives, were flabbergasted, and some of them even argued that this was a hoax.

I spent the next few months demonstrating magnetic levitation to colleagues and visitors, as well as trying to make a 'non-boffin' illustration for this beautiful phenomenon. Out of the many objects that we had floating inside the magnet, it was the image of a levitating frog (Figure 1) that started the media hype. More importantly, though, behind all the media noise, this image found its way into many textbooks. However quirky, it has become a beautiful symbol of ever-present diamagnetism, which is no longer perceived to be extremely feeble. Sometimes I am stopped at conferences by people exclaiming "I know you! Sorry, it is not about graphene. I start my lectures with showing your frog. Students always want to learn how it could fly." The frog story, with some intricate physics behind the stability of diamagnetic levitation, is described in my review in *Physics Today* [11].

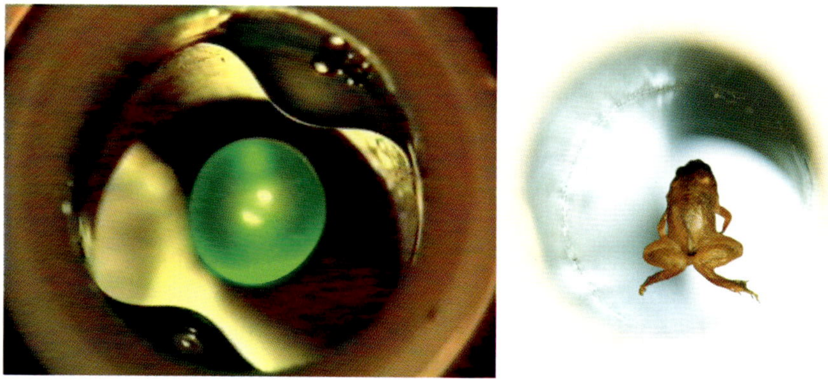

Figure 1. Levitating moments in Nijmegen. Left – Ball of water (about 5 cm in diameter) freely floats inside the vertical bore of an electromagnet. Right – The frog that learned to fly. This image continues to serve as a symbol showing that magnetism of 'nonmagnetic things', including humans, is not so negligible. This experiment earned Michael Berry and me the 2000 Ig Nobel Prize. We were asked first whether we dared to accept this prize, and I take pride in our sense of humour and self-deprecation that we did.

FRIDAY NIGHT EXPERIMENTS

The levitation experience was both interesting and addictive. It taught me the important lesson that poking in directions far away from my immediate area of expertise could lead to interesting results, even if the initial ideas were extremely basic. This in turn influenced my research style, as I started making similar exploratory detours that somehow acquired the name 'Friday night experiments'. The term is of course inaccurate. No serious work can be accomplished in just one night. It usually requires many months of lateral thinking and digging through irrelevant literature without any clear idea in sight. Eventually, you get a feeling – rather than an idea – about what could be interesting to explore. Next, you give it a try, and normally you fail. Then, you may or may not try again. In any case, at some moment you must decide (and this is the most difficult part) whether to continue further efforts or cut losses and start thinking of another experiment. All this happens against the backdrop of your main research and occupies only a small part of your time and brain.

Already in Nijmegen, I started using lateral ideas as under- and post-graduate projects, and students were always excited to buy a pig in a poke. Kostya Novoselov, who came to Nijmegen as a PhD student in 1999, took part in many of these projects. They never lasted for more than a few months, in order not to jeopardise a thesis or career progression. Although the enthusiasm inevitably vanished towards the end, when the predictable failures materialised, some students later confided that those exploratory detours were invaluable experiences.

Most surprisingly, failures sometimes failed to materialise. Gecko tape is one such example. Accidentally or not, I read a paper describing the mechanism behind the amazing climbing ability of geckos [12]. The physics is rather straightforward. Gecko's toes are covered with tiny hairs. Each hair attaches to the opposite surface with a minute van der Waals force (in the nN range), but billions of hairs work together to create a formidable attraction sufficient to keep geckos attached to any surface, even a glass ceiling. In particular, my attention was attracted by the spatial scale of their hairs. They were submicron in diameter, the standard size in research on mesoscopic physics. After toying with the idea for a year or so, Sergey Dubonos and I came up with procedures to make a material that mimicked a gecko's hairy feet. He fabricated a square cm of this tape, and it exhibited notable adhesion [13]. Unfortunately, the material did not work as well as a gecko's feet, deteriorating completely after a couple of attachments. Still, it was an important proof-of-concept experiment that inspired further work in the field. Hopefully, one day someone will develop a way to replicate the hierarchical structure of gecko's setae and its self-cleaning mechanism. Then gecko tape can go on sale.

BETTER TO BE WRONG THAN BORING

While preparing for my lecture in Stockholm, I compiled a list of my Friday night experiments. Only then did I realise a stunning fact. There were two dozen or so experiments over a period of approximately fifteen years and, as expected, most of them failed miserably. But there were three hits: levitation, gecko tape and graphene. This implies an extraordinary success rate: more than 10%. Moreover, there were probably near-misses, too. For example, I once read a paper [14] about giant diamagnetism in FeGeSeAs alloys, which was interpreted as a sign of high-*T* superconductivity. I asked Lamarches for samples and got them. Kostya and I employed ballistic Hall magnetometry to check for giant diamagnetism but found nothing, even at 1 K. This happened in 2003, well before the discovery of iron pnictide superconductivity, and I still wonder whether there were any small inclusions of a superconducting material which we missed with our approach. Another miss was an attempt to detect "heartbeats" of individual living cells. The idea was to use 2DEG Hall crosses as ultrasensitive electrometers to detect electrical signals due to physiological activity of individual cells. Even though no heartbeats were detected while a cell was alive, our sensor recorded huge voltage spikes at its "last gasp" when the cell was treated with excess alcohol [15]. Now I attribute this near-miss to the unwise use of yeast, a very dormant microorganism. Four years later, similar experiments were done using embryonic heart cells and – what a surprise – graphene sensors, and they were successful in detecting such bioelectrical activity [16].

Frankly, I do not believe that the above success rate can be explained by my lateral ideas being particularly good. More likely, this tells us that poking in new directions, even randomly, is more rewarding than is generally perceived. We are probably digging too deep within established areas, leaving plenty of unexplored stuff under the surface, just one poke away. When one dares to try, rewards are not guaranteed, but at least it is an adventure.

THE MANCUNIAN WAY

By 2000, with mesoscopic superconductivity, diamagnetic levitation and four *Nature* papers under my belt, I was well placed to apply for a full professorship. Colleagues were rather surprised when I chose the University of Manchester, declining a number of seemingly more prestigious offers. The reason was simple. Mike Moore, chairman of the search committee, knew my wife Irina when she was a very successful postdoc in Bristol rather than my co-author and a part-time teaching lab technician in Nijmegen. He suggested that Irina could apply for the lectureship that was there to support the professorship. After six years in the Netherlands, the idea that a husband and wife could officially work together had not even crossed my mind. This was the decisive factor. We appreciated not only the possibility of sorting out our dual career problems but also felt touched that our future colleagues cared. We have never regretted the move.

So in early 2001, I took charge of several dilapidated rooms storing ancient equipment of no value, and a start-up grant of £100K. There were no central facilities that I could exploit, except for a helium liquefier. No problem. I followed the same routine as in Nijmegen, combining help from other places, especially Sergey Dubonos. The lab started shaping up surprisingly quickly. Within half a year, I received my first grant of £500K, which allowed us to acquire essential equipment. Despite being consumed with our one year old daughter, Irina also got her starting grant a few months later. We invited Kostya to join us as a research fellow (he continued to be officially registered in Nijmegen as a PhD student until 2004 when he defended his thesis there). And our group started generating results that led to more grants that in turn led to more results.

By 2003 we published several good-quality papers including *Nature, Nature Materials* and *Phys. Rev. Letters,* and we continued beefing up the laboratory with new equipment. Moreover, thanks to a grant of £1.4M (research infrastructure funding scheme masterminded by the then science minister David Sainsbury), Ernie Hill from the Department of Computer Sciences and I managed to set up the Manchester Centre for Mesoscience and Nanotechnology. Instead of pouring the windfall money into bricks-and-mortar, we utilised the existing clean room areas (~250 m^2) in Computer Sciences. Those rooms contained obsolete equipment, and it was thrown away and replaced with state-of-the-art microfabrication facilities, including a new electron-beam lithography system. The fact that Ernie and I are most proud of is that many groups around the world have more expensive facilities but our Centre has continuously, since 2003, been producing new structures and devices. We do not have a posh horse here that is for show, but rather a draft horse that has been working really hard.

Whenever I describe this experience to my colleagues abroad, they find it difficult to believe that it is possible to establish a fully functional laboratory and a microfabrication facility in less than three years and without an astronomical start-up grant. If not for my own experience, I would not believe it either. Things progressed unbelievably quickly. The University was supportive, but my greatest thanks are reserved specifically for the responsive mode of the UK Engineering and Physical Sciences Research Council (EPSRC). The funding system is democratic and non-xenophobic. Your position in an academic hierarchy or an old-boys network counts for little. Also, 'visionary ideas' and grand promises to 'address social and economic needs' play little role when it comes to the peer review. In truth, the responsive mode distributes its money on the basis of a recent track record, whatever that means in different subjects, and the funding normally goes to researchers who work both efficiently and hard. Of course, no system is perfect, and one can always hope for a better one. However, paraphrasing Winston Churchill, the UK has the worst research funding system, except for all the others that I am aware of.

THREE LITTLE CLOUDS

As our laboratory and Nanotech Centre were shaping up, I got some spare time for thinking of new research detours. Gecko tape and the failed attempts with yeast and quasi-pnictides took place during that time. Also, Serge Morozov, a senior fellow from Chernogolovka, who later became a regular visitor and invaluable collaborator, wasted his first two visits on studying magnetic water. In the autumn of 2002, our first Manchester PhD student, Da Jiang, arrived, and I needed to invent a PhD project for him. It was clear that for the first few months he needed to spend his time learning English and getting acquainted with the lab. Accordingly, as a starter, I suggested to him a new lateral experiment. It was to make films of graphite 'as thin as possible' and, if successful, I promised we would then study their 'mesoscopic' properties. Recently, trying to analyse how this idea emerged, I recalled three badly shaped thought clouds.

One cloud was a concept of 'metallic electronics'. If an external electric field is applied to a metal, the number of charge carriers near its surface changes, so that one may expect that its surface properties change, too. This is how modern semiconductor electronics works. Why not use a metal instead of silicon? As an undergraduate student, I wanted to use electric field effect (EFE) and X-ray analysis to induce and detect changes in the lattice constant. It was naïve because simple estimates show that the effect would be negligible. Indeed, no dielectric allows fields much higher than 1V/nm, which translates into maximum changes in charge carrier concentration n at the metal surface of about 10^{14} per cm^2. In comparison, a typical metal (e.g., Au) contains $\sim10^{23}$ electrons per cm^3 and, even for a 1 nm thick film, this yields relative changes in n and conductivity of ~1%, leaving aside much smaller changes in the lattice constant.

Previously, many researchers aspired to detect the field effect in metals. The first mention is as far back as 1902, shortly after the discovery of the electron. J. J. Thomson (1906 Nobel Prize in Physics) suggested to Charles Mott, the father of Nevill Mott (1977 Nobel Prize in Physics), to look for the EFE in a thin metal film, but nothing was found [17]. The first attempt to measure the EFE in a metal was recorded in scientific literature in 1906 [18]. Instead of a normal metal, one could also think of semimetals such as bismuth, graphite or antimony which have a lot fewer carriers. Over the last century, many researchers used Bi films ($n \sim10^{18}$ cm^{-3}) but observed only small changes in their conductivity [19,20]. Aware of this research area and with experience in GaAlAs heterostructures, I was continuously, albeit casually, looking for other candidates, especially ultra-thin films of superconductors in which the field effect can be amplified in proximity to the superconducting transition [21,22]. In Nijmegen, my enthusiasm was once sparked by learning about nm-thick Al films grown by MBE on top of GaAlAs heterostructures but, after estimating possible effects, I decided that the chances of success were so poor it was not worth trying.

Carbon nanotubes were the second cloud hanging around in the late

1990s and early 2000s. Those were the years when nanotubes were at the peak of their glory. Living in the Netherlands, I heard talks by Cees Dekker and Leo Kouwenhoven and read papers by Thomas Ebbesen, Paul McEuen, Sumio Iijima, Pheadon Avouris and others. Each time, those exceptionally nice results inevitably triggered thoughts about entering this research area. But I was too late and needed to find a different perspective, away from the stampede.

As for the third cloud, I read a review of Millie Dresselhaus about intercalated graphite compounds [23], which clearly showed that, even after many decades, graphite was still a material little understood, especially in terms of its electronic properties. This influential review prompted me to look further into graphite literature. In doing so, I encountered papers by Pablo Esquinazi and Yakov Kopelevich, who reported ferromagnetism, superconductivity and a metal-insulator transition, all in the same good old graphite and at room T [24,25]. Those provocative papers left me with a distinct feeling that graphite was very much worth having a careful look at.

The three thought clouds (and maybe some more that I cannot recall) somehow merged into Da's project. I reckoned that if we were to succeed in making thin films of graphite, instead of Bi, they could exhibit some electric field effect and/or some other interesting properties resembling those of carbon nanotubes. In the worst-case scenario, our mesoscopic samples would be monocrystals and this could help to clarify those controversies about graphite. Why not try to poke in this direction for a few months, I thought.

LEGEND OF SCOTCH TAPE

To make thin graphite films, I provided Da with a tablet of pyrolytic graphite, which was several mm thick and an inch in diameter, and suggested using a polishing machine. We had a fancy one that allowed submicron flatness. A few months later, Da declared that he had reached the ultimate thickness and showed me a tiny speck of graphite at the bottom of a Petri dish. I looked at it in an optical microscope and, by focusing on the top and bottom surfaces, estimated that the speck was ~10 μm thick. Too thick, I thought and suggested trying a finer polishing liquid. However, it turned out that Da had polished away the whole tablet to obtain this one speck. It was actually my fault: Da successfully finished his PhD later, but at that time he was just a fresh foreign student with a huge language barrier. Moreover, by mistake I gave him high-density graphite instead of highly-oriented pyrolytic graphite (HOPG) as was intended. The former does not shed as easily as HOPG.

Oleg Shklyarevskii, a senior fellow from Kharkov, Ukraine was working nearby and had to listen to the typical flow of my teasing remarks, this time about polishing a mountain to get one grain of sand. Oleg was an expert in scanning tunnelling microscopy (STM) and worked on a project that later turned out to be another bad 'Friday night' idea of mine. He interjected by bringing over a piece of cellotape with graphite flakes attached to it. Allegedly, he just fished out the tape from a litter bin. Indeed, HOPG is the standard reference sample for STM, where a fresh surface of graphite is normally prepared by removing a top layer with sticky tape. We used this technique for years but never looked carefully at what was thrown away along with the tape. I looked in the microscope at the remnants of graphite (Figure 2) and found pieces much thinner than Da's speck. Only then did I realise how silly it was of me to suggest the polishing machine. Polishing was dead, long live Scotch tape!

This moment was not a breakthrough yet, but things started to look promising and required more people to get involved. Oleg did not volunteer to take on yet another project but Kostya did. 'Volunteer' is probably not the right word. Everyone in our lab has always been welcome to move around and participate in whatever project they want. At that time, Kostya was working on a nicely-moving project on ferromagnetism [26]. He was also our 'caretaker' when things went wrong, especially with measuring equipment. As for me, at that time I used to spend a few hours a day in the lab preparing samples, doing measurements and analysing results. It was only after 2006 that I turned into a paper-writing machine combined with a data analyser. I have always loved the latter but hated to write papers. Unfortunately, no lab can survive without its Shakespeare.

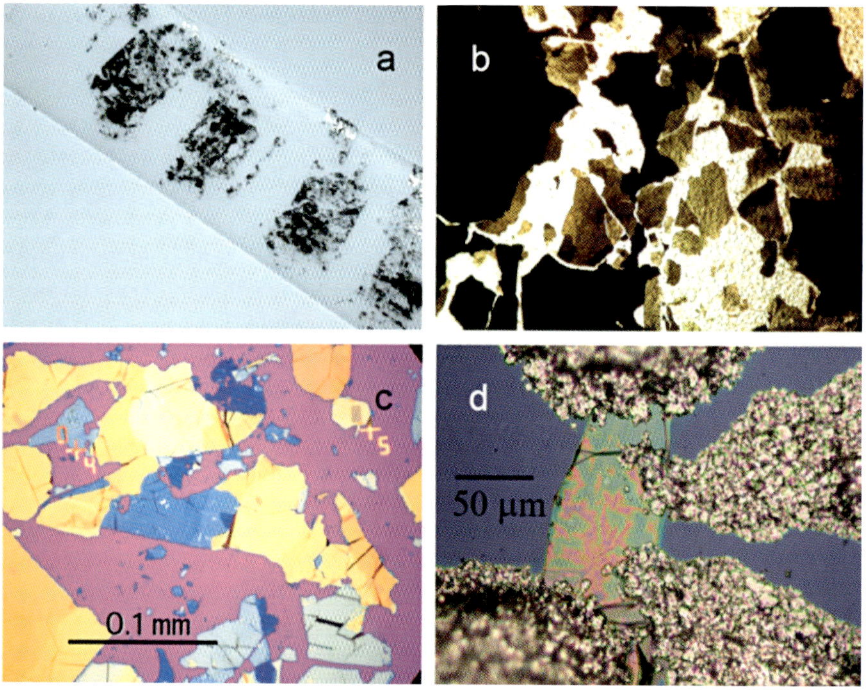

Figure 2. In hindsight, thin crystals of graphite are easy to obtain. **a** – Remnants of HOPG left attached to Scotch tape. **b** – Some of the crystals are optically transparent if viewed in an optical microscope or just with a magnifying glass. **c** – If placed on an oxidised Si wafer, transparent crystals give rise to various shades of blue. **d** – One of our very first devices made by using 'a shoestring and sealing wax': in this case, tweezers, a toothpick and silver paint.

Kostya and I decided to check out the electrical properties of the graphite flakes found on the cellotape and, to this end, he started transferring them onto glass slides, initially by using just tweezers. A few days later and keeping in mind the initial motivation, I brought in oxidised Si wafers in order to use them as substrates and detect the EFE. This delivered an unexpected bonus. Placing thin graphite fragments onto those wafers allowed us to observe interference colours that indicated that some of the fragments were optically transparent. Moreover, the colours provided us with a very intuitive way of judging which flakes were thin (Figure 2c). We quickly found that some of them were just a few nm thick. This was our first real breakthrough.

EUREKA MOMENT

In graphene literature and especially in popular articles, a strong emphasis is placed on the Scotch tape technique, and it is hailed for allowing the isolation and identification of ultra-thin graphite films and graphene. For me, this was an important development but still not a Eureka moment. Our goal always was to find some exciting physics rather than just observing ultra-thin films in a microscope.

Within a couple of days after Oleg prompted the use of Scotch tape, Kostya was already using silver paint to make electrical contacts to graphite platelets transferred from the Scotch tape. To our surprise, they turned out to be highly conductive and even the painted contacts exhibited a reasonably low resistance. The electronic properties could be studied, but we felt it was too early to put the ugly-looking devices (see Fig. 2d) in a cryostat for proper measurements. As a next step, we applied voltage, first, through the glass slides and, a bit later, to the Si wafer, using it as a back gate to check for the field effect. Figure 2 shows a photograph of one of our first devices. The central part is a graphite crystal that is ~20 nm thick, and its lateral size is comparable to the diameter of a human hair. To transfer the crystal by tweezers from the tape and then make four such closely-spaced contacts by using just a toothpick and silver paint is the highest level of experimental skill. These days, not many researchers have fingers green enough to make such samples. I challenge readers to test their own skills against this benchmark!

The very first hand-made device on glass exhibited a clear EFE such that its resistance could be changed by several per cent. It may sound small and of marginal importance but, aware of how hard it was previously to detect any EFE at all, I was truly shocked. If those ugly devices made by hand from relatively big and thick platelets already showed some field effect, what could happen, I thought, if we were to use our thinnest crystallites and apply the full arsenal of microfabrication facilities? There was a click in my head that we had stumbled onto something really exciting. This was my Eureka moment.

What followed was no longer a random walk. From this point, it was only logical to continue along the same path by improving procedures for cleaving and finding thinner and thinner crystals and making better and better devices, which we did. It was both painstaking and incredibly rapid, depending on one's viewpoint. It took several months until we learned how to identify monolayers by using optical and atomic force microscopy. On the microfabrication side, we started using electron-beam lithography to define proper Hall bar devices and started making contacts by metal evaporation rather than silver painting. The microfabrication development was led by Dubonos, aided by his PhD student Anatoly Firsov. Initially, they employed facilities in Chernogolovka but, when our new postdoc Yuan Zhang got fully acquainted with the recently-installed lithography system at our Nanotech Centre, the process really speeded up.

The move from multilayers to monolayers and from hand-made to lithography devices was conceptually simple but never straightforward. We took numerous detours and wasted much effort on ideas that only led us into dead ends. An example of grand plans that never worked out was the idea to plasma-etch graphite mesas in the form of Hall bars which, after cleavage, would provide readily shaped devices, or so I thought. Later, we had to return to the unprocessed graphite. The teething problems we experienced at that time can also be illustrated by the fact that initially we believed that Si wafers should have a very precise thickness of oxide (within several nm) to allow hunting for monolayers. These days we can find graphene on practically any substrate. Crystal sizes also went up from a few microns to nearly a millimetre, just by tinkering with procedures and using different sources of graphite.

The most essential part of our 2004 report [27] was the electrical measurements, and this required a lot of work. For several months, Kostya and Serge Morozov were measuring full time, and I was around as well, discussing and analysing raw data, often as soon as they appeared on the screen. The feedback to our microfabrication guys was almost instantaneous. As always in the case of encountering a new system where one does not know what to expect, we had to be particularly careful in those first experiments. We disregarded any curve, unless it was reproducible for many devices and, to avoid any premature conclusions, we studied more than 50 ultra-thin devices. Those were years of hard work compressed into just a few months, but we were excited as every new device got better and better, and we could work 24x7, which typically meant fourteen hour days and no breaks for the weekends.

Finally, by the end of 2003, we got a reliable experimental picture ready for publication. Between that moment and the end of my timeline when the *Science* paper was accepted in September 2004, there is a lengthy gap. Those nine months were consumed by excruciating efforts to publish the results in a high-profile journal. We continuously added data and polished the presentation. Irina's help was invaluable in this time-consuming process, which can be fully appreciated only by those readers who ever published in such glossy journals. First, we submitted the manuscript to *Nature*. It was rejected and, when further information requested by referees was added, rejected again. According to one referee, our report did "not constitute a sufficient scientific advance." *Science* referees were more generous (or more knowledgeable?), and the presentation was better polished by that time. In hindsight, I should have saved the time and nerves by submitting to a second-tier journal, even though we all felt that the results were ground-breaking. Readers aspiring to get published in those glossy magazines and having their papers recently rejected can use this story to cheer up: Their papers may also be prize winning!

DEFIANT EXISTENCE

One of the most surprising results of our *Science* report was the observation that, after being isolated, atomic planes remained continuous and conductive under ambient conditions. Even with hindsight, there are many reasons to be surprised.

First, for many decades researchers studied ultra-thin films, and their collective experience proves that continuous monolayers are practically impossible to make (see, e.g., [28,29]). Try to evaporate a metal film a few nm in thickness, and you will find it discontinuous. The material coagulates into tiny islands. This process, called island growth, is universal and driven by the fact that a system tries to minimise its surface energy. Even by using epitaxial substrates that provide an interaction working against the surface energy contribution and cooling them down to liquid-helium T, which prevents migration of deposited atoms, it is hard to find the right conditions to create continuous nm thick films, let alone monolayers [28,29].

The second reason to be surprised is that theory unequivocally tells us that an isolated graphene sheet should be thermodynamically unstable. Calculations show that 'graphene is the least stable [carbon] structure until about 6000 atoms' [30]. Until ~24,000 atoms (that is, a flat sheet with a characteristic size of ~25 nm), various 3D configurations are energetically more favourable than the 2D geometry [30,31]. For larger sizes, theory shows again that a graphene sheet is unstable but now with respect to scrolling. The latter conclusion is based on considering competing contributions from the bending and surface energies [32,33]. These calculations are specific to carbon, but the underlying physics is conceptually connected to the surface energy mechanism that leads to island growth.

Third, 2D crystals cannot be grown in isolation, without an epitaxial substrate that provides an additional atomic bonding. This follows from the Landau-Peierls argument that shows that the density of thermal fluctuations for a 2D crystal in the 3D space diverges with temperature [1]. Although the divergence is only logarithmic, crystal growth normally requires high T such that atoms become sufficiently mobile. This also implies a softer lattice with little shear rigidity. The combination of the two conditions sets a limit on possible sizes L of 2D atomic crystals. One can estimate L as $\sim a\exp(E/T_G)$ where $a\sim 1$Å is the lattice spacing, $E\sim 1$eV the atomic bond energy and T_G the growth temperature. This consideration should not be applied to graphene at room T, which would yield astronomical sizes. T_G is usually comparable to the bond energy, which renders the disorder-generating mechanism irrelevant at much lower *T*. Note that, in principle, self-assembly may allow growth of graphene at room *T* but, so far, this has been achieved only for nm-sized graphene sheets [34].

The fourth and probably the most important reason to be surprised is that graphene remains stable under ambient conditions. Surfaces of materials can react with air and moisture, and monolayer graphene has not one but two surfaces, making it more reactive. Surface science research involves ultra-high vacuum facilities and, often, liquid-helium *T* to keep surfaces stable and away from reactive species. For example, gold is one of the most inert materials in nature but, even for Au, it is hard to avoid its near-surface layer being partially oxidised in air. What then are the chances for a monolayer exposed to ambient conditions to remain unaffected?

Graphene flouts all the above considerations. It is instructive to analyse how. First, any existing method of obtaining graphene starts with 3D rather than 2D growth. Graphene sheets are initially formed either within the bulk or on top of an epitaxial substrate, which quenches the diverging thermal fluctuations. The interaction can be relatively weak, as in the case of graphene grown on graphite [35], but it is always present. This allows graphene to dodge the Landau-Peierls argument and, also, to avoid coagulation into islands and 3D carbon structures. Second, if graphene is cleaved or released from a substrate, the process is normally carried out at room *T* so that energy barriers remain sufficiently high. This allows atomic planes to persist in an isolated, non-scrolled form without any substrate [36], even though this is energetically unfavourable. If placed on a substrate, the van der Waals interaction may also be sufficient to prevent a graphene sheet from scrolling. Third, graphite is even more chemically inert than gold. Although graphene is more reactive than graphite and weakly reacts with air and pollutants at room *T*, this does not destroy its crystal lattice and high conductivity [37,38]. It requires *T* twice as high as room *T* to irreversibly damage graphene in air. Our ambient conditions appear fortuitous enough for the graphene lattice to survive.

REQUIEM FOR BRILLIANT IDEAS

Science literature is full of brilliant ideas that did not work. Searching the literature for those is not a good idea at all. At a start of a new project, a couple of decent reviews usually do the job of making sure that one does not reinvent the wheel. The alternative can be truly detrimental. I have met many promising researchers who later failed to live up to their promise because they wasted their time on searching literature, instead of spending it on searching for new phenomena. What's more, after months of literature search, they inevitably came to the same conclusion: Everything they planned had been done before. Therefore, they saw no reason to try their own ideas and, consequently, began a new literature search. One should realise that ideas are never new. However brilliant, every idea is always based on previous knowledge and, with so many smart people around, the odds are that someone somewhere had already thought of something similar before. This should not be used as an excuse for not trying because local circumstances vary and, moreover, facilities change with time. New technologies offer a reasonable chance that old failed ideas may work unpredictably well the next time round.

In 2002–2003, the merged thought clouds that I would not even call a brilliant idea were sufficient to instigate the project. They also provided us with an Ariadne's thread that helped with choosing specific directions. Literature search was done in due course, after we roughly scouted the new area and especially when the results were being prepared for publication. In addition to the literature relevant to the thought clouds, our *Science* paper cited the challenges of obtaining isolated 2D crystals, their thermodynamic instability, the observation of nanoscrolls and papers on epitaxial growth. Those references were important to show the experimental progress we achieved. The first review of earlier literature was done in our 2007 progress article [1]. Since then, I updated my conference presentations whenever a historically important paper came to light. This is the first opportunity to update the history chapter in writing by adding several new references. Furthermore, my recent call for further historical insights [39] was answered by a number of researchers and, for completeness, I want to acknowledge their early ideas and contributions, too.

GRAPHENE INCARNATIONS

Looking back at graphene history, we should probably start with an observation by the British chemist Benjamin Brodie [40]. In 1859, by exposing graphite to strong acids, he obtained what he called 'carbonic acid' (Figure 3a). Brodie believed that he discovered 'graphon', a new form of carbon with a molecular weight of 33. Today we know that he observed a suspension of tiny crystals of graphene oxide, that is, graphene sheets densely covered with hydroxyl and epoxide groups [41]. Over the next century, there were quite a few papers describing the laminated structure of graphite oxide, but the next crucial step in graphene history was the proof that this 'carbonic acid' consisted of floating atomic planes. In 1948, G. Ruess and F. Vogt used transmission electron microscopy (TEM) and, after drying a droplet of a graphene-oxide suspension on a TEM grid, they observed creased flakes down to a few nm in thickness [42]. These studies were continued by the group of Ulrich Hofmann. In 1962, he and Hanns-Peter Boehm looked for the thinnest possible fragments of reduced graphite oxide and identified some of them as monolayers [43] (Figure 3b).

This remarkable observation received little attention until 2009–2010. I have to mention that the 1962 identification relied on a relative TEM contrast, an approach that would not stand today's scrutiny because the contrast strongly depends on focusing conditions [44]. For example, Rahul Nair and I tried but predictably failed to distinguish between monolayers and somewhat thicker flakes by using only their TEM contrast. Graphene monolayers were unambiguously identified in TEM only forty years after the 1962 paper by counting the number of folding lines [45–47]. Nonetheless, the Boehm-Hofmann work should, in my opinion, stand as the first observation of graphene because monolayers should have been present among the residue, and the idea was correct. Furthermore, it was Boehm and his colleagues who in 1986 introduced the term graphene, deriving it from the combination of the word 'graphite' and the suffix that refers to polycyclic aromatic hydrocarbons [48].

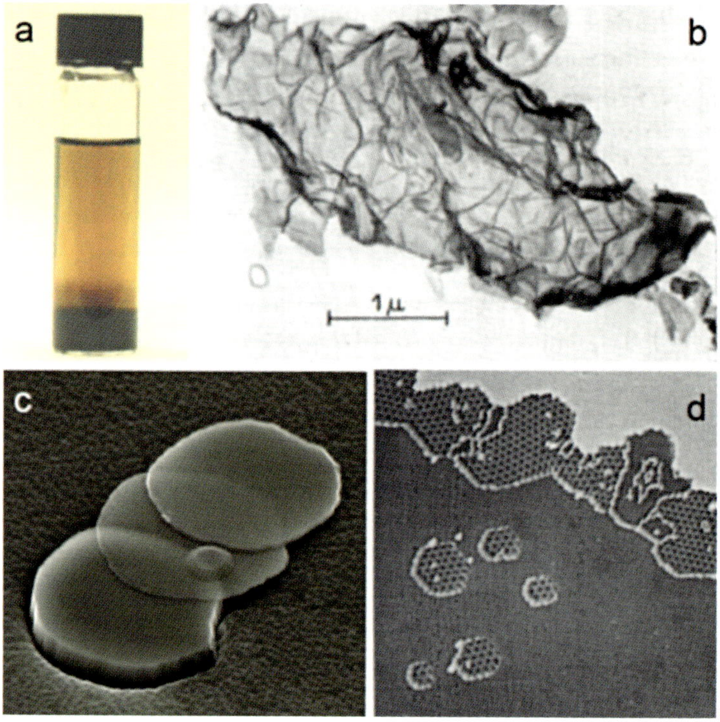

Figure 3. Prehistory of graphene. **a** – Graphene as probably seen by Brodie 150 years ago. Graphite oxide at the bottom of the container dissolves in water making the yellow suspension of floating graphene flakes. **b** – TEM image of ultra-thin graphitic flakes from the early 1960s (copied with permission from ref. [43]). **c** – Scanning electron microscopy (SEM) image of thin graphite platelets produced by cleavage (similar to images reported in ref. [60]). **d** – STM of graphene grown on Pt (copied with permission from ref. [53]). The image is 100x100 nm^2 in size. The hexagonal superstructure has a period of ~22 Å and appears due to the interaction of graphene with the metal substrate.

In addition to the TEM observations, another important line in pre-2004 graphene research was its epitaxial growth. Ultra-thin graphitic films and, sometimes, even monolayers were grown on metal substrates [49–53], insulating carbides [54–57] and graphite [35] (see Fig. 3d). The first papers I am aware of go back to 1970 when John Grant reported graphitic films on Ru and Rh [49] and Blakely *et al.* on Ni [50]. Epitaxial growth on insulating substrates was first demonstrated by van Bommel *et al.* in 1975 [54] whereas

Chuhei Oshima found other carbides allowing graphene growth (for example, TiC) [55]. The grown films were usually analysed by surface science techniques that average over large areas and say little about the film's continuity and quality. Occasionaly, STM was also used for visualisation and local analysis.

Even more relevant were earlier attempts to obtain ultra-thin films of graphite by cleavage, similar to what we did in 2003. In 1990, Heinrich Kurz's group reported 'peeling optically thin layers with transparent tape' (read Scotch tape), which were then used to study carrier dynamics in graphite [58]. In 1995, Thomas Ebbesen and Hidefumi Hiura described few-nm-thick 'origami' visualised by atomic force microscopy (AFM) on top of HOPG [59]. Rod Ruoff also photographed thin graphite platelets in SEM [60] (Fig. 3c). In 2003, monolayers were reported by Yang Gan who used STM for their cleavage on top of HOPG [61].

Finally, there were electrical studies of thin graphite films. Between 1997 and 2000, Yoshiko Ohashi succeeded in cleaving crystals down to ~20 nm in thickness, studied their electrical properties including Shubnikov – de Haas oscillations and, quite remarkably, observed the electric field effect with resistivity changes of up to 8% [62,63]. Also, Ebbesen's group succeeded in growth of micron-sized graphitic disks with thickness down to 60 layers and measured their electrical properties [64].

As for theory, let me make only a short note (for more references, see [1,65]). Theoretically, graphene ('a monolayer of graphite') was around since 1947 when Phil Wallace first calculated its band structure as a starting point to understanding the electronic properties of bulk graphite [66]. Gordon Semenoff and Duncan Haldane realised that graphene could provide a nice condensed-matter analogue of (2+1)-dimensional quantum electrodynamics [67,68] and, since then, the material served as a toy model to address various questions of QED (see, e.g., [69,70]). Many of the theories became relevant to experiment well before 2004, when electronic properties of carbon nanotubes (rolled-up graphene ribbons) were investigated. A large amount of important theoretical work on graphene was done by Tsuneya Ando, and Millie Dresselhaus and co-workers (see, e.g., [71–73]).

To complete the history of graphene, let me also acknowledge some earlier ideas. Thomas Ebbesen and Hidefumi Hiura envisaged a possibility of graphene-based nanoelectronics in 1995 (as an example, they referred to epitaxial graphene grown on TiC) [59]. In patent literature, speculations about "field effect transistors employing pyrolytic graphite" go back as far as 1970 [74]. Also, it was pointed out to me by Rod Ruoff and Reginald Little that their pre-2004 papers discussed possibilities and mentioned an intention of obtaining isolated monolayers [60,75]. Finally, the layered structure of graphite was known since early days of X-ray crystallography, and researchers certainly have been aware of graphite being a deck of weakly bonded graphene planes for an even longer time. This property has been widely used to create a variety of intercalated graphite compounds [23] and, of course, to make drawings. After all, we now know that isolated monolayers can be found in every pencil trace, if one searches carefully enough in an optical microscope [2]. Graphene has literally been before our eyes and under our noses for many centuries but was never recognised for what it really is.

Πλανήτη GRAPHENE

The reader may find some of the cited ideas and historical papers irrelevant, but I tried my best to avoid any pre-2004 result, especially experimental, being overlooked. All the mentioned studies poked in the right direction, but there were no big surprises to spark a graphene gold rush. This is probably because the earlier experiments had one thing in common. They were observational. They observed ultra-thin graphitic films, and occasionaly even monolayers without reporting any of graphene's distinguishing properties. The very few electrical and optical measurements cited above were done using thin films of graphite and could not assess the physics that graphene has brought to the fore since 2004.

Our *Science* paper provided a clear watershed. Of course, the article reported the isolation of graphene crystals large enough to do all sorts of measurements, beyond the observation in an electron or scanning probe microscope. The method of graphene isolation and identification it described was so straightforward and accessible that even schoolchildren could probably do it. This was important but, if we were to stop there, just with the observations, our work would only add to the previous literature and, I believe, disappear into oblivion. It is not the observation and isolation of graphene but its electronic properties that took researchers by surprise. Our measurements delivered news well beyond the Scotch tape technique, which persuaded many researchers to join in the graphene rush.

First, the 2004 paper reported an ambipolar electric field effect, in which resistivity changed by a factor of ~100. This is thousands of times more than the few per cent changes observed previously for any metallic system and amounted to a qualitative difference. To appreciate the exquisiteness of this observation, imagine a nanometre thick Au film. No matter what you do with such a film by physical means, it will remain a normal metal with the same properties. In contrast, properties of graphene can be altered by simply varying the gate voltage. We can tune graphene from a state close to a normal metal with electrons in concentration $\sim 10^{21}$ cm^{-3} to a metal with a similar concentration of holes, all the way through a 'semiconducting' state with few charge carriers.

Even more remarkably, our devices exhibited an astonishing electronic quality. Graphene was completely unprotected from the environment, as it was placed on a microscopically rough substrate and covered from both sides with adsorbates and a polymer residue. Still, electrons could travel submicron distances without scattering, flouting all the elements outside. This level of electronic quality is completely counterintuitive. It contradicts the common wisdom that surface science requires ultra-high vacuum and, even then, thin films become progressively poorer in quality as their thickness decreases. Even with hindsight, such electronic quality is mystifying and, in fact, not fully understood so far.

In semiconductor physics, electronic quality is described in terms of charge carrier mobility μ. Our *Science* paper reported graphene with room-T μ ≈10,000 cm^2/Vs (as of 2010, μ can be 10 and 100 times higher at room and low T, respectively [76,77]). For a general reader, 10,000 may sound like just another number. To explain its significance, let us imagine that in 2004 we made devices from, for example, reduced graphene oxide, which exhibits μ ~1cm^2/Vs due to its irreversibly damaged crystal lattice [78]. In our second paper on graphene [79], we reported 2D dichalcogenides with equally low μ. Since then, there has been little interest in them. The reported ballistic transport over submicron distances was essential to spark interest in graphene and to allow the observation of many quantum effects reported both in 2004 and later. This would have been impossible if graphene exhibited μ below several 1,000 cm^2/Vs.

If not for graphene's high quality and tunability, there would be no new physics and, therefore, no graphene boom. In this respect, graphene history has something in common with that of solar planets. Ancient Greeks observed them and called them wandering stars, πλανήτες. After the physics behind this wandering was discovered, people started perceiving planets quite differently from πλανήτες. Similarly, during the last six years people discovered what graphene really is, which completely changed the earlier perception. Our *Science* paper offered the first glimpse of graphene in its new avatar as a high quality 2D electronic system and beyond.

MAGIC OF FLAT CARBON

What is this new incarnation? For me, 2004 was only the starting point for the unveiling of many unique properties of graphene. Since then, we have demonstrated that charge carriers in graphene are massless fermions described by a Dirac-like equation rather than by the standard Schrödinger equation [80]. In bilayer graphene, electrons receive yet another makeup as massive Dirac fermions [81]. These properties were unveiled by the observation of two new types of the integer quantum Hall effect, which corresponded to the two types of Dirac fermions [1,65]. We also found that graphene remained metallic in the limit of no charge carriers, even when just a few electrons remained present in a micron-sized device [1,77]. Our experiments have revealed that graphene exhibits a universal optical conductivity of $\pi e^2/2h$, such that its visible opacity is just $\pi\alpha$, where α is the fine structure constant [82]. We suggested that the phenomenon of Klein tunnelling, which was known in relativistic quantum physics for many decades but assumed non-observable, could be probed using graphene devices [83]. Several groups later demonstrated this experimentally. We were lucky to be slightly quicker than others in showing that bilayer graphene was a tuneable-gap semiconductor [84] and that graphene could be carved into devices on a true nm scale [85]. We demonstrated sensors capable of detecting individual molecules, more sensitive than any sensor before [38]. We suggested that strain in graphene creates pseudo-magnetic fields that alter its electronic properties [86] and later discussed the possibility of creating uniform pseudo-fields and observation of the quantum Hall effect without an external magnetic field [87]. Pseudo-magnetic fields in excess of 400 T were reported experimentally half a year later. We made the first step into graphene chemistry by experimentally introducing its derivatives, graphane and stoichiometric fluorographene [88,89]. This is not even an exhaustive list of the nice phenomena that we and our collaborators found in graphene and, of course, many other researchers reported many other beautiful discoveries that propelled graphene into its new status of a system that can nearly deliver magic.

ODE TO ONE

After reading about the beautiful properties of graphene, the reader may wonder why many atomic layers stacked on top of each other, as in graphite, do not exhibit similar properties. Of course, any graphitic derivative has something in common with its parent, but in the case of graphene, differences between the parent and descendants are fundamental. To appreciate this, let us simplify the task and compare graphene with its bilayer. The crucial distinctions are already there.

First, graphene exhibits record stiffness and mechanical strength [90]. As for its bilayer, this strength is jeopardised by the possibility that the two layers will slide relative to each other. This leads to a principal difference if, for example, graphene or any thicker platelets are used in composite materials. Second, graphene chemistry is different depending on whether one or both surfaces of a monolayer are exposed. For example, atomic hydrogen cannot bind to graphene from one side but makes a stoichiometric compound (graphane) if both surfaces are exposed. This makes graphene much more reactive than its bilayer. Third, an electric field is screened in graphite at distances of about the interlayer separation, and the electric screening becomes important even for a bilayer. For multilayer graphene, the electric field can dope no more than a couple of near-surface atomic planes, leaving the bulk unaffected. This makes it naïve to speculate about the use of graphitic multilayers in active electronics. Fourth, charge carriers in a monolayer are massless Dirac fermions whereas they are massive in a graphene bilayer. This leads to essential differences in many electronic properties including Shubnikov-de Haas oscillations, quantum Hall effect, Klein tunnelling and so on. The Sorites paradox refers to a moment when a heap is no longer a heap if the grains are removed one by one. For graphene, even its bilayer is so different that two already make a heap.

TO COLLEAGUES AND FRIENDS

Our *Science* report was a collective effort, and I would again – on behalf of Kostya and myself – like to thank all the other contributors. Serge Morozov was and remains our 'multitasking measurement machine' working 24x7 when in Manchester. His electrical measurement skills are unmatched, and I know that any curve he brings in is completely reliable and no questions are ever asked whether this and that was checked and crosschecked. Da Jiang was around from the very start, and it is unfortunate that I had to take the project away from him because it was beyond the scope of a single new PhD student. Sergey Dubonos and Yuan Zhang were the ones who made all the devices without which our work would obviously have been impossible. I utterly regret that our life trajectories have later diverged and, especially, that Sergey has switched from microfabrication technology to goat farming. I also acknowledge help of Anatoly Firsov in making those devices. Irina Grigorieva helped with scanning electron microscopy but, more importantly, with writing up the 2004 manuscript.

Figure 4. Those who made our first graphene paper possible but did not get the Prize.

The end of my timeline was only a start for further hard work involving many collaborators. Our rapid progress would be impossible without Misha Katsnelson who provided us with all the theoretical help an experimentalist can only dream of. Since 2006, I have been enjoying collaboration with other great theory guys including Antonio Castro Neto, Paco Guinea, Nuno Peres, Volodya Fal'ko, Leonid Levitov, Allan MacDonald, Dima Abanin, Tim Wehling and their co-workers. In particular, I want to acknowledge many illuminating discussions and banter over dinners with Antonio and Paco. As for experimentalists, the list is longer and includes Philip Kim, Ernie Hill, Andrea Ferrari, Eva Andrei, Alexey Kuzmenko, Uschi Bangert, Sasha Grigorenko, Uli Zeitler, Jannik Meyer, Marek Potemskii and many of their colleagues.

Philip deserves special praise. In August 2004, before our *Science* paper was published, his group submitted another important paper [91]. His report described electronic properties of ultra-thin graphite platelets (down to ~35 layers). Except for the thicker devices, Philip's group followed the same route as our now-celebrated paper. How close he was can be judged from the fact that, after adopting the Scotch tape technique, Philip started studying monolayers in early 2005. This allowed him to catch up quickly and, in mid-2005, our two groups submitted independent reports that appeared back-to-back in *Nature*, both describing the all-important observation of Dirac fermions in monolayer graphene [80,92]. Later, I had the pleasure of closely working with Philip on two joint papers, for *Science* and *Scientific American*. For me personally, those back-to-back *Nature* papers signified a watershed. People within the large semiconducting community no longer rumoured that 'the results were as difficult to reproduce as those by Hendrik Schön', and friends no longer stopped me in corridors with 'be more careful; you know …' I owe Philip a great deal for this, and many people heard me saying – before and after the Nobel Prize – that I would be honoured to share it with him.

Last but not least, let me acknowledge many bright young, and not so young, colleagues: Peter Blake, Rahul Nair, Roman Gorbachev, Leonid Ponomarenko, Fred Schedin, Daniel Elias, Sasha Mayorov, Rui Yang, Vasyl Kravets, Zhenhua Ni, Wencai Ren, Rashid Jalil, Ibtsam Riaz, Soeren Neubeck, Tariq Mohiuddin and Tim Booth. They were PhD students and postdocs here in Manchester over the last six years and, as always, I avoid using the feudal word 'my'.

Finally, I acknowledge the financial support of EPSRC in its best, that is, the responsive mode. This Nobel Prize would be absolutely impossible without this mode. Let me also thank the Royal Society and the Leverhulme Trust for reducing my teaching loads, which allowed me to focus on the project. I have also received funding from the Office of Naval Research and the Air Force Office of Scientific Research, which helped us to run even faster. The Körber Foundation is gratefully acknowledged for its 2009 award. However, I can offer no nice words for the EU Framework programmes which, except for the European Research Council, can be praised only by Europhobes for discrediting the whole idea of an effectively working Europe.

REFERENCES

1. A. K. Geim, K. S. Novoselov. *Nature Mater.* **6**, 183 (2007).
2. A. K. Geim, P. Kim. *Sci. Am.* **298**, 90 (2008).
3. A. K. Geim. *Science* **324**, 1530 (2009).
4. A. K. Geim. *JETP Lett.* **50**, 389 (1989).
5. S. J. Bending, K. von Klitzing, K. Ploog. *Phys. Rev. Lett.* **65**, 1060 (1990).
6. A. K. Geim, P. C. Main, P. H. Beton, P. Streda, L. Eaves, C. D. Wilkinson, S. P. Beaumont. *Phys. Rev. Lett.* **67,** 3014 (1991).
7. A. K. Geim, P. C. Main, P. H. Beton, L. Eaves, C. D. W. Wilkinson, S. P. Beaumont. *Phys. Rev. Lett.* **69**, 1248 (1992).
8. A. K. Geim, S. V. Dubonos, J. G. S. Lok, I. V. Grigorieva, J. C. Maan, L. T. Hansen, P. E. Lindelof. *Appl. Phys. Lett.* **71**, 2379 (1997).
9. I. V. Grigorieva. *Supercond. Sci. Technol.* **7**, 161 (1994).
10. J. S. Baker, S. J. Judd. *Water Research* **30**, 247 (1996).
11. A. Geim. *Phys. Today* **51**, No. 9, 36 (1998).
12. K. Autumn, Y. A. Lang, S. T. Ksieh, W. Zesch, W. P. Chan, T. W. Kenny, R. Fearing, R. J. Full. *Nature* **405**, 681 (2000).
13. A. K. Geim, S. V. Dubonos, I. V. Grigorieva, K. S. Novoselov, A. A. Zhukov, S. Y. Shapoval. *Nature Mater.* **2**, 461 (2003).
14. G. Lamarche, F. Lamarche, A. M. Lamarche, *Europhys. Lett.* **53**, 378 (2001).
15. I. I. Barbolina, K. S. Novoselov, S. V. Morozov, S. V. Dubonos, M. Missous, A. O. Volkov, D. A. Christian, I. V. Grigorieva, A. K. Geim. *Appl. Phys. Lett.* **88**, 013901 (2006).
16. T. Cohen-Karni, Q. Qing, Q. Li, Y. Fang, C. M. Lieber. *Nano Lett.* **10**, 1098 (2010).
17. Nevill Mott, *A Life in Science*, Taylor & Francis, 1986.
18. E. Bose, *Phys. Z.* **7**, 373 (1906).
19. A. V. Butenko, Dm. Shvarts, V. Sandomirsky, Y. Schlesinger, R. Rosenbaum. *J. Appl. Phys.* **88**, 2634 (2000).
20. V. T. Petrashov, V. N. Antonov, B. Nilsson. *J. Phys.– Cond. Mat.* **3**, 9705 (1991).
21. R. E. Glover, M. D. Sherrill. *Phys. Rev. Lett.* **5,** 248 (1960).
22. C. H. Ahn *et al. Rev. Mod. Phys.* **78**, 1185 (2006).
23. M. S. Dresselhaus, G. Dresselhaus. *Adv. Phys.* **30**, 139 (1981).
24. Y. Kopelevich, P. Esquinazi, J. H. S. Torres, S. Moehlecke. *J. Low Temp. Phys.* **119**, 691 (2000).
25. H. Kempa, P. Esquinazi, Y. Kopelevich. *Phys. Rev. B* **65**, 241101 (2002).
26. K. S. Novoselov, A. K. Geim, S. V. Dubonos, E.W. Hill, I. V. Grigorieva. *Nature* **426**, 812 (2003).
27. K. S. Novoselov, A. K. Geim, S. V. Morozov, D. Jiang, Y. Zhang, S. V. Dubonos, I. V. Grigorieva, A. A. Firsov. *Science* **306**, 666 (2004).
28. J. A. Venables, G. D. T. Spiller, M. Hanbucken. *Rep. Prog. Phys.* **47**, 399 (1984).
29. J. W. Evans, P. A. Thiel, M. C. Bartelt. *Sur. Sci. Rep.* **61**, 1 (2006).
30. O. A. Shenderova, V. V. Zhirnov, D. W. Brenner, *Crit. Rev. Solid State Mater. Sci.* **27** 227 (2002).
31. D. Tománek, W. Zhong, E. Krastev. *Phys. Rev. B* **48**, 15461 (1993).
32. R. Seton. *Carbon* **34**, 69 (1996).
33. S. Braga, V. R. Coluci, S. B. Legoas, R. Giro, D. S. Galvão, R. H. Baughman. *Nano Lett.* **4**, 881 (2004).
34. C. D. Simpson, J. D. Brand, A. J. Berresheim, L. Przybilla, H. J. Rader, K Mullen. *Chem. Eur. J.* **8**, 1424 (2002).
35. A. Affoune, B. Prasad, H. Sato, T. Enoki, Y. Kaburagi, Y. Hishiyama, *Chem. Phys. Lett.* **348**, 17 (2001).
36. T. J. Booth, P. Blake, R. R. Nair, D. Jiang, E. W. Hill, U. Bangert, A. Bleloch, M. Gass, K. S. Novoselov, M. I. Katsnelson, A. K. Geim. *Nano Lett.* **8**, 2442 (2008).

37. S. Ryu, L. Liu, S. Berciaud, Y. J. Yu, H. Liu, P. Kim, G. W. Flynn, L. E. Brus. *Nano Lett.* **10**, 4944 (2010).
38. F. Schedin, A. K. Geim, S. V. Morozov, E.W. Hill, P. Blake, M. I. Katsnelson, K. S. Novoselov. *Nature Mater.* **6**, 652 (2007).
39. *Letter to the Editor. "October 22, 2004: Discovery of Graphene"* www.aps.org/publications/apsnews/201001/letters.cfm
40. B. C. Brodie. *Philos. Trans. R. Soc. London* **149**, 249 (1859).
41. D. R. Dreyer, S. Park, C. W. Bielawski, R. S. Ruoff. *Chem. Soc. Rev.* **39**, 228 (2010).
42. G. Ruess, F. Vogt. *Monatshefte Chem.* **78**, 222 (1948).
43. H. P. Boehm, A. Clauss, G. O. Fischer, U. Hofmann. *Z. anorg. allgem. Chem.* **316**, 119 (1962).
44. J. C. Meyer, A. K. Geim, M. I. Katsnelson, K. S. Novoselov, D. Obergfell, S. Roth, C. Girit, A. Zettl. *Solid State Commun.* **143**, 101 (2007).
45. H. Shioyama. *J. Mat. Sci. Lett.* **20**, 499 (2001).
46. L. M. Viculis, J. J. Mack, R. B. Kaner. *Science* **299**, 1361 (2003).
47. S. Horiuchi, T. Gotou, M. Fujiwara, T. Asaka, T. Yokosawa, Y. Matsui. *Appl. Phys. Lett.* **84**, 2403 (2004).
48. H. P. Boehm, R. Setton, E. Stumpp. *Carbon* **24**, 241 (1986).
49. J. T. Grant, T. W. Haas. *Surf. Sci.* **21**, 76 (1970).
50. J. M. Blakely, J. S. Kim, H. C. Potter. *J. Appl. Phys.* **41**, 2693 (1970). 51. R. Rosei, M. De Crescenzi, F. Sette, C. Quaresima, A. Savois, P. Perfetti. *Phys. Rev. B* **28**, 1161 (1983).
52. C. F. McConville, D. P. Woodruff, S. D. Kevan, M. Weinert, J. W. Davenport. *Phys. Rev. B* **34**, 2199 (1986).
53. T. A. Land, T. Michely, R. J. Behm, J. C. Hemminger, G. Comsa. *Surf. Sci.* **264**, 261 (1992).
54. A. J. van Bommel, J. E. Crombeen, A. van Tooren. *Surf. Sci.* **48**, 463 (1975).
55. A. Nagashima, K. Nuka, K. Satoh, H. Itoh, T. Ichinokawa, C. Oshima, S. Otani. *Surf. Sci.* **287**, 609 (1993).
56. M. Terai, N. Hasegawa, M. Okusawa, S. Otani, C. Oshima. *Appl. Surf. Sci.* **130**, 876 (1998).
57. I. Forbeaux, J. M. Themlin, J. M. Debever. *Phys. Rev. B* **58**, 16396 (1998).
58. K. Seibert, G. C. Cho, W. Kütt, H. Kurz, D. H. Reitze, J. I. Dadap, H. Ahn, M. C. Downer, A. M. Malvezzi. *Phys. Rev. B* **42**, 2842 (1990).
59. T. W. Ebbesen, H. Hiura, *Adv. Mater.* **7**, 582 (1995).
60. X. Lu, M. Yu, H. Huang, R. S. Ruoff. *Nanotechnology* **10**, 269 (1999).
61. Y. Gan, W. Chu, L. Qiao, *Surf. Sci.* **539**, 120 (2003).
62. Y. Ohashi, T. Koizumi, T. Yoshikawa, T. Hironaka, K. Shiiki. *TANSO* **180**, 235 (1997).
63. Y. Ohashi, T. Hironaka, T. Kubo, K. Shiiki. *TANSO* **195**, 410 (2000).
64. E. Dujardin, T. Thio, H. Lezec, T. W. Ebbesen. *Appl. Phys. Lett.* **79**, 2474 (2001).
65. A. H. Castro Neto, F. Guinea, N. M. R. Peres, K. S. Novoselov, A. K. Geim. *Rev. Mod. Phys.* **81**, 109 (2009).
66. P. R. Wallace. *Phys. Rev.* **71**, 622 (1947).
67. G. W. Semenoff. *Phys. Rev. Lett.* **53**, 2449 (1984).
68. F. D. M. Haldane. *Phys. Rev. Lett.* **61**, 2015 (1988).
69. J. Gonzales, F. Guinea, M. A. H. Vozmediano. *Phys. Rev. B* **59**, 2474 (1999).
70. E. V. Gorbar, V. P. Gusynin, V. A. Miransky, I. A. Shovkovy. *Phys. Rev. B* **66**, 045108 (2002).
71. T. Ando, T. Nakanishi, R. Saito. *J. Phys. Soc. Jpn.* **67**, 2857 (1998).
72. Y. Zheng, T. Ando. *Phys. Rev. B* **65**, 245420 (2002).
73. R. Saito, M. Fujita, G. Dresselhaus, M. S Dresselhaus. *Appl. Phys. Lett.* **60**, 2204 (1992).
74. H. J. Teuschler. *Method of producing isolated field effect transistors employing pyrolytic graphite.* US Patent 3,522,649.
75. R. B. Little. *J. Cluster Sci.* **14**, 135 (2003).

76. C. R. Dean *et al. Nature Nano* **5**, 722 (2010).
77. E. V. Castro, H. Ochoa, M. I. Katsnelson, R. V. Gorbachev, D. C. Elias, K. S. Novoselov, A. K. Geim, F. Guinea. *Phys. Rev. Lett.* **105**, 266601 (2010).
78. K. Erickson, R. Erni, Z. Lee, N. Alem, W. Gannett, A. Zettl. *Adv. Mat.* **22**, 4467 (2010).
79. K. S. Novoselov, D. Jiang, F. Schedin, T. J. Booth, V. V. Khotkevich, S. V. Morozov, and A. K. Geim. *Proc. Natl. Acad. Sci. U.S.A.* **102**, 10451 (2005).
80. K. S. Novoselov, A. K. Geim, S. M. Morozov, M. I. Katsnelson, I. V. Grigorieva, S. V. Dubonos, A. A. Firsov. *Nature* **438**, 197 (2005).
81. K. S. Novoselov, E. McCann, S. V. Morozov, V. I. Fal'ko, M. I. Katsnelson, U. Zeitler, D. Jiang, F. Schedin, A. K. Geim. *Nature Phys.* **2**, 177 (2006).
82. R. R. Nair, P. Blake, A. N. Grigorenko, K. S. Novoselov, T. J. Booth, T. Stauber, N. M. R. Peres, A. K. Geim. *Science* **320**, 1308 (2008).
83. M. I. Katsnelson, K. S. Novoselov, A. K. Geim. *Nature Phys.* **2**, 620 (2006).
84. E.V. Castro, K. S. Novoselov, S. V. Morozov, N. M. R. Peres, J. M. B. Lopes dos Santos, J. Nilsson, F. Guinea, A. K. Geim, A. H. Castro Neto. *Phys. Rev. Lett.* **99**, 216802 (2007).
85. L. A. Ponomarenko, F. Schedin, M. I. Katsnelson, R. Yang, E. W. Hill, K. S. Novoselov, A. K. Geim. *Science* **320**, 356 (2008).
86. S. V. Morozov, K. S. Novoselov, M. I. Katsnelson, F. Schedin, L. A. Ponomarenko, D. Jiang, A. K. Geim. *Phys. Rev. Lett.* **97**, 016801 (2006).
87. F. Guinea, M. I. Katsnelson, A. K. Geim. *Nature Phys.* **6**, 30 (2010).
88. D. C. Elias *et al. Science* **323**, 610 (2009).
89. R. R. Nair *et al. Small* **6**, 2877 (2010).
90. C. Lee, X. Wei, J. W. Kysar, J. Hone. *Science* **321**, 385 (2008).
91. Y. Zhang, J. P. Small, M. E. S. Amori, P. Kim. *Phys. Rev. Lett.* **94**, 176803 (2005).
92. Y. B. Zhang, Y. W. Tan, H. L. Stormer, P. Kim. *Nature* **438**, 201 (2005).

Portrait photo of Professor Geim by photographer Ulla Montan.

KONSTANTIN S. NOVOSELOV

"Now, here, you see, it takes all the running you can do, to keep in the same place. If you want to get somewhere else, you must run at least twice as fast as that!"

Lewis Carroll, *Through the Looking Glass*

"Everything's got a moral..."

Lewis Carroll, *Alice's Adventures in Wonderland*

I was born in 1974 in Nizhnii Tagil, a middle-sized industrial city in the Ural Mountains in Russia. My mother, Tatiana Novoselova, was an English teacher at my school (though, in spite of all her efforts, I only started to speak, not even proper, but *any* English after I moved to the Netherlands), and my father, Sergey Novoselov, was an engineer at the local Factory.

The Factory – a huge enterprise the size of the city itself – was central to our life, even at the most basic level: every morning there would be a whistle loud enough to wake people several miles away at 7.00 am, two at 7.30 am to get people out of their homes, three at 8.00 am as a signal to start working and another at 4.30 pm when the workers could go home. It produced railway carriages and tanks, including the legendary T-34 (it was moved from the occupation zone of Kharkov during the Second World War), a fact I was very proud of despite the trouble it brought to our family (my granddad Gleb Komarov, a tank test-driver who was evacuated from Kharkov with the factory, lost his legs in an accident in his tank in 1944).

Having such high-technology industry in the vicinity meant there were large numbers of highly qualified engineers and specialist technicians around, and inevitably, our hobbies were rather technical as well. So, along with cross-country skiing, I was seriously into carting, mainly due to my father, who was himself into auto sports, and many parts of the cars were produced or modified by our own hands. Through this hobby, I learned bits of lathing, milling and welding, skills which I also put to use during summer placements at the factory.

I had always been quite technical. When I was eight, my father gave me a German model railway, and the part I used most was its variable DC power source, which came in handy in experiments from electrolysis to building electromagnets. With my parents working full time, and my seven-years-younger sister Elena in the nursery, I had a few hours after school each day

Figure 1. Myself (steering) and my friend Dima Zamiatin racing on ice track (approx. 1980).

to do 'research' such as looking for gunpowder recipes or casting metals and then cleaning up the kitchen afterwards.

The load on our kitchen was significantly reduced when I reached the higher grades, and my passion for such experiments was supported by my physics teacher, Ljudmila Rastorgueva, who allowed me free rein with the equipment in our school physics laboratory. She also, together with my maths teachers Valentina Filippova and Ljudmila Bashmakova, introduced me to the Distance Learning School of the Moscow Institute of Physics and Technology (Phystech), as well as pushing me to participate in physics and maths Olympiads at various levels. Other great sources of information and encouragement at this time were the monthly journal *Quant*, a series of fantastic books by the same publisher and translated texts by Martin Gardner. But, it would be wrong to suggest that I limited myself to physics and maths literature; quite a keen reader, my school-time favourites included Pasternak, Pushkin, Jack London, H.G. Wells, Jerome K. Jerome, Lewis Carroll and Mark Twain (though my tastes changed dramatically over time).

My participation in the Distance Learning School and Olympiads made entering Phystech in 1991 fairly straightforward. I chose the Faculty of Physical and Quantum Electronics and experienced an amazing and bizarre combination of the highest standards of education and rather tough living conditions. The curriculum was also quite intense, especially during the third year when one could easily spend ten straight hours a day in the lectures,

Figure 2. At this moment we were "heavily" involved in construction business in Moscow (1993). Left to right: Danil Melnikov, Konstantin Antipin, Kostya Novoselov, Vladimir Ryabenko, Sergey Lemekhov with Taras Taran pressing the button on the camera.

tutorials and research labs. But with our courses given by the leading actively-working scientists, we felt privileged and extremely proud to study there.

The Phystech students formed a very close and friendly community, and these connections helped us survive the turbulent times of 1991–95 in Russia. I remember, during one of the blackouts which were unfortunately very regular (especially during winter), Sasha Zhuromskii reading something from Tolkien using the last candle we could find and a good dozen people hanging around on the double-decker beds in our hostel room, which was small even for the four of us who lived there. Another source of entertainment, despite a continuous shortage of money on all of our parts, were regular visits to the Bolshoi Theatre, where we traded work as *claqueurs* (paid applauders) for a chance to see the performances.

Of course, there were quite a few temptations outside science and many alternative paths to take. In 1993, I participated in the October Putsch in Moscow, and it was quite an experience. I still feel lucky that they refused to give me a gun, despite my strong insistence. As a consequence, I decided that my revolutions would be in physics, definitely not politics – in fact I decided to stay as far away as possible from any politics at all. My romance with business was somewhat longer lasting: for about three years I was heavily involved with a construction company in parallel with my study (luckily the work was mostly during my summer breaks) and for some time it was good fun to learn another profession, meet new people and earn good money. But, after

Figure 3. At the military camp (1996). The first lesson in the camp is to learn how to put the foot-binding properly (wearing boots require foot-binding rather than socks). Left to right: Kostya Novoselov, Alexander Fedichkin, Sergey Zhukov, Yurii Kapin, Misha Meleshkevich, Alexey Sobolev.

a while, I got bored, and when the question of science or business arose, I chose science; it is absolutely impossible to do 'part-time science', and I feel lucky that I realised that quite early in my career.

Phystech is rather different from other Russian universities. Science in the country is traditionally concentrated in research institutes and Phystech uses these as so-called 'bases', where students can follow specialised courses and get involved in research projects. Typically students spend about a day a week on a 'base' during their third year, with the proportion reaching 100% by their sixth year. My first base was Astrophysica, the State Research Centre originally focused on research into powerful laser systems and their use in military applications, but within a year I had decided that it was not what I wanted and moved to Chernogolovka's Institute of Microelectronics Technology.

Chernogolovka is a very small town in the middle of a forest about sixty km east of Moscow, with 20,000 people and a dozen research institutes. I loved everything about it: the place itself looked amazing, especially during the winter (I would have to walk through the forest for a good half hour each day to get to the institute), the people were enthusiastic and passionate about science, and the range of courses we were offered was excellent. In addition, the lectures were given by the leading scientists at the Institute of Solid State Physics, the Landau Institute for Theoretical Physics and the Institute of Microelectronics Technology: Vsevolod Gantmakher, Vladislav Timofeev and Mikhail Trunin to name just three.

At Chernogolovka, I started to learn microelectronics technology (now it would be happily called 'nanotechnology') from Sergey Dubonos and worked with Zhenia Vdovin, Yura Khanin and Sergey Morozov on tunnelling spectroscopy [1] in the laboratory of the late Yura Dubrovskii. I learned so much from these people, from basic human communication skills to the most complicated experimental techniques. I remember that I so envied the skills with which Yura Khanin and Zhenia Vdovin handled the most miniscule samples that I asked a good friend of mine, Marina Dvinianina, to get me a cut-throat razor to develop the steadiness of my hands when shaving – it was quite a painful and bloody experience, but I soon achieved a less dangerous level of expertise.

In 1997, I was doing my PhD in the same lab [2] (and still shaving with the same razor, which I use to this day), when I got an opportunity to go to Nijmegen in the Netherlands to work with Andre Geim. Andre already had a reputation for being an innovative and creative experimentalist, so I didn't think twice. During the spring of 1999, I spent a couple of months in Nijmegen as a probation period, where I did everything possible to disappoint Andre, once forgetting to close the lid on the helium dewar (which has never happened to me before or since) and using a 'u' instead of an 'a' in the phrase "*last* opportunity" when Andre asked me to write to a journalist for him from his e-mail account. Yet, despite all my "efforts" to sabotage my chances, I started my PhD with Andre in Professor Jan Kees Maan's high magnetic field laboratory in August 1999.

Figure 4. High magnetic field installation in Nijmegen is a huge enterprise and requires rather radical moves if something needs to be fixed (2000).

This was quite a different experience for me. The laboratory was large, international, had a huge variety of projects running [3–4] simultaneously, and always had visitors coming in for measurements on the high magnetic field installation. It certainly broadened my horizons in terms of science but unfortunately not with regards to Dutch: with our community being so international (my closest friends were Igor Shkliarevskii, Fabio Pulizzi and Cecilia Possanzini) we spoke quite a weird dialect of English, with a smattering of Italian, French, Dutch and Russian words and grammar lifted from Guy Ritchie's movies *Lock, Stock…* and *Snatch* (kindly and patiently explained to us by A. Keen and A. Quinn).

Figure 5. Back to the beginning. At the first ever graphite mines (Seathwaite, near Keswick, Lake District, England, 2009). After reaching Great Gable in nice, blasting Lake District weather, we finally got to the mines (just 50 meters away from the main road) to find plenty of graphite in the mine hillock.

In 2001, as many people finished their PhDs and postdocs, the community started to break up. Andre himself moved to Manchester early that year, and I didn't hesitate for a moment when he invited me to join him, even though it meant leaving my PhD unfinished for the second time in a row. When I arrived in Manchester, it was to an empty room with one lock-in (still working), a turbo-pump (still there) and Sergey Morozov (still around), measuring 'magnetic water' while on a short visit to Andre. It was my third lab in less than three years and a different experience again: everything had to be built from scratch both in there and in the clean rooms (though I had less involvement with the latter), but this did allow for plenty of fun as everything was bespoke to our specific requirements.

Despite the fact that the lab then included only Andre, Irina Grigorieva, myself and a couple of other postdocs and visitors, the number of projects (in comparison to what I was exposed to in Nijmegen) hasn't dropped. Probably even otherwise, besides mainstream projects like Irina's cryogenic Bitter decoration [5] and mine and Andre's domain wall motion in garnets [6], we were all involved in several others, including the mentioned 'magnetic water', mesoscopic superconductivity [7], gecko tape [8], scanning tunnelling microscopy (STM) with a gate electrode… the list was endless. In between the projects, I also managed to convince Irina Barbolina to marry me (over the phone, she was in Nijmegen at the time), and, with a PhD in microbiology, she joined the group for a few months, helping with bio-related experiments. We all enjoyed watching the turbulent life of yeast and other micro-organisms under a microscope during a project we dubbed "the last fart of a living cell" [9].

One of our projects, initiated by Andre, was an attempt to make a metallic field effect transistor. The choice of material, quite naturally, fell to graphite, mostly due to its low carrier concentration. I will skip giving a detailed description of the first stage of the project as Andre Geim describes the process in his lecture, but I thought I would just mention that we thought we would have to drop it altogether when Andre's PhD student Da Jiang enthusiastically polished a piece of very expensive graphite into dust. The unexpected solution to the problem came from the scanning tunnelling microscopy project, which was led by Oleg Shkliarevskii.

At that time, I was doing very long measurements on domain walls, with magnetic field sweeps easily taking a day or more, so I was often hanging around the cold STM. Oleg was doing the first scans and showed me the way he cleaned graphite, by cleaving it with Scotch tape. Using Scotch tape (with residual flakes on it) taken literally from the dustbin, it took me less than an hour to produce a device which immediately demonstrated some miserable field effect; but, however small the effect – it was clear we had stumbled upon something very big (though I doubt at that time I realised how far it would go). We got onto it, and within a few months we had our first graphene device (sample ZYH-K51) [10].

The results we were getting were quite puzzling though, and I admit we got a great deal of help from theorists. I was organiszing our group seminars at the time and invited Dima Khmelnitsky from Cambridge to visit us. The seminar was due to start at 3 pm and Dima arrived at 7 in the evening, arguing that he calculated it should have taken him 3.5 hours to reach Manchester. Obviously, the seminar had to be cancelled, but we were able to spend the rest of the evening chatting about our recent results. When he heard about graphite, Dima immediately told us about the linear spectrum and pointed out (off the top of his head) that the Landau level quantisation for such a spectrum was considered in the problem to the paragraph on "Dirac equation for an electron in external field", on page 148 of *Quantum Electrodynamics* (book four of the Russian edition of the Landau and Lifshitz course) – information which came just in time. Another piece of useful information came from Dima the very next morning, when he called me to confirm that his theory was correct and that he had reached Cambridge within 3.5 hours after having left Manchester at 4 am.

By 2004, my first postdoc was coming to an end, and I was actively searching for funding to allow me to continue in Manchester. I was granted a fellowship from the Leverhulme Trust, but they pointed out something unfortunate in the fine print: the recipient must have a PhD. I started running around like a headless chicken, searching for a body which would give me a degree within three months – with my project finishing soon and close to being kicked out of the country. I was extremely close to buying a so called 'life experience' PhD on the internet, but things came together, and I was awarded a PhD from Nijmegen – though even then things were touch-and-go as the company I was flying with went bust and my passport got stuck at the Foreign Office. Finally, after months of havoc, I was able to proudly phone the Leverhulme Trust to ask where to send my certificate. "We don't want to see it," they replied, "If you say you've got it, you've got it." I was absolutely charmed with this kind of attitude.

In 2009, Irina's and my bio-mechanical experiments paid off, and we produced a pair of amazingly good-looking samples: Sophia and Victoria. I sincerely hope that they will continue to work for many, many, many years.

Finally, the promised moral. Four of us (Andre Geim, Volodia Falko, Boris Altshuler and I) were sitting in a seminar room in Lancaster discussing our recent experiments on weak localisation in graphene. Volodia was telling us that it was unlikely that there was no weak localisation at all, and that we should measure better, and we were insisting that those were the facts and encouraging everyone to discuss the real physical situation and try to understand it. The truth, as usual, appeared to be somewhere in the middle, but the discussion became rather heated and even personal (as it often does between Andre and Volodia). Eventually Boris jumped up, ran away and brought back a poorly copied paper by Stark. It is really a bizarre reading on "The Pragmatic and the Dogmatic Spirit in Physics"[11]. It starts with, probably, the most concise description of how science should be done, and you are ready to sign under every single word until you turn the page, where… Well, let's say Volodia turned out to be a bad guy (in the illustrious company of Einstein, Schrodinger and Heisenberg).

The moral is that it is impossible to learn the spirit of science from a textbook or article. They may be able to teach us physics and chemistry and many other disciplines at university, but it is up to us to develop a gut feeling for how best to 'do science'. I am extremely lucky to have worked with and learned from Andre Geim, who is highly innovative and broad in his perspective but, at the same time, very truthful and critical of himself, with manic attention to details. It is so easy to lose sight of the bigger picture underpinning the details or get carried away with your 'beautiful theory' and stop paying attention to the facts; Andre is a master of finding the narrow path between these extremes, and, if there is one thing I am proud of in my life, it is that I have learned a little of this style.

REFERENCES

1. Y. Khanin *et al.*, *Japanese Journal of Applied Physics* Part 1-Regular Papers Short Notes & Review Papers **37**, 3245 (1998).
2. K. S. Novoselov *et al.*, *Europhys. Lett.* **52**, 660 (2000).
3. A. K. Geim *et al.*, *Nature* **407**, 55 (2000).
4. K. S. Novoselov *et al.*, *Phys. Rev.* B **65**, 233312 (2002).
5. I. V. Grigorieva *et al.*, *Phys. Rev. Lett.* **96**, 077005 (2006).
6. K. S. Novoselov *et al.*, *Nature* **426**, 812 (2003).
7. I. V. Grigorieva *et al.*, *Phys. Rev. Lett.* **92**, 237001 (2004).
8. A. K. Geim *et al.*, *Nature Mater.* **2**, 461 (2003).
9. I. I. Barbolina *et al.*, *Appl. Phys. Lett.* **88**, 013901 (2006).
10. K. S. Novoselov *et al.*, *Science* **306**, 666 (2004).
11. J. Stark, *Nature* **141**, 770 (1938).

GRAPHENE: MATERIALS IN THE FLATLAND

Nobel Lecture, December 8, 2010

by

KONSTANTIN S. NOVOSELOV

School of Physics and Astronomy, The University of Manchester, Oxford Road, Manchester M13 9PL, UK.

"I call our world Flatland…"
Edwin Abbott Abbott, *Flatland. A Romance of Many Dimensions*

Much like the world described in Abbott's Flatland, *graphene is a two-dimensional object. And, as 'Flatland' is "a romance of many dimensions", graphene is much more than just a flat crystal. It possesses a number of unusual properties which are often unique or superior to those in other materials. In this brief lecture I would like to explain the reason for my (and many other people's) fascination with this material, and invite the reader to share some of the excitement I have experienced while researching it.*

GRAPHENE AND ITS UNUSUAL PROPERTIES

Graphene is a two-dimensional (2D) atomic crystal which consists of carbon atoms arranged in a hexagonal lattice (Fig. 1). Although sporadic attempts to study it can be traced back to 1859 [1], active and focused investigation of this material started only a few years ago, after a simple and effective way to produce relatively large isolated graphene samples was found [2–3]. The original 'Scotch tape method' [2–3] appeared to be so simple and effective that this area of science grew extremely quickly, and now hundreds of laboratories around the world deal with different aspects of graphene research. Also known as the micromechanical cleavage technique, the Scotch tape method has a low barrier to entry in that it does not require large investments or complicated equipment, which has helped considerably to broaden the geography of graphene science.

Figure 1. The crystal structure of graphene – carbon atoms arranged in a honeycomb lattice.

Another source of graphene's widespread popularity is that it appeals to researchers from a myriad of different backgrounds. It is the first example of 2D atomic crystals, whose properties from the thermodynamics point of view are significantly different from those of 3D objects. It is also a novel electronic system with unprecedented characteristics [4]. It can be thought of as a giant molecule which is available for chemical modification [5-6] and is promising for applications [7–8] ranging from electronics [7–9] to composite materials [8, 10–11]. These factors allow for true multi- and cross-disciplinary research. Thanks to these attributes, within seven years of the first isolation of graphene we have accumulated as many results and approached the problem from as many different perspectives as other areas of science would more commonly achieve over several decades.

The major draw to people in the field, though, is graphene's unique properties, each of which seems to be superior to its rivals. This material is the first 2D atomic crystal ever known to us [3]; the thinnest object ever obtained; the world's strongest material [12]; its charge carriers are massless Dirac fermions [7, 13–14]; it is extremely electrically [15] and thermally [16] conductive; very elastic; and impermeable to any molecules [17] – the list goes on. Even a simple inventory of graphene's superlative qualities would require several pages, and new entries are being added on a monthly basis.

As it is not possible to give a comprehensive overview of all of graphene's properties in one lecture, I will limit myself to just three, which, in my opinion, give the best possible impression of graphene: (i) it is the first example of 2D atomic crystals, (ii) it demonstrates unique electronic properties, thanks to charge carriers which mimic massless relativistic particles, and (iii) it has promise for a number of applications.

I) TWO-DIMENSIONAL CRYSTALS

Stability of 2D crystals

Intuitively one can easily discern the difference between two- and three-dimensional objects: restrict the size or motion of an object to its width and length and forget (or reduce to zero) its height and you will arrive in 'flatland'. The consequences of subtracting one (or more) dimensions from our 3D world are often severe and dramatic. To give just a few examples: there are no knots in 2D space; the probability of reaching any point in d-dimensional space by random walking is exactly unity, for d=1 and d=2 and smaller than 1 in higher dimensions [18]; the problem of bosons with repulsive potential in 1D is exactly equivalent to that of fermions, since particles cannot penetrate each other and cannot be swapped [19–20] (the Tonks–Girardeau gas and fermionisation of bosons in 1D problem); and it is impossible to have thermodynamic equilibrium between different phases in 1D systems [21].

Many of the peculiar properties that one can expect in 2D systems are present due to so-called 'logarithmic divergences', with the most well-known example being the weak localisation quantum corrections to the conductivity in 2D. In particular, a series of works by Peierls [22–23], Landau [21, 24], Mermin [25] and Wagner [26] demonstrated the theoretical impossibility of long-range ordering (crystallographic or magnetic) in 2D at any finite temperatures. The stability of 2D crystals (here the theory has to be expanded to take flexural phonons or out-of-plane displacements into account [27–29]) is a simple consequence of divergences at low k-vectors, when the integration of the atomic displacements is taken over the whole 2D k-space.

It is important to mention that such instabilities are the result of thermal fluctuations and disappear at $T = 0$. Also, strictly speaking, at any finite temperature the fluctuations diverge only for infinitely large 2D crystals ($k \to 0$); as the divergences are weak (logarithmic), crystals of limited sizes might exhibit infinitely small fluctuations at least at low temperatures.

These fluctuations place a fundamental restriction on the existence and synthesis of low-dimensional crystals. Growth or synthesis generally requires elevated temperatures – at which only crystallites of very limited size can be stable in the flat form – and, as the bending rigidity of such crystals is usually low, they would generally crumple and fold easily and form 3D structures (which might also help in reducing the energy of unsaturated dangling bonds at the perimeter). Therefore the largest flat molecule synthesised to date is C_{222} [30], and the method used to create it is the *low (room) temperature* cyclodehydrogenation of a 3D precursor molecule.

A possible way around the problem of 2D crystals' instability is synthesis as part of a 3D structure, with subsequent extraction of the 2D part of the system at low temperatures (in fact, such a strategy is the basis of all methods of graphene synthesising available to date) [7]. The fluctuations, which might diverge at high temperatures, are quenched during the synthesis due to the interaction with the 3D matrix (substrate) whenever the extraction of 2D crystals can be done at low temperatures where the fluctuations are suppressed.

Graphene creation

The simplest implementation of this method for graphitic materials is to use bulk graphite and exfoliate it into individual planes. Graphite is a layered material and can be considered as a stack of individual graphene layers. High-quality graphite typically requires growth temperatures of above 3000K, but exfoliation can be done at room temperatures – an order of magnitude lower than the growth temperatures. In fact, many of us have performed this procedure numerous times while using pencils, as drawing with a pencil relies on exfoliation of graphite (though not up to the monolayer limit, which would be practically invisible to the naked eye).

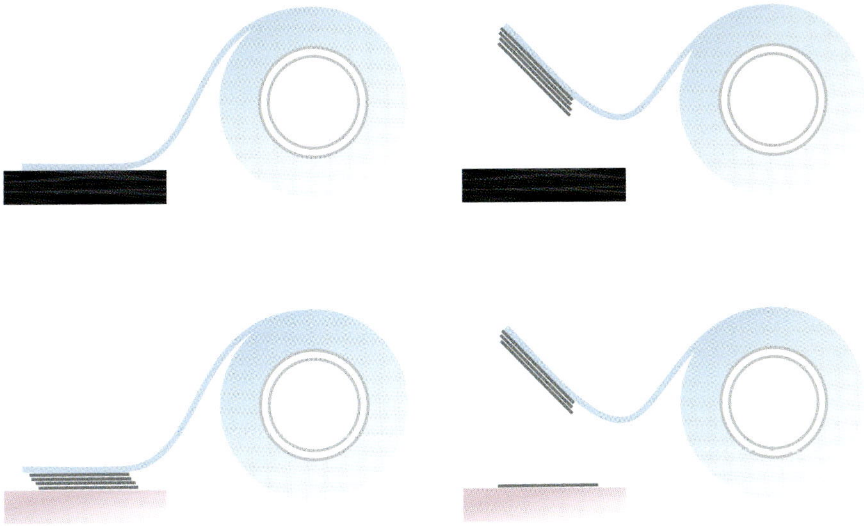

Figure 2. The micromechanical cleavage technique ('Scotch tape' method) for producing graphene. *Top row:* Adhesive tape is used to cleave the top few layers of graphite from a bulk crystal of the material. *Bottom left:* The tape with graphitic flakes is then pressed against the substrate of choice. *Bottom right:* Some flakes stay on the substrate, even on removal of the tape.

Graphite exfoliation techniques slightly more elaborate than writing with a pencil have been attempted by several groups [31–37] and thin graphitic films obtained. But even graphitic films only 20 layers thick would generally behave similarly to bulk graphite, so the real breakthrough came when monolayer films of graphene, large enough to be studied by conventional techniques, were obtained [2–3]. The technique used in those cases is known as the micromechanical cleavage or 'Scotch-tape' method (Fig. 2). The top layer of the high-quality graphite crystal is removed by a piece of adhesive tape, which – with its graphitic crystallites – is then pressed against the substrate of choice. If the adhesion of the bottom graphene layer to the substrate is stronger than that between the layers of graphene, a layer of graphene can be transferred onto the surface of the substrate, producing extremely high-quality graphene crystallites via an amazingly simple procedure. In principle, this technique works with practically any surface which has reasonable adhesion to graphene.

Figure 3. Thin graphitic flakes on a surface of Si/SiO_2 wafer (300 nm of SiO_2, purple colour). The different colours correspond to flakes of differing thicknesses, from ~100 nm (the pale yellow ones) to a few nanometres (a few graphene layers – the most purple ones). The scale is given by the distance between the lithography marks (200 μm).

However, especially in the first experiments, the yield of the process was extremely low, and one would have to scan macroscopically large areas to find a micrometer-sized graphene flake (Fig. 3). Needless to say, this search is a practically impossible task for conventional microscopy methods like Atomic Force Microscopy or Scanning Electron Microscopy; realistically only optical microscopy, which relies on the high-sensitivity, speed and processing power of the human eye and brain, can do the job. So it came as a pleasant surprise that monolayers of graphite on some substrates (Si/SiO_2 with a 300 nm SiO_2 layer, for instance) can produce an optical contrast of up to 15% for some wavelengths of incoming light. The phenomenon is now well understood [38–39] and made Si/SiO_2 with an oxide layer either 100 nm or 300 nm thick the substrate of choice for a number of years for most experimental groups relying on the micromechanical cleavage method of graphene production.

Similar techniques (growth at high temperatures as a part of 3D system, with subsequent extraction of the 2D part at low temperatures) have been used in other graphene preparation methods. Probably the closest to the micromechanical exfoliation method is chemical exfoliation, which can be traced back to the original work of Professor Brodie[1] who treated graphite with acids and arrived at 'Graphon' (or graphite oxide as we now know it). Graphite oxide can be thought of as graphite intercalated with oxygen and hydroxyl groups, which makes it a hydrophilic material and easily dispersible in water. This technique produces extremely thin, sometimes even monolayer, flakes of this material [40–44], which can then subsequently be reduced, producing low quality graphene [5, 10, 45–47].

One can imagine an even simpler path for chemical exfoliation. Although graphene is hydrophobic, it can be dispersed in other, mostly organic, solvents [48–49]. By repeating the exfoliation and purification (centrifugation) process several times one can obtain 50% and higher fractions of graphene in suspension.

There are also well-known graphene-growing recipes from surface science. Catalytic cracking of hydrocarbons, or precipitation of dissolved carbon on a metal surface with subsequent graphitisation, has long been known to produce high-quality graphene layers [50–57]. A similar process is the graphitisation of excess carbon atoms after sublimation of silicon from the surface of silicon carbide [58–59]. One should note that it is also true in these cases that graphene must be a part of the 3D structure, as the underlying substrate aids in quenching the diverging fluctuations at high temperatures.

Other 2D crystals

Thus far we have been talking mainly about graphene, but the 2D materials family is of course not limited to carbonic crystals, although similar problems are faced when attempts are made to synthesise other 2D materials. At least two possible methods of obtaining other 2D crystals come immediately to mind.

One possibility is to apply the same recipes we saw working for graphene to other chemical compounds.

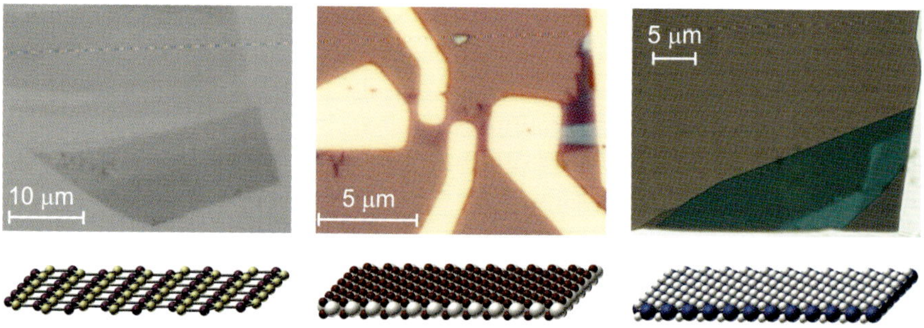

Figure 4. Optical micrographs of various 2D crystals (top) and their crystal structures (bottom). *Left:* Flakes of boron-nitride on top of an oxidised Si wafer (290 nm of SiO_2, the image taken using narrow band yellow filter, λ =590 nm). The central crystal is a monolayer. *Centre:* A device prepared from mono- and bi-layer $NbSe_2$ crystals on an oxidised Si wafer (290 nm of SiO_2). *Right:* Flakes of MoS_2 on top of an oxidised Si wafer (90 nm of SiO_2). The piece at the bottom-right corner is a monolayer. *Colour coding for the crystal structures at the bottom:* yellow spheres = boron, purple = nitrogen; large white spheres = niobium, red = selenide, blue = molybdenum, small white spheres = sulfur.

Micromechanical or chemical exfoliation can be successfully applied to other layered materials (Fig. 4) such as $Bi_2Sr_2CaCu_2O_x$ [3], $NbSe_2$ [3], BN [3], MoS_2 [3, 60], Bi_2Te_3 [61] and other *dichalcogenides,* and epitaxial growth has been applied to grow monolayers of boron-nitride [62–63]. As with graphene, the crystal quality of the obtained monolayer samples is very high. Many of the 2D materials conduct and even demonstrate field effects (changes of the resistance with gating). The properties of the obtained 2D materials might be very different from those of their 3D precursors. For example, the overlapping between the valence and conduction bands in graphene is exactly zero while it is finite in graphite [2], and a monolayer of molybdenum disulphide is a direct band semiconductor while the bulk material has got an indirect band gap [60].

A second approach is to start with an existing 2D crystal and modify it chemically to obtain a new material. One can think of graphene, for instance, as a giant molecule. All the atoms of this molecule are, in principle, accessible for chemical reaction (as opposed to the 3D case, where atoms in the interior of the crystal cannot participate in such reactions).

Graphene, due to the versatility of carbon atoms, is a particularly good candidate for such modification. Depending on the environment, the electron configuration of a carbon atom (which has four electrons in the outer shell) might change dramatically, allowing it to bond to two, three or four other atoms. Bonding between the carbon atoms is exceptionally strong (the strongest materials on Earth are all carbon-based), whereas bonding to other species, though stable, can be changed by chemical reactions. To give an example of such versatility: a backbone of two carbon atoms each can accept one, two or three hydrogen atoms, forming ethyne (also known as acetylene), ethene (a.k.a. ethylene) or ethane respectively. It is possible to convert any one of those into another by adding or removing hydrogen, thus changing the electron configuration of carbon atoms between so-called sp, sp^2 and sp^3 hybridisations.

Carbon atoms in graphene are sp^2-hybridised, meaning that only three electrons form the strong σ-bonds and the fourth has a communal use forming the so-called π-bonds. So, graphene is a zero-overlap semimetal and conducts electricity very well (in contrast to diamond, where each carbon atom is in sp^3-hybridisation and therefore has four neighbours. In that case all four electrons in the outer shell are involved in forming σ-bonds, so a huge gap appears in the electronic band structure, making diamond an insulator). The versatility of carbon atoms, then, gives us an idea of how to create novel 2D crystals: one can attach something to carbon atoms, creating a new material with a different chemical composition and exciting properties.

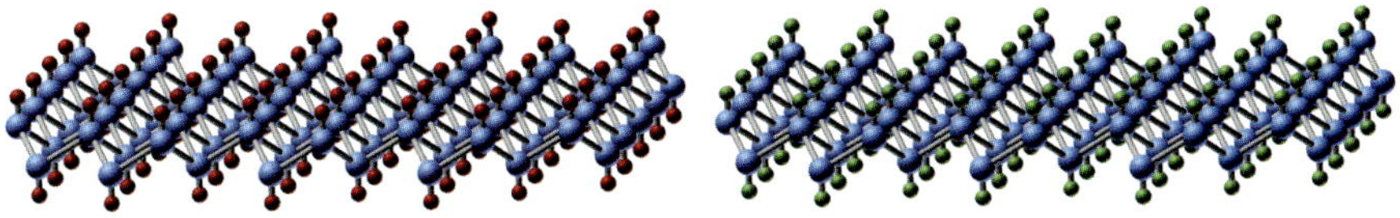

Figure 5. Chemically modified graphene. One can add different species (like hydrogen or fluorine) to graphene scaffolding. Carbon atoms are represented by blue spheres.

A wide variety of chemicals can be attached to graphene. So far only two crystallographically-ordered chemical modifications of graphene have been predicted and achieved: graphane (when one hydrogen atom is attached to each of the carbon atoms) [64–65] and fluorographene (Fig. 5) [66–69]. Both derivatives are insulators (exhibiting large band-gaps) of very high crystallographic quality and very stable at ambient temperatures (though it should be mentioned that fluorographene generally exhibits more robust properties, probably due to stronger C-F bonding in comparison to C-H). Graphane and fluorographene open the floodgates for the chemical modification of graphene and for the appearance of novel two-dimensional atomic crystals with predetermined properties. It would be interesting to see other derivatives, probably filling the gap between graphene and graphane in terms of electrical conductance (the size of the band gap).

Clearly then, the importance of graphene is not only that it has unique properties but also that it has paved the way for, and promoted interest in, the isolation and synthesis of many other 2D materials. We can now talk about a whole new class of materials, 2D atomic crystals, and already have examples with a large variety of properties (from large band-gap insulators to the very best conductors, the extremely mechanically strong to the soft and fragile, and the chemically active to the very inert). Further, many of the properties of these 2D materials are very different from those of their 3D counterparts. Given that, even after seven years' intensive research, graphene still regularly delivers surprises, it seems reasonable to expect a huge influx of breathtakingly interesting results from the field of 2D atomic crystals.

Out to Spaceland: 2D-based heterostructures

As mentioned earlier, the properties of 2D crystals can be very different from their 3D counterparts. Even bilayer graphene [70–71] (two graphene layers stacked on top of each other in special, so-called Bernal- or A–B, stacking), is remarkably different from graphene. The latter is a zero-overlap semi-metal, with linear dispersion relations whenever the bands are parabolic in bilayer graphene, and a gap can be opened in the spectrum if the symmetry between the layers is broken (say, by applying electric field between the layers) [72–75]. However, the properties of multilayered materials depend not only on the number of layers [2, 76] but also on how those layers are stacked. For instance, in the case of graphite, consider Bernal stacking versus rhombohedral versus hexagonal versus turbostratic, and, in bilayer, a small rotation between the individual layers leads to the appearance of van Hove singularities at low energies [77–80].

As we have full control over the 2D crystals, we can also create stacks of these crystals according to our requirements. Here, we are not merely talking about stacks of the same material: we can combine several different 2D crystals in one stack. Insulating, conducting, probably superconducting and magnetic layers can all be combined in one layered material as we wish, the properties of such heterostructures depending on the stacking order and easily tuneable.

Thus a completely new world of 'materials on demand' is opening up to us. Because the pool of the original 2D crystals is very rich, the properties of such heterostructures can cover a huge parameter space, combining characteristics which previously we would not even dare to think of being found together in one material.

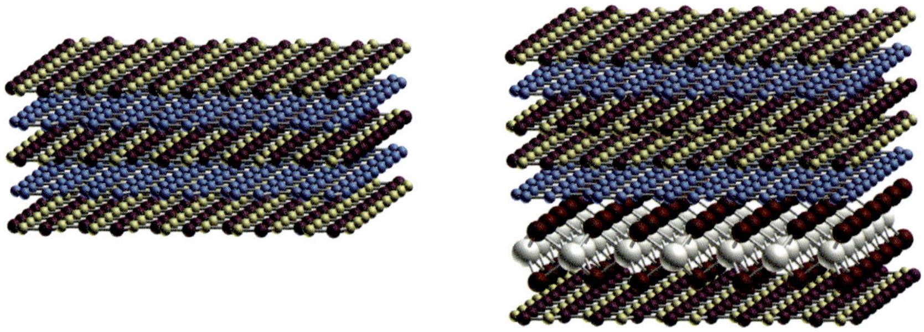

Figure 6. 2D-based heterostructures. Blue spheres = carbon atoms, yellow = boron, purple = nitrogen, white = niobium, and red = selenide.

The first members of this huge family are already there. By combining (alternating) monolayers of insulating boron-nitride and graphene, one can get weakly coupled graphene layers whose coupling would depend on the number of BN layers between the graphene planes (Fig. 6). The level of interaction between the graphene planes ranges from tunnelling (for single or double BN layers in between) to purely Coulomb (for thicker BN spacers).

Figure 7. The low-energy band structure of graphene.

II) CHIRAL QUASIPARTICLES AND THE ELECTRONIC PROPERTIES OF GRAPHENE

Linear dispersion relation and chirality

What really makes graphene special is its electronic properties. Graphene is a zero-overlap semimetal, with valence and conduction bands touching at two points (K and K') of the Brillouin zone [81–83] (Fig. 7). This is a consequence of the hexagonal symmetry of graphene's lattice (which is not one of the Bravais lattices): it has two atoms per unit cell and can be conceptualised as two interpenetrating triangular lattices. The p_z-orbitals from the carbon atoms hybridise to form π and π^* bands, whose crossing at the K and K' points guarantees a gapless spectrum with linear dispersion relation. Thus, the absence of a gap between the valence and the conduction bands in graphene makes for a very robust phenomenon and is a consequence of the symmetry between the sub-lattices [83] (in boron-nitride, where the symmetry between the sub-lattices is broken [one consists of boron, another of nitrogen], a large gap is opened in the electronic spectrum [84]).

Linear dispersion relation already makes graphene special, but there is more to it than that. States in the valence and conduction bands are essentially described by the same spinor wave function, so electrons and holes are linked via charge conjugation. This link implies that quasiparticles in graphene obey chiral symmetry, similar to that which exists between particles and antiparticles in quantum electrodynamics (QED). This analogy between relativistic particles and quasiparticles in graphene is extremely useful and often leads to interesting interpretations of many phenomena observed in experiments [85].

The Klein paradox

Probably the most striking result of the quasiparticles' chiral symmetry is the prediction [86] and observation [87–88] of the Klein paradox in graphene (to explore it, the p-n junction is a natural venue [89]). The paradox refers [90–91] to the enhanced tunnelling probability of a relativistic particle, which approaches unity as the height of the potential barrier exceeds $2m_0c^2$ (where m_0 is the rest mass of the particle and c is the speed of light) and is exactly 1 for massless particles. It can be seen as a result of suppressed back-scattering (massless relativistic particles, like photons, always move with constant velocity – the speed of light – whereas backscattering requires velocity to become zero at the turning point) or as particle-antiparticle pair production and annihilation due to the Schwinger [92] mechanism in the areas of high electric field.

The Klein paradox for chiral quasiparticles in graphene leads to perfect electron-to-hole conversion at a potential barrier, and an equal to unity probability of tunnelling through such a barrier at least for the normal incidence [86, 89, 93–94]. This guarantees the absence of localisation [95–96] and finite minimum conductivity [7], even in relatively disordered graphene, which – in the limit of nominally zero carrier concentration – splits into electron-hole puddles [97]. The absence of backscattering, which leads to the Klein paradox, also ensures that quantum (interference) corrections to the conductivity are positive (at least if we neglect the inter-valley scattering and the effect of trigonal warping) [4, 98], resulting in weak anti-localisation which has indeed been observed experimentally [99–100].

Half-integer quantum Hall effect

The charge conjugation symmetry between electrons and holes also guarantees that there should always be an energy level exactly at $E = 0$. In the magnetic field this symmetry results in a sequence of the Landau levels as $E_n = \pm\sqrt{2e\hbar v^2 B(n + \frac{1}{2} \pm \frac{1}{2})}$ (here e is an electron charge, $\hbar$ the Planck constant, B the magnetic field, v the Fermi velocity and $n = 0,1,2...$), rather different from that for normal massive particles. The ± 1/2 term is related to the chirality of the quasiparticles and ensures the existence of two energy levels (one electron- and one hole-like) at exactly zero energy, each with degeneracy two times smaller than that of all the other Landau levels [101–107].

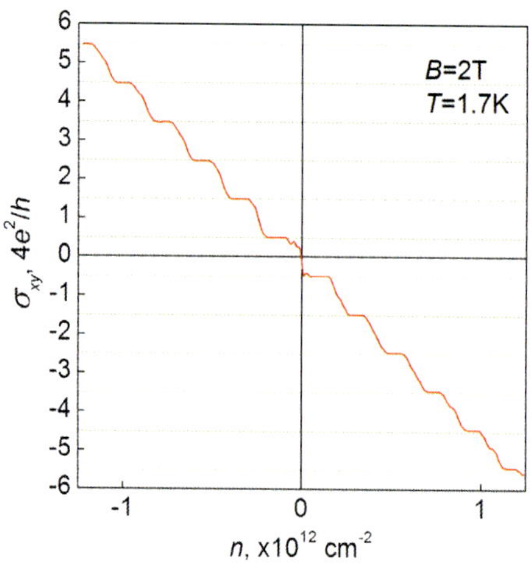

Figure 8. Hall conductivity as a function of the carrier concentration.

Experimentally such a ladder of Landau levels exhibits itself in the observation of a 'half-integer' quantum Hall effect [13–14] (Fig. 8). The two times smaller degeneracy of the zero Landau level is revealed by the $\pm \frac{1}{2}(4e^2/h)$ plateaus in Hall conductivity at filling factors $\pm$ 2. Furthermore, due to the linear dispersion relation and the relatively high value of the Fermi velocity ($v \approx 10^6$ m/s), the separation between the zero and first Landau levels is unusually large (it exceeds the room temperature even in a modest magnetic field of 1T). This, in conjunction with the low broadening of the zero Landau level [108], makes it possible to observe the quantum Hall effect even at room temperatures [109]. This is rather exciting news for people working in metrology, as it should allow much simpler realisation of quantum resistance standard (no need for ultra-low temperatures), an idea which has recently been supported by several experiments [110–111].

Effect of mechanical deformation

It is important to remember that graphene is not just another 2D electronic system, similar to electrons on the surface of silicon MOSFET or in 2D quantum wells in GaAs/AlGaAs heterostructures. Graphene is a truly 2D atomic crystal as well as having electronic properties as in 2D. Essentially the thinnest possible fabric, it can be easily deformed mechanically and can be stretched [12, 17, 112–116], compressed [117], folded [80, 118], rippled [119] – even torn into pieces [120]. Needless to say, each of these mechanical manipulations would result in strong changes to the electronic structure [121–123].

Furthermore, it can be demonstrated that strain is equivalent to the local magnetic field (of opposite directions for quasiparticles in the K and K' valleys, to preserve the time-reversal symmetry) – a phenomenon used to explain the suppression of the weak localisation [99, 124] and additional broadening of all but zeroth Landau levels [108]. In principle, one can imagine engineering strain distribution of a special geometry so that the electronic band-structure would be modified as if a constant magnetic field was being applied to a particular area of the sample [125–126]. Since graphene is mechanically strong and very elastic [12], the strains applied (and thus the pseudomagnetic fields which would be generated) can be extremely large, resulting in the opening of sizeable gaps in the electronic spectrum [127]. This allows us to talk about a completely new and unexplored direction in electronics: strain engineering of electronic structure [123] and valleytronics [128–130].

Graphene optics

Can one expect anything interesting from the optical properties of graphene? Rather counterintuitively, despite being only one atom thick, graphene absorbs quite a large fraction of light. In the infrared limit the absorption coefficient is exactly $\pi\alpha \approx 2.3\%$ (where $\alpha = e^2/\hbar c$ is the fine structure constant), and the corrections to this number in the visible range of the spectrum are less than 3% [131–134]. Such a significant absorption coefficient makes it possible to see graphene without the use of a microscope; thus, one can observe (literally) the most fundamental constant of this universe with the naked eye. At higher frequencies the absorption becomes even larger, reaching 10% due to the presence of the van Hove singularities at the zone edge [135–136].

By changing the carrier concentration, one can shift the position of the Fermi level and change graphene's optical absorption due to Pauli blocking [133]. Since the density of states in graphene is relatively low (at least in the vicinity of the Dirac point), even electrostatic gating can be enough to shift the Fermi level as high as few hundred meV [137], so Pauli blocking happens in the visible range of the spectrum. By executing such strong gating on several tens of graphene layers in series, it would be possible to control the light transmission in such structures to a large extent, an observation which might be promising for novel photonic devices.

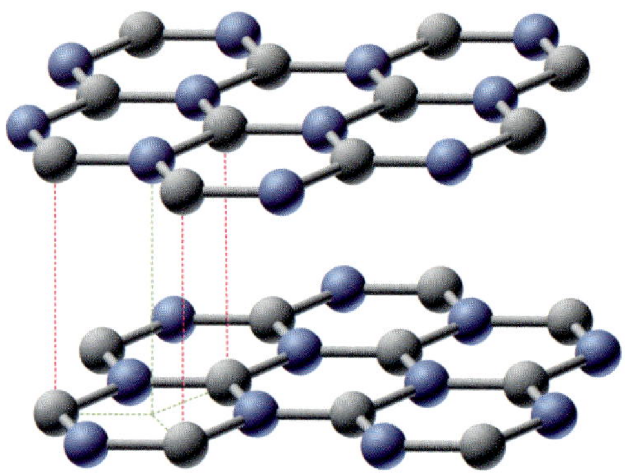

Figure 9. Crystal structure of bilayer graphene.

Bilayer graphene

Although the addition of one layer on top of graphene is all that is needed to arrive at bilayer graphene, the properties of the latter are not simply twice those of the monolayer crystal; this is one of those cases where 'one plus one is greater than two'. Bilayer graphene is remarkably different – sometimes even richer in its properties than its monolayer cousin – and fully deserves to be called a different material in its own right.

Two graphene layers, when placed together, do not like to lie exactly one on top of each other with each atom having a couterpart in the adjacent layer (unlike boron-nitride, which does exactly that). Instead, bilayer graphene is mostly found [78] in so-called A–B or Bernal stacking [138] (named after the famous British scientist John Desmond Bernal, one of the founders of X-ray crystallography, who determined the structure of graphite in 1924). In such an arrangement, only half of the carbon atoms have a neighbour in another layer and the other half do not (and so are projected right into the middle of the hexagon) (Fig. 9). The quantum-mechanical hopping integral between the interacting atoms (generally called γ_1) is in the order of 300 meV, which gives rise to a pair of high-energy electronic sub-bands [70–71, 139]. The offset from the zero energy (the position of the Fermi level in undoped bilayer graphene) is exactly γ_1, so these sub-bands do not contribute to electronic transport unless a very high level of doping is achieved (though these sub-bands can be easily observed in optical experiments [140–141]) (Fig. 10).

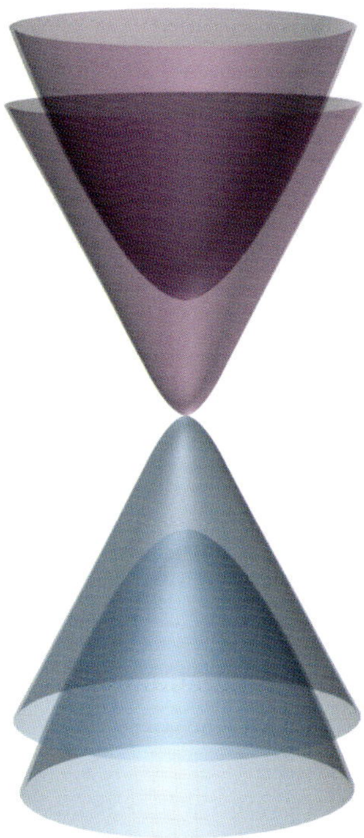

Figure 10. The band structure of bilayer graphene in the vicinity of the K point.

The non-interacting atoms give rise to low-energy bands which are still crossing at zero energy (as in graphene), but are parabolic, Fig. 10. The symmetry between the layers is analogous to the sub-lattice symmetry in monolayer graphene, and it guarantees the chiral symmetry between electrons and holes. Thus we have a new type of quasiparticle in graphene – massive chiral fermions – which does not have an analogy in QED [70–71]. Similarly to graphene, the chirality reveals itself in the unusual quantum Hall effect. The sequence of the Landau levels in the magnetic field is now $E_N = \pm\hbar\omega_C\sqrt{N(N-1)}$; here $\omega_c = eB/m^*$ is the cyclotron frequency and $m^* = \gamma_1/2v^2$ is the cyclotron mass. It is easy to see that two Landau levels exist at zero energy ($N = 0$ and $N = 1$), which again ensures a peculiar sequence of the Hall plateaus and metallic behaviour in the limit of zero filling factor (at least if we neglect the many-body effects) [70–71].

As has been said, the chiral symmetry in mono- and bi-layer graphene is protected by the symmetry between the sub-lattices. In the case of graphene it is rather difficult to break this symmetry – one would have to diligently apply a certain potential to atoms which belong to one sub-lattice only, while applying different potential to another sub-lattice – but in bilayer graphene it is possible to do just that. By applying a gate voltage or by chemically doping from only one side, we can discriminate between the layers and thus between the sub-lattices (breaking the inversion symmetry). This results in lifting the chiral symmetry and opening a gap in the spectrum. Both strategies have been implemented in experiments and yielded a rather striking result: a gap as large as 0.5 eV could be opened [72–74, 140–143]. Thus bilayer graphene is a rare case of a material where the band-gap can be directly controlled by (and its size directly propotional to) the electric field applied across the layers.

As the quality of bilayer graphene samples improves [144–146], we will see more and more interesting properties in them. Topological transitions at low electron energies [146], excitonic effects [147] and topological one-dimensional states [130] are among those which can be expected.

III) GRAPHENE APPLICATIONS

Fans of the American sitcom *The Big Bang Theory* (splendidly produced by Chuck Lorre) might recall the episode 'The Einstein Approximation' in which string theorist Sheldon Cooper tries to resolve 'the graphene problem': why do quasiparticles in graphene behave like massless Dirac fermions? (Fig. 11) The whole crew is great as usual, particularly actor Jim Parsons' grotesquely brilliant depiction of the tough but enjoyable process of searching for a solution to a scientific problem. It is also probably the best episode from a physics point of view (thanks to Chuck Lorre, other writers and the scientific advisor David Saltzberg), as – unusually – the whole plot hinges on the scientific problem, rather than this serving merely to link its parts (the only other example I can recall of such an episode is the one about the paper on supersolids). I'd like to think that the reason for this is the simple and appealing physics of graphene, which is Sheldon sophisticated, Penny beautiful, Raj exotic, Leonard practical and Howard intrusive. On the day the episode was shot, Professor Saltzberg wrote in his blog, "...graphene has captured the imagination of physicists with its potential applications," and, in fact, graphene applications are already here.

The point of this paragraph is not to advertise *The Big Bang Theory* but to demonstrate the kind of applications we are expecting from graphene. The fact that one of the first practical uses of this material was not in a high-expectation, predictable field like transistors or photonics, but instead in the entertainment industry indicates its great potential and versatility. Indeed, in graphene we have a unique combination of properties, which are not seen together anywhere else: conductivity and transparency, mechanical strength and elasticity. Graphene can successfully replace many materials in a great

Figure 11. Dr Sheldon Cooper (Jim Parsons) "…either isolating the terms of his formula and examining them individually or looking for the alligator that swallowed his hand after Peter Pan cut it off." From *The Big Bang Theory*, series 3, episode 14 'The Einstein Approximation'. Photo: Sonja Flemming/CBS ©2010 CBS Broadcasting Inc.

number of existing applications, but I would also like to see things going in the other direction, with the unique combination of properties found in graphene inspiring completely new applications.

Graphene support

I would like to start by talking about relatively simple graphene devices for quite a minute market: graphene support to study biological and other samples in transmission electron microscopy (TEM) [148–150]. It is appealing for the simple reason that graphene membranes are already available on the market and are sold by several companies in both Europe and the US.

Graphene is an ultimately thin, ultimately conductive, ultimately mechanically strong and crystallographically ordered material, and it would be strongly beneficial to use it as a support for nano-objects when observing them in TEM. Its mechanical strength provides rigidity and ease in sample preparation, and it has a very high radiation damage threshold (in the order of 80 keV). High conductivity eliminates the problem of charging of the support. As it is only one atom thick (and also made of a very light element), graphene ensures the highest possible contrast (one can only go higher in contrast if suspending the object). Finally, because it is highly crystallographically ordered, graphene produces few diffraction spots, and those that do appear can be easily filtered out, leaving the image completely unperturbed by the presence of support. Although graphene is already quite compatible with biomolecules, it could also be functionalised to achieve a certain surface potential (for example, changed from being hydrophobic to hydrophilic). Chemical modification of graphene is already well developed, but there are still a large number of opportunities available in this area [5–6].

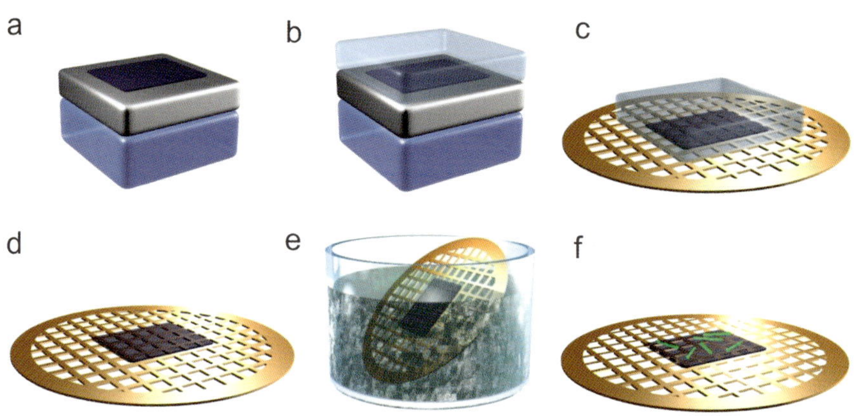

Figure 12. Production of graphene membranes for TEM support application. Graphene, grown on metal (a) is covered with a layer of plastic (b). The sacrificial metal layer is etched away and graphene on plastic is fished on a standard TEM grid (c). Upon the removal of the plastic layer (d) the graphene membrane can be exposed to a solution of biomolecules (e) which adsorb on the surface of graphene (f) and can be studied in a TEM.

Initially, free-standing graphene membranes were produced from exfoliated graphene [151–152] and required several lithography steps during their manufacture. With the availability of chemical vapor deposition (CVD) – grown graphene [153–154], the technique became dramatically simpler, enabling production on an industrial scale. Graphene, epitaxially grown on the surface of a metal (either by precipitation of the dissolved carbon upon cooling or by direct catalytic cracking of the hydrocarbons on the hot surface of the metal, with subsequent graphitisation), is covered by a sacrificial layer of plastic. The underlying metal is then removed by etching and the plastic film (with graphene attached) can be transferred onto practically any surface. It can be placed, for example, on a metal grid with holes typically the size of few microns, where – upon the removal of the sacrificial plastic film – a free standing graphene membrane is formed (Fig. 12). This entire process is very reproducible and can result in a large total area of graphene membranes.

Transparent conductive coating

Another area, which should benefit significantly from the availability of CVD-grown graphene, is transparent conductive coating. Graphene is unusually optically active [131–133, 135] and for a monolayer absorbs a rather large fraction of incoming light (2.3%), but this is still significantly smaller than the typical absorption coefficient which could be achieved with a more traditional transparent conductive coating materials [155]. In conjunction with its low electrical resistivity, high chemical stability and mechanical strength, this absorption coefficient makes graphene an attractive material for optoelectronic devices.

Transparent conductors are an essential part of many optical devices, from solar cells to liquid crystal displays and touch screens. Traditionally metal oxides or thin metallic films have been used for these purposes [155], but with existing technologies often complicated (thin metallic films require anti-reflection coating for example) and expensive (often using noble or rare metals), there has been an ongoing search for new types of conductive thin films. Furthermore, many of the widely used metal oxides exhibit non-uniform absorption across the visible spectrum and are chemically unstable; the commonly used indium tin oxide (ITO, In_2O_3:Sn), for instance, is known to inject oxygen and indium ions into the active media of a device.

Graphene avoids all of these disadvantages. Moreover, it has recently been demonstrated that large areas of graphene can be grown by the CVD method [115, 153–154] and transferred onto practically any surface. Initial prototype devices (solar cells and liquid crystal displays), which use graphene as a transparent conductive coating have already been created [48, 156].

Graphene transistors

Even the very first graphene field effect transistors demonstrated remarkable quality: prepared using rather humble methods in poorly controlled environments, they still showed reasonably high quasiparticle mobility (up to 20,000 $cm^2/V{\cdot}s$) [2–3] (Fig. 13). Although the role of different scattering mechanisms is still debated [95, 107, 157–168] the careful elimination of Coulomb and resonant scatterers, as well as ripples, has allowed the achievement of mobilities of over 10^6 $cm^2/V{\cdot}s$ at low temperatures [169] in free-standing devices [144–145], and offers hope that values above 10^5 $cm^2/V{\cdot}s$ can be achieved even at ambient temperatures [170].

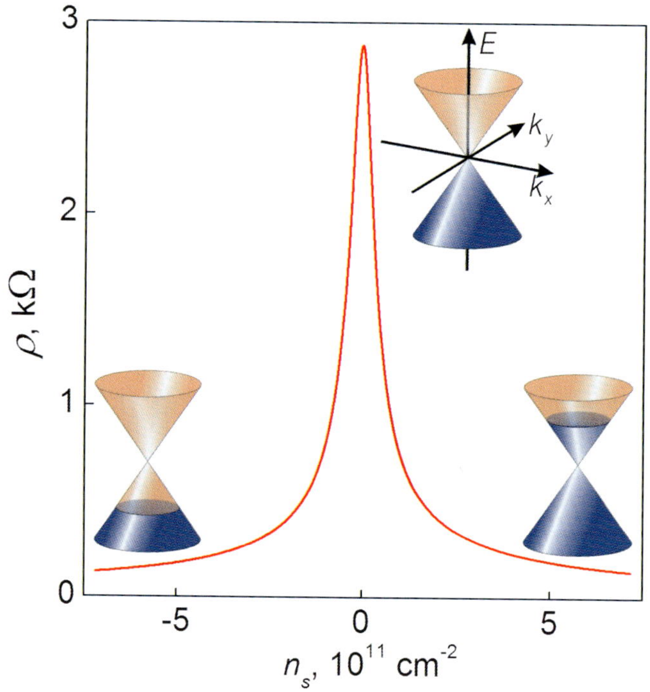

Figure 13. Ambipolar field effect in graphene.

Such characteristics make graphene field effect transistors extremely promising for high-frequency applications [171–172]. Additional benefits are also emerging from very favourable electrostatics of 2D films and high Fermi velocity (which is important in ballistic regimes). Even when using graphene with rather modest mobility (~10^3 $cm^2/V{\cdot}s$) current gain has been achieved at frequencies as high as 100 GHz for 240 nm gate length transistors (power gain for similar devices was achieved below 14 GHz) [173], which is better than for Si metal–oxide-semiconductor FETs of the same gate length.

The situation is not as bright for integrated circuits. The Klein paradox ensures a finite minimum conductivity for graphene (of the order of $4\ e^2/h$) even within the limit of nominally zero carrier concentration. This is definitely too high for applications in logic elements, as it leads to high 'leakage' current in the 'off' state and limits the possible on/off ratio of such transistors to about 10^3 even in very favourable circumstances.

There are several possible tricks one can play to increase the on/off ratio of graphene transistors, however. One is to utilise low-dimensional graphene nanostructures such as graphene nanoribbons [174], quantum dots [175] and single electron transistors [175–176] where a band gap can be engineered due to quantum confinement or Coulomb blockade. The smallest quantum dots (a few nanometres in size) demonstrate a significant gap in the order of few hundred meV, which is enough for such transistors to achieve an on/off ratio in the order of 10^5 even at room temperatures [175]. The strong carbon-carbon bonds ensure the mechanical and chemical stability of such devices, which also can pass a significant current without diminishing their properties. Basically, we can think of it as top-down molecular electronics – one nanometre-sized graphene quantum dot contains only $\sim 10^2$ atoms. The major problem with implementing such quantum dots would be the limits of modern lithographic techniques, which do not currently allow true nanometre resolution. Also, one would have to control the roughness and chemistry of the edges with atomic precision, which is also beyond the capabilities of modern technology.

Although modern microelectronics relies on lithographic techniques, one can imagine using other approaches to form nanostructures, which would eventually allow reproduction of fine details far beyond the resolution of lithography. One promising method would be the use of the self-organisation properties of chemical reactions. Graphene nanostructures could be formed, for instance, by fluorination of the supposed-to-be-insulating parts. Partial fluorination or hydrogenation can result in the formation of self-organised structures on a graphene surface [177–178], which, in principle, could be used to modify its transport and optical properties.

The other possible way to open a gap in the spectrum of quasiparticles in graphene is to use chemically modified graphene [65–69], where the π-electrons are participating in the covalent bond with foreign atoms attached to the carbon scaffolding. One could also use bilayer graphene, as a gap can be opened by applying a potential difference between the two layers [72–74, 140–143]. An on/off ratio of 2000 has recently been achieved in dual-gated devices at low temperatures [179].

Graphene composites

The unique combination of graphene's electronic, chemical, mechanical and optical properties can be utilised in full in composite materials. It is also relatively easy to prepare graphene for such an application: one can either use the direct chemical exfoliation of graphene [48–49], which allows a rather high yield of graphene flakes in a number of organic solvents or go through an oxidation process to prepare graphite oxide – which can be easily exfoliated in water – with subsequent reduction in a number of reducing media [10].

The strongest and simultaneously one of the stiffest known materials, with a Young's modulus of 1 TPa, graphene is an ideal candidate for use as a reinforcement in high-performance composites [11]. There is a huge advantage in its being exactly one atom thick: it cannot cleave, giving it the maximum possible strength in the out-of-plane direction. Its high aspect ratio also allows graphene to act as an ideal stopper for crack propagation. As for interaction with the matrix – the central issue for all nanocomposite fillers like carbon fibre or carbon nanotubes – chemical modification of the surface or edges may significantly strengthen the interface between the graphene and a polymer.

Using chemical derivatives of graphene would not only broaden the range of possible matrices but also widen the functionality of the possible composites. Given that the mechanical strength of fluorographene is only slightly smaller than that of pristine graphene [66], one can obtain composites with similar mechanical properties but a range of other characteristics, from optically transparent to opaque, and from electrically conductive to insulating.

We should also benefit strongly from the possibility of optically monitoring the strain in graphene [11, 112–114]. The Raman spectrum of graphene [180–182] changes significantly with strain, so even mechanical distortions of a fraction of a percent can be easily detected. As the stress transfer between graphene and polymer is reasonably good[11], and graphene's Raman signal is very strong (the strongest peaks in the Raman spectrum of graphene are due to either single or double resonant processes), one can easily detect stress build-up in the composite material under investigation by monitoring the position of the Raman peaks.

Other applications

It is impossible to review all the potential applications of graphene in one lecture: with practically all the properties of this 2D crystal superior to those in other materials, and the combination of these properties unique, we are limited only by our own imaginations. In terms of electronic properties it is worth mentioning ultimately-sensitive gas detectors [183] (graphene, being surface without bulk, can detect a single foreign molecule attached to it) and variable quantum capacitors [184–185]. In photonics, ultrafast photodetectors [186] (utilising the high mobility and high Fermi velocity of quasiparticles in graphene) and extremely efficient mode lockers [187] must be noted. Additionally, its unprecedented mechanical strength and high crystallographic quality allow one to use graphene to provide the perfect gas barrier [17] and strain gauges [115].

ACKNOWLEDGEMENTS

The huge diversity of graphene properties researched and experiments undertaken was only possible thanks to a large and friendly community, which is still growing day by day. I would really like to thank each and every member for their interaction, for teaching me new techniques and for the excitement I feel every morning opening *cond-mat* (often tinged with disappointment, admittedly, that I haven't done it first!). Unfortunately it is impossible to name everybody, so I will limit myself to just my closest collaborators and the leaders of the groups.

I am most grateful to Andre Geim, who has been a teacher, colleague and friend to me over the years. Universities teach us physics, maths, chemistry and tens of other disciplines, but each of us has to work out for himself how to do science. I feel extremely lucky to have worked for all these years alongside such a fantastic and dedicated scientist and researcher.

I have also learned a lot from my colleagues in the condensed matter group, the Centre of Mesoscience & Nanotechnology and Chernogolovka: Irina Grigorieva, Ernie Hill, Sasha Grigorenko, Fred Schedin, Alexander Zhukov, Yuan Zhang, Cinzia Casiraghi, Ursel Bangert, Ian Kinloch, Bob Young, Helen Gleeson, Stan Gillot, Mark Sellers, Oleg Shkliarevskii, Yurij Dubrovskii, Zhenia Vdovin, Yurii Khanin, Sergey Dubonos and Vsevolod Gantmacher. Special appreciation goes to Sergey Morozov, one of the key figures in graphene research and a fantastic friend.

It is impossible to overstate the contribution of our research students and postdocs: their ingenious, creative, and active research often opening new directions of study. I would like to mention Peter Blake, Rahul Nair, Da Jiang, Leonid Ponomarenko, Daniel Elias, Roman Gorbachev, Sasha Mayorov, Tolik Firsov, Soeren Neubeck, Irina Barbolina, Zhenhua Ni, Ibtsam Riaz, Rahul Jalil, Tariq Mohiuddin, Rui Yang, Tim Booth, Liam Britnell, Sveta Anissimova, Frank Freitag, Vasil Kravets, Paul Brimicombe, Margherita Sepioni and Thanasis Georgiou.

The theoretical support provided by a number of condensed matter theorists around the globe was most valuable, and I only hope that the process was reciprocal and that we guided (and misguided) each other equally. The long list of theorists who contributed strongly to our research includes (but is not limited to): Misha Katsnelson, Antonio Castro Neto, Paco Guinea, Nuno Peres, Volodia Falko, Ed McCann, Leonid Levitov, Dima Abanin, Tim Wehling, Allan MacDonald, Sasha Mirlin, Sankar Das Sarma.

Finally, as I have mentioned already, we benefited enormously from collaboration and competition with and information from the other experimental groups. Primarily I must mention Philip Kim – a great physicist and a very good collaborator. Other experimentalists whose results influenced us include (unfortunately only partial list) Andrea Ferrari, Eva Andrei, Jannik Meyer, Alexey Kuzmenko, Uli Zeitler, Jan Kees Maan, Jos Giesbers, Robin Nicholas, Michal Fuhrer, Tatiana Latychevskaia, Mildred Dresselhaus, Alberto Morpurgo, Lieven Vandersypen, Klaus Ensslin and Jonathan Coleman.

REFERENCES

1. B. C. Brodie, *Philos. T. Roy. Soc.* **149**, 249 (1859).
2. K. S. Novoselov *et al.*, *Science* **306**, 666 (2004).
3. K. S. Novoselov *et al.*, *P. Natl. Acad. Sci. USA* **102**, 10451 (2005).
4. A. H. Castro Neto *et al.*, *Rev. Mod. Phys.* **81**, 109 (2009).
5. R. Ruoff, *Nature Nanotechnology* **3**, 10 (2008).
6. K. P. Loh *et al.*, *J. Mater. Chem.* **20**, 2277 (2010).
7. A. K. Geim, and K. S. Novoselov, *Nature Mater.* **6**, 183 (2007).
8. A. K. Geim, *Science* **324**, 1530 (2009).
9. F. Schwierz, *Nature Nanotech.* **5**, 487 (2010).
10. S. Stankovich *et al.*, *Nature* **442**, 282 (2006).
11. L. Gong et al., *Adv. Mater.* **22**, 2694 (2010).
12. C. Lee et al., *Science* **321**, 385 (2008).
13. K. S. Novoselov *et al.*, *Nature* **438**, 197 (2005).
14. Y. B. Zhang *et al.*, *Nature* **438**, 201 (2005).
15. I. Meric *et al.*, *Nature Nanotech.* **3**, 654 (2008).
16. A. A. Balandin *et al.*, *Nano Lett.* **8**, 902 (2008).
17. J. S. Bunch *et al.*, *Nano Lett.* **8**, 2458 (2008).
18. G. Pólya, *Mat. Ann.* **84**, 149 (1921).
19. M. Girardeau, *Journal of Mathematical Physics* **1**, 516 (1960).
20. E. H. Lieb, and W. Liniger, *Phys. Rev.* **130**, 1605 (1963).
21. L. D. Landau, and E. M. Lifshitz, *Statistical Physics, Part I* (Pergamon, Oxford, UK, 1980), Vol. 1.
22. R. E. Peierls, *Helv. Phys. Acta* **7**, 81 (1934).
23. R. E. Peierls, *Ann. I. H. Poincare* **5**, 177 (1935).
24. L. D. Landau, *Phys. Z. Sowjetunion* **11**, 26 (1937).
25. N. D. Mermin, *Phys. Rev.* **176**, 250 (1968).
26. N. D. Mermin, and H. Wagner, *Phys. Rev. Lett.* **17**, 1133 (1966).
27. P. Chaikin, and T. C. Lubensky, *Introduction to Condensed Matter Physics* (Cambridge University Press, Cambridge, 1995).
28. D. Nelson, D. R. Piran, and S. Weinberg, *Statistical Mechanics of Membranes and Surfaces* (World Scientific, Singapore, 2004).
29. A. Fasolino, J. H. Los, and M. I. Katsnelson, *Nature Mater.* **6**, 858 (2007).
30. C. D. Simpson *et al.*, *Chem.-Eur. J.* **8**, 1424 (2002).
31. S. Mizushima, Y. Fujibayashi, and K. Shiiki, *J. Phys. Soc. Jpn.* **30**, 299 (1971).
32. Y. Fujibayashi, *J. Phys. Soc. Jpn.* **34**, 989 (1973).
33. Y. Fujibayashi, and S. Mizushima, *J. Phys. Soc. Jpn.* **34**, 281 (1973).
34. T. W. Ebbesen, and H. Hiura, *Adv. Mater.* **7**, 582 (1995).
35. Y. Ohashi *et al.*, *Tanso* **180**, 235 (1997).
36. X. K. Lu *et al.*, *Appl. Phys. Lett.* **75**, 193 (1999).
37. Y. Gan, W. Y. Chu, and L. J. Qiao, *Surf. Sci.* **539**, 120 (2003).
38. P. Blake *et al.*, *Appl. Phys. Lett.* **91**, 063124 (2007).
39. D. S. L. Abergel, A. Russell, and V. I. Fal'ko, *Appl. Phys. Lett.* **91**, 063125 (2007).
40. G. Ruess, and F. Vogt, *Monatsh. Chem.* **78**, 222 (1948).
41. H. P. Boehm, A. Clauss, and U. Hofmann, *J. Chim. Phys.* PCB **58**, 141 (1961).
42. H. P. Boehm *et al.*, *Z. Naturforsch. Pt. B* **B 17**, 150 (1962).
43. H. P. Boehm *et al.*, *Z. Anorg. Allg. Chem.* **316**, 119 (1962).
44. S. Horiuchi *et al.*, *Appl. Phys. Lett.* **84**, 2403 (2004).
45. D. A. Dikin *et al.*, *Nature* **448**, 457 (2007).
46. S. Park, and R. S. Ruoff, *Nature Nanotech.* **4**, 217 (2009).
47. C. Gomez-Navarro *et al.*, *Nano Lett.* **7**, 3499 (2007).
48. P. Blake *et al.*, *Nano Lett.* **8**, 1704 (2008).
49. Y. Hernandez *et al.*, *Nature Nanotech.* **3**, 563 (2008).
50. J. T. Grant, and T. W. Haas, *Surf. Sci.* **21**, 76 (1970).

51. N. R. Gall *et al.*, *Fiz. Tverd. Tela* **27**, 2351 (1985).
52. N. R. Gall *et al.*, *Surf. Sci.* **191**, 185 (1987).
53. N. R. Gall, E. V. Rutkov, and A. Y. Tontegode, *Int. J. Mod. Phys. B* **11**, 1865 (1997).
54. A. Nagashima *et al.*, *Surf. Sci.* **291**, 93 (1993).
55. I. Forbeaux *et al.*, *Surf. Rev. Lett.* **5**, 193 (1998).
56. A. M. Affoune *et al.*, *Chem. Phys. Lett.* **348**, 17 (2001).
57. K. Harigaya, and T. Enoki, *Chem. Phys. Lett.* **351**, 128 (2002).
58. A. J. van Bommel, J. E. Crombeen, and A. van Tooren, *Surf. Sci.* **48**, 463 (1975).
59. C. Berger *et al.*, *J. Phys. Chem. B* **108**, 19912 (2004).
60. K. F. Mak *et al.*, *Phys. Rev. Lett.* **105**, 136805 (2010).
61. D. Teweldebrhan, V. Goyal, and A. A. Balandin, *Nano Lett.* **10**, 1209 (2010).
62. A. Nagashima *et al.*, *Phys. Rev. Lett.* **75**, 3918 (1995).
63. A. Nagashima *et al.*, *Surf. Sci.* **357**, 307 (1996).
64. J. O. Sofo, A. S. Chaudhari, and G. D. Barber, *Phys. Rev. B* **75**, 153401 (2007).
65. D. C. Elias *et al.*, *Science* **323**, 610 (2009).
66. R. R. Nair *et al.*, *Small* **6**, 2877 (2010).
67. K. A. Worsley *et al.*, *Chem. Phys. Lett.* **445**, 51 (2007).
68. S. H. Cheng *et al.*, *Phys. Rev. B* **81**, 205435 (2010).
69. F. Withers, M. Dubois, and A. K. Savchenko, *Phys. Rev. B* **82**, 073403 (2010).
70. K. S. Novoselov *et al.*, *Nature Phys.* **2**, 177 (2006).
71. E. McCann, and V. I. Fal'ko, *Phys. Rev. Lett.* **96**, 086805 (2006).
72. E. V. Castro *et al.*, *Phys. Rev. Lett.* **99**, 216802 (2007).
73. J. B. Oostinga *et al.*, *Nature Mater.* **7**, 151 (2008).
74. T. Ohta *et al.*, *Science* **313**, 951 (2006).
75. E. McCann, *Phys. Rev. B* **74**, 161403 (2006).
76. S. V. Morozov *et al.*, *Phys. Rev. B* **72**, 201401 (2005).
77. M. Aoki, and H. Amawashi, *Solid State Commun.* **142**, 123 (2007).
78. K. F. Mak, J. Shan, and T. F. Heinz, *Phys. Rev. Lett.* **104**, 176404 (2010).
79. J. dos Santos, N. M. R. Peres, and A. H. Castro, *Phys. Rev. Lett.* **99**, 256802 (2007).
80. G. H. Li *et al.*, *Nature Phys.* **6**, 109 (2010).
81. P. R. Wallace, *Phys. Rev.* **71**, 622 (1947).
82. J. W. McClure, *Phys. Rev.* **108**, 612 (1957).
83. J. C. Slonczewski, and P. R. Weiss, *Phys. Rev.* **109**, 272 (1958).
84. K. Novoselov, *Nature Mater.* **6**, 720 (2007).
85. A. C. Neto, F. Guinea, and N. M. R. Peres, *Physics World* **19**, 33 (2006).
86. M. I. Katsnelson, K. S. Novoselov, and A. K. Geim, *Nature Phys.* **2**, 620 (2006).
87. A. F. Young, and P. Kim, *Nature Phys.* **5**, 222 (2009).
88. R. V. Gorbachev *et al.*, *Nano Lett.* **8**, 1995 (2008).
89. V. V. Cheianov, and V. I. Fal'ko, *Phys. Rev. B* **74**, 041403 (2006).
90. O. Klein, Z. Phys. A – *Hadron Nucl.* **53**, 157 (1929).
91. F. Sauter, Z. Phys. A – *Handron Nucl.* **73**, 547 (1932).
92. J. Schwinger, *Phys. Rev.* **82**, 664 (1951).
93. V. V. Cheianov, V. Fal'ko, and B. L. Altshuler, *Science* **315**, 1252 (2007).
94. C. W. J. Beenakker, *Rev. Mod. Phys.* **80**, 1337 (2008).
95. P. M. Ostrovsky, I. V. Gornyi, and A. D. Mirlin, *Phys. Rev. B* **74**, 235443 (2006).
96. J. H. Bardarson *et al.*, *Phys. Rev. Lett.* **99**, 106801 (2007).
97. J. Martin *et al.*, *Nature Phys.* **4**, 144 (2008).
98. E. McCann *et al.*, *Phys. Rev. Lett.* **97**, 146805 (2006).
99. S. V. Morozov *et al.*, *Phys. Rev. Lett.* **97**, 016801 (2006).
100. F. V. Tikhonenko *et al.*, *Phys. Rev. Lett.* **100**, 056802 (2008).
101. J. W. McClure, *Phys. Rev.* **119**, 606 (1960).
102. G. W. Semenoff, *Phys. Rev. Lett.* **53**, 2449 (1984).
103. N. H. Shon, and T. Ando, *J. Phys. Soc. Jpn.* **67**, 2421 (1998).
104. F. D. M. Haldane, *Phys. Rev. Lett.* **61**, 2015 (1988).

105. Y. Zheng, and T. Ando, *Phys. Rev. B* **65**, 245420 (2002).
106. V. P. Gusynin, and S. G. Sharapov, *Phys. Rev. B* **71**, 125124 (2005).
107. N. M. R. Peres, F. Guinea, and A. H. C. Neto, *Phys. Rev. B* **73**, 125411 (2006).
108. A. J. M. Giesbers *et al., Phys. Rev. Lett.* **99**, 206803 (2007).
109. K. S. Novoselov *et al., Science* **315**, 1379 (2007).
110. A. J. M. Giesbers *et al., Appl. Phys. Lett.* **93**, 222109 (2008).
111. A. Tzalenchuk *et al., Nature Nanotech.* **5**, 186 (2010).
112. Z. H. Ni *et al., ACS Nano* **2**, 2301 (2008).
113. T. M. G. Mohiuddin *et al., Phys. Rev. B* **79**, 205433 (2009).
114. M. Y. Huang *et al., P. Natl. Acad. Sci. USA* **106**, 7304 (2009).
115. K. S. Kim *et al., Nature* **457**, 706 (2009).
116. J. S. Bunch *et al., Science* **315**, 490 (2007).
117. G. Tsoukleri *et al., Small* **5**, 2397 (2009).
118. H. Hiura *et al., Nature* **367**, 148 (1994).
119. W. Z. Bao *et al., Nature Nanotech.* **4**, 562 (2009).
120. D. Sen *et al., Small* **6**, 1108 (2010).
121. E. A. Kim, and A. H. C. Neto, *Europhys. Lett.* **84**, 57007 (2008).
122. V. M. Pereira, A. H. C. Neto, and N. M. R. Peres, *Phys. Rev. B* **80**, 045401 (2009).
123. V. M. Pereira, and A. H. C. Neto, *Phys. Rev. Lett.* **103**, 046801 (2009).
124. A. F. Morpurgo, and F. Guinea, *Phys. Rev. Lett.* **97**, 196804 (2006).
125. F. Guinea, M. I. Katsnelson, and A. K. Geim, *Nature Phys.* **6**, 30 (2010).
126. F. Guinea *et al., Phys. Rev. B* **81**, 035408 (2010).
127. N. Levy *et al., Science* **329**, 544 (2010).
128. T. Low, and F. Guinea, *Nano Lett.* **10**, 3551 (2010).
129. A. Rycerz, J. Tworzydlo, and C. W. J. Beenakker, *Nature Phys.* **3**, 172 (2007).
130. I. Martin, Y. M. Blanter, and A. F. Morpurgo, *Phys. Rev. Lett.* **100**, 196804 (2008).
131. A. B. Kuzmenko *et al., Phys. Rev. Lett.* **100**, 117401 (2008).
132. R. R. Nair *et al., Science* **320**, 1308 (2008).
133. Z. Q. Li *et al., Nature Phys.* **4**, 532 (2008).
134. K. F. Mak *et al., Phys. Rev. Lett.* **101**, 196405 (2008).
135. V. G. Kravets *et al., Phys. Rev. B* **81**, 155413 (2010).
136. L. Yang *et al., Phys. Rev. Lett.* **103**, 186802 (2009).
137. A. Pachoud *et al., Europhys. Lett.* **92**, 27001 (2010).
138. J. D. Bernal, *P. Roy. Soc. A* **106**, 749 (1924).
139. M. S. Dresselhaus, and G. Dresselhaus, *Adv. Phys.* **51**, 1 (2002).
140. A. B. Kuzmenko *et al., Phys. Rev. B* **80**, 165406 (2009).
141. A. B. Kuzmenko *et al., Phys. Rev. B* **79**, 115441 (2009).
142. E. V. Castro *et al., J. Phys. – Condens. Mat.* **22**, 175503 (2010).
143. Y. B. Zhang *et al., Nature* **459**, 820 (2009).
144. X. Du *et al., Nature Nanotech.* **3**, 491 (2008).
145. K. I. Bolotin *et al., Solid State Commun.* **146**, 351 (2008).
146. B. E. Feldman, J. Martin, and A. Yacoby, *Nature Phys.* **5**, 889 (2009).
147. H. K. Min *et al., Phys. Rev. B* **78**, 121401 (2008).
148. R. S. Pantelic *et al., J. Struct. Biol.* **170**, 152 (2010).
149. N. R. Wilson *et al., ACS Nano* **3**, 2547 (2009).
150. R. R. Nair *et al., Appl. Phys. Lett.* **97**, 153102 (2010).
151. J. C. Meyer *et al., Nature* **446**, 60 (2007).
152. J. C. Meyer *et al., Solid State Commun.* **143**, 101 (2007).
153. A. Reina *et al., Nano Lett.* **9**, 30 (2009).
154. X. S. Li *et al., Science* **324**, 1312 (2009).
155. C. G. Granqvist, *Solar Energy Materials and Solar Cells* **91**, 1529 (2007).
156. X. Wang, L. J. Zhi, and K. Mullen, *Nano Lett.* **8**, 323 (2008).
157. T. Ando, *J. Phys. Soc. Jpn.* **75**, 074716 (2006).
158. T. Ando, and T. Nakanishi, *J. Phys. Soc. Jpn.* **67**, 1704 (1998).

159. S. V. Morozov *et al.*, *Phys. Rev. Lett.* **100**, 016602 (2008).
160. L. A. Ponomarenko *et al.*, *Phys. Rev. Lett.* **102**, 206603 (2009).
161. J. H. Chen *et al.*, *Nature Phys.* **4**, 377 (2008).
162. Z. H. Ni *et al.*, *Nano Lett.* **10**, 3868 (2010).
163. T. Stauber, N. M. R. Peres, and F. Guinea, *Phys. Rev. B* **76**, 205423 (2007).
164. V. V. Cheianov, and V. I. Fal'ko, *Phys. Rev. Lett.* **97**, 226801 (2006).
165. K. Nomura, and A. H. MacDonald, *Phys. Rev. Lett.* **96**, 256602 (2006).
166. K. Nomura, and A. H. MacDonald, *Phys. Rev. Lett.* **98**, 076602 (2007).
167. S. Adam *et al.*, *P. Natl. Acad. Sci. USA* **104**, 18392 (2007).
168. M. I. Katsnelson, and A. K. Geim, *Philos. Trans. R. Soc. A – Math. Phys. Eng. Sci.* **366**, 195 (2008).
169. E. V. Castro *et al.*, *arxiv.org, arXiv*:1008.2522v1 (2010).
170. C. R. Dean *et al.*, *Nature Nanotech.* **5**, 722 (2010).
171. J. S. Moon *et al.*, *IEEE Electron Device Letters* **30**, 650 (2009).
172. Y. M. Lin *et al.*, *Nano Lett.* **9**, 422 (2009).
173. Y. M. Lin *et al.*, *Science* **327**, 662 (2010).
174. M. Y. Han *et al.*, *Phys. Rev. Lett.* **98**, 206805 (2007).
175. L. A. Ponomarenko *et al.*, *Science* **320**, 356 (2008).
176. C. Stampfer *et al.*, *Appl. Phys. Lett.* **92**, 012102 (2008).
177. L. A. Chernozatonskii, P. B. Sorokin, and J. W. Bruning, *Appl. Phys. Lett.* **91**, 183103 (2007).
178. A. V. Shytov, D. A. Abanin, and L. S. Levitov, *Phys. Rev. Lett.* **103**, 016806 (2009).
179. F. N. Xia *et al.*, *Nano Lett.* **10**, 715 (2010).
180. A. C. Ferrari *et al.*, *Phys. Rev. Lett.* **97**, 187401 (2006).
181. A. C. Ferrari, *Solid State Commun.* **143**, 47 (2007).
182. L. M. Malard *et al.*, *Physics Reports-Review Section of Physics Letters* **473**, 51 (2009).
183. F. Schedin *et al.*, *Nature Mater.* **6**, 652 (2007).
184. Z. Chen, and J. Appenzeller, in *Ieee International Electron Devices Meeting 2008, Technical Digest* (2008), pp. 509.
185. L. A. Ponomarenko *et al.*, *Phys. Rev. Lett.* **105**, 136801 (2010).
186. F. N. Xia *et al.*, *Nature Nanotech.* **4**, 839 (2009).
187. H. Zhang *et al.*, *Optics Express* **17**, 17630 (2009).

Portrait photo of Professor Novoselov by photographer Ulla Montan.